# Gallium Nitride (GaN)

## Physics, Devices, and Technology

# Devices, Circuits, and Systems

**Series Editor**
*Krzysztof Iniewski*
Emerging Technologies CMOS Inc.
Vancouver, British Columbia, Canada

## PUBLISHED TITLES:

## PUBLISHED TITLES:

## PUBLISHED TITLES:

## FORTHCOMING TITLES:

# Gallium Nitride (GaN)

## Physics, Devices, and Technology

Edited by
## Farid Medjdoub
University of Lille (IEMN-CNRS), France

## Krzysztof Iniewski MANAGING EDITOR
Emerging Technologies CMOS Inc.
Vancouver, British Columbia, Canada

CRC Press
Taylor & Francis Group
Boca Raton London New York

CRC Press is an imprint of the
Taylor & Francis Group, an **informa** business

# Contents

**Chapter 9** Gallium Nitride–Based Interband Tunnel Junctions........................299

*Siddharth Rajan, Sriram Krishnamoorthy, and Fatih Akyol*

**Chapter 10** Trapping and Degradation Mechanisms in GaN-Based HEMTs.....327

*Matteo Meneghini, Gaudenzio Meneghesso, and Enrico Zanoni*

# Preface

Gallium nitride (GaN) is a binary III/V direct-bandgap semiconductor commonly used in bright-light-emitting diodes since the 1990s. Its wide bandgap of 3.4 eV and high electron mobility affords it special properties for applications in optoelectronic, high-power, and high-frequency devices. GaN is the material that makes violet (405 nm) laser diodes possible, without use of nonlinear optical frequency-doubling.

Its sensitivity to ionizing radiation is low, making it a suitable material for solar cell arrays for satellites. Military and space applications could also benefit as devices have shown stability in radiation environments. Because GaN transistors can operate at higher temperature and at higher voltages than gallium arsenide (GaAs) or silicon (Si) transistors, they make ideal power amplifiers at microwave frequencies.

This book, *Gallium Nitride (GaN): Physics, Devices, and Technology*, will cover large aspects of GaN from the fundamental physics of this emerging material to the fabrication of devices and circuits that are already foreseen to replace standard silicon- or GaAs-based materials in many applications.

**Farid Medjdoub**
IEMN-CNRS

# Editors

**Farid Medjdoub** is a CNRS senior scientist at IEMN in France. He earned his PhD in electrical engineering from the University of Lille in 2004. Then, he moved to the University of Ulm in Germany as research associate before joining IMEC as a senior scientist.

Dr. Medjdoub's research interests are the design, fabrication, and characterization of innovative GaN-based devices. He is author and co-author of more than 100 articles in this field. He holds several patents deriving from his research. He serves as a reviewer for IEEE journals and is a TPC member in several conferences. He is also part of the French observatory of wide-bandgap devices.

**Krzysztof (Kris) Iniewski** is managing R&D at Redlen Technologies Inc., a start-up company in Vancouver, Canada. Redlen's revolutionary production process for advanced semiconductor materials enables a new generation of more accurate, all-digital, radiation-based imaging solutions. Kris is also a founder of Emerging Technologies CMOS Inc. (www.etcmos.com), an organization of high-tech events covering communications, microsystems, optoelectronics, and sensors. In his career, Dr. Iniewski held numerous faculty and management positions at the University of Toronto, University of Alberta, SFU, and PMC-Sierra Inc. He has published over 100 research articles in international journals and conferences. He holds 18 international patents granted in the U.S.A., Canada, France, Germany, and Japan. He is a frequent invited speaker and has consulted for multiple organizations internationally. He has written and edited several books for CRC Press, Cambridge University Press, IEEE Press, Wiley, McGraw-Hill, Artech House, and Springer. His personal goal is to contribute to healthy living and sustainability through innovative engineering solutions. In his leisurely time, Kris can be found hiking, sailing, skiing, or biking in beautiful British Columbia. He can be reached at kris.iniewski@gmail.com.

# Contributors

**Fatih Akyol**
Department of Electrical and Computer
    Engineering
The Ohio State University
Columbus, Ohio, United States

**Subramaniam Arulkumaran**
Microsystems, Temasek Laboratories
Nanyang Technological University
Nanyang, Singapore

**Mark Beeler**
CEA-Grenoble
Grenoble, France

**Yvon Cordier**
Centre de Recherche sur l'Hétéro-
    Epitaxie et ses Applications (CRHEA)
Nice, France

**Ezgi Dogmus**
IEMN-CNRS
Lille, France

**Keisuke Shinohara**
Teledyne Scientific Company
Malibu, California, United States

**Sriram Krishnamoorthy**
Department of Electrical and Computer
    Engineering
The Ohio State University
Columbus, Ohio, United States

**Ching-Ting Lee**
Department of Electrical Engineering
National Cheng Kung University
Taiwan, Republic of China

**Hsin-Ying Lee**
Department of Photonics
National Cheng Kung University
Taiwan, Republic of China

**Li-Ren Lou**
Department of Physics
University of Science and Technology
    of China
Hefei, China

**Farid Medjdoub**
IEMN-CNRS
Lille, France

**Gaudenzio Meneghesso**
Department of Information
    Engineering
University of Padova
Padova, France

**Matteo Meneghini**
Department of Information
    Engineering
University of Padova
Padova, Italy

**Eva Monroy**
CEA-Grenoble
Grenoble, France

**Geok Ing Ng**
School of EEE
Nanyang Technological University
Nanyang, Singapore

**Rüdiger Quay**
IAF
Freiburg, Germany

**Siddharth Rajan**
Department of Electrical and Computer
    Engineering
Department of Material Science and
    Engineering
The Ohio State University
Columbus, Ohio, United States

**Joachim Würfl**
Ferdinand-Braun-Institut (FBH)
Leibniz Institut für
  Höchstfrequenztechnik
Berlin, Germany

**Enrico Zanoni**
Department of Information
  Engineering
University of Padova
Padova, Italy

# 1 GaN High-Voltage Power Devices

*Joachim Würfl*

## CONTENTS

## 1.1  INTRODUCTION

This section discusses GaN power switching devices and provides a general comparison to GaN microwave power devices. In the literature, many overview articles on GaN device technologies and applications are available [1–9].

### 1.1.1  Advantages of GaN Power Devices

AlGaN/GaN high-electron-mobility transistors (HEMTs) are promising candidates for switching power transistors because of their high off-state breakdown strength combined with excellent channel conductivity at on state. These features are consequences of the particular physical properties of GaN in combination with its heterostructure material AlGaN. Table 1.1 compares some of the physical properties of GaN to other competitive device families such as Si and SiC. One of the most important features is the breakdown strength of the material. Compared to Si, this parameter is 10 times higher for GaN. This means that 10 times the voltage can be applied to GaN devices for a given device dimension compared to Si devices. If the device is switched on, the remaining on-state resistance $R_{on}$ defines device losses at this condition. As the specific on-state resistance $R_{on}$ scales with the length of the device drift region necessary to maintain a given breakdown voltage, the more compact GaN devices feature much lower on-state resistances as possible with Si devices. Additionally, due to the electron transport properties of GaN/AlGaN HEMTs, the specific on-state resistance is almost two orders of magnitude lower compared to Si power devices with the same voltage rating [10]. Thus, GaN devices enable high breakdown voltages and high current levels simultaneously and feature small semiconductor areas. This additionally translates to high switching frequencies at high power levels [11]. According to current state of the art, the static on-state resistance of GaN power devices outperforms silicon power electronic devices and is close to the performance of silicon carbide (SiC) devices [12]. Several figures of merit have been defined in the past that describe these dependencies. They can be used to compare different device families in a more quantitative manner. For example, the resistive losses of the devices are reflected by the so-called Baliga figure of merit (BFM) [13], switching losses by the Baliga high-frequency figure of merit (BHFFM) [14], and microwave performance by the Johnson's figure of merit (JFM) [15]. In Table 1.1, these figures of merit have been normalized to Si performance and are shown for comparison. It can be seen that GaN devices easily outperform existing Si and SiC devices.

**TABLE 1.1**

**Comparison of Physical Parameters of Epitaxial GaN Layers as well as Bulk Ga Crystals against Si and 4H-SiC**

| | Si | 4H-SiC | GaN (Epitaxial) | GaN (Bulk) |
|---|---|---|---|---|
| Band gap energy, $E_g$ (eV) | 1.1 ind. | 3.26 ind. | 3.42 dir. | 3.42 dir. |
| Electric breakdown field, $E_{crit}$ ($10^6$ V/cm) | 0.3 | 2.2 | 2 | 3.3 |
| Relative dielectric constant, $\varepsilon_r$ | 11.9 | 10.1 | 9 | 9 |
| Thermal conductivity, $k$ (W/K·cm) | 1.5 | 4.9 | 1.3 | 2.3 |
| Electron mobility, $\mu_e$ ($cm^2$/V·s) | 1350 | 900 | 1150 (2000)[a] | 1150 (2000)[a] |
| Saturation velocity, $v_{sat}$ ($10^7$ cm/s) | 1.0 | 2.0 | 3 | 3 |
| BFM$_{Si}$, $\varepsilon_r\mu_e E_{crit}^3$ | 1 | 223 | 190 (330)[a] | 850 (1480)[a] |
| BHFFM$_{Si}$, $\mu_e E_{crit}^2$ | 1 | 45 | 36 (63)[a] | 98 (170)[a] |
| JFM$_{Si}$, $v_{sat}E_{crit}/2$ | 1 | 215 | 400 | 1090 |
| Maximum estimated operation temperature, $T_{max}$ (°C) | 200 | 500 | 700 | 700 |

*Note:* [a]As most of the GaN power transistors realized today rely on the two-dimensional electron gas (2DEG) properties at the GaN/AlGaN interface, the 2DEG-related data are given in parentheses wherever applicable.

dir., direct band gap; ind., indirect band gap.

The specific on-state resistance of any semiconductor devices used for power electronic applications is one of the key performance parameters. An ideal power switching device should have a high breakdown voltage to cope with the off-state condition, whereas its on-state resistance should be as small as possible. This request translates into the trade-off situation depicted in Figure 1.1. High-breakdown devices require extended drift regions, for example, the distance between gate and drain needs to be sufficiently large. This means that as breakdown voltage increases, more and more resistive elements are added to the device, thus increasing the on-state resistance. Clearly, those device technologies are advantageous that allow for a high blocking voltage at comparably low drift region extensions. As shown in Figure 1.1, the optimum on-state resistance for each device family follows a particular theoretical line, which has been calculated for unipolar devices. In fact, as power switching devices are operated in the linear (mobility-controlled) region at on state, the on-state resistance increases with breakdown voltage practically in a quadratic dependency [16]. Accordingly, GaN devices have an advantage over Si devices by theoretically at least three orders of magnitude. They even outperform SiC devices. The data shown in Figure 1.1 represent the static value of $R_{on}$. It is shown in Section 1.3 that this value may increase during dynamic switching.

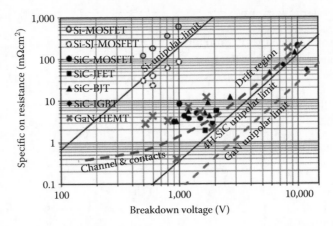

**FIGURE 1.1**  Comparison of specific on-state resistance $R_{on} \times A$ versus breakdown voltage for different device families. The straight lines represent theoretical values of $R_{on} \times A$ for unipolar devices. The channel & contacts represents limitations due to channel and contact resistance in lateral field-effect transistors and the drift region in vertical SiC devices. (From Kaminski, N., and O. Hilt, "GaN device physics for electrical engineers," ECPE GaN and SiC User Forum, Proceedings, Birmingham, United Kingdom, 2011 [17].)

In GaN HEMTs, the lateral device design with a comparably small total gate width for a given drain current combined with a low area-specific on-state resistance leads to very low device capacitances and a particularly low gate charge, $Q_g$. The product $Q_g \times R_{on}$, a figure of merit for switching efficiency, is at least 10 times smaller than that for Si-based devices (for illustration and further information, see Figure 1.26). The lower switching losses of GaN devices allow higher switching frequencies on the system level. As conversion frequency increases, the size of passive components such as inductors and capacitor can be decreased. This means that volume and weight of such systems can be reduced significantly, which, in turn, opens up new possibilities for system applications.

In contrast to Si and SiC devices, the majority of GaN devices for power electronics are realized on foreign substrates such as Si and SiC. The root cause of this relates to the difficulty to fabricate larger diameter freestanding GaN substrates. Therefore, historically, the development of GaN devices started on foreign substrates with a lattice constant very close to that of bulk GaN, such as SiC. Table 1.2 gives a comparison of different substrate types for GaN epitaxial growth of layers suitable for GaN power electronics. Although GaN on SiC substrates are generally advantageous for high power devices because of the high thermal conductivity of SiC, the growth on Si substrates is more promising for power applications because of the scalability of wafer diameters to 200 mm and more and the comparably low price level. Clearly, in terms of material quality the epitaxial growth on GaN substrate would be the best choice. Freestanding GaN substrates with low dislocation density (~$10^4$/ cm$^2$) even show a much better thermal conductivity compared to epi-GaN layers on foreign substrates (2.3 vs. 1.3 W/K·cm as shown in Table 1.1). However, technologies toward an increase of wafer diameter beyond 100 mm are still not available and a matter of ongoing research. Freestanding GaN substrates are expected to be very

**TABLE 1.2**
**Comparison of Substrates Suitable for GaN Power Electronic Devices**

|  | n-Type SiC | si-SiC | GaN Bulk | Si |
|---|---|---|---|---|
| Lattice mismatch (%) | 3.1 | 3.1 | 0 | 17 |
| Thermal conductivity (W/K·cm) | 4 | 4 | 2.3 | 1.5 |
| Thermal expansion coefficient ($10^{-6}$/K) at room temperature/1000 K | 4.5 | 4.5 | 3.72/5.45 | 2.4/4.4 |
| Availability/price | Medium | High | Very high | Low |
| Remarks | High thermal conductivity Requires thick buffer layers to accommodate vertical voltage drop | High thermal conductivity Highly isolating (thin buffer layers possible) Interesting for GaN power electronics | Limitations in wafer size Very interesting for true vertical GaN devices | Large lattice mismatch requires special technological solutions Requires thick buffer layers to accommodate vertical voltage drop Can be scaled to large wafer diameters (200 mm) Will dominate GaN power electronics |

attractive for true vertical GaN devices (see Section 1.5). For microwave applications, the growth on semi-insulating SiC (si-SiC) substrates has matured to a standard; to a lesser extent, GaN on Si devices are also feasible. si-SiC is preferred due to its inherently good electrical isolation properties giving low signal losses and due to its high thermal conductivity. Also, power devices grown on si-SiC wafers are very attractive as they benefit from the superior heat-sinking properties of SiC compared to Si. However, the price level of si-SiC wafers is so high that, besides very special applications, it is expected that the volume production of GaN power devices will be realized on Si substrates in future.

## 1.1.2 GaN Power Switching versus GaN Microwave Devices

Traditionally, the development of GaN transistors has been motivated by their impressive microwave capabilities compared to competing device families. This technology enabled highly efficient small-volume microwave devices and power amplifiers with a couple of hundred watts of output power in the low-GHz frequency range (e.g., S band at 2 GHz) and tens of watts in the 30-GHz range (Ka band) and still watt-level power levels in the W band [18]. The lower frequency devices are operating at a drain bias of typically 50 V, which means that their safe breakdown voltage is optimized to a level of around 200 V. In contrast, GaN power electronic devices are intended to safely operate at the 600-V or even the 1200-V node. This means that the breakdown voltage should be much higher to guarantee a safe margin of operation. Thus, the development of GaN power switching devices required novel technological concepts such as modified epitaxial buffer structures, field plates, and passivation layers. These technologies are described in Section 1.2.

The operation conditions of power devices significantly differ from those of standard microwave operation principles, as shown in Figure 1.2. Analogue microwave amplifiers (e.g., operated in class AB) operate around a fixed bias point; the dynamic load line (e.g., the time-dependent position of the bias point in the output I/V characteristics) makes use of the full I/V characteristic and can have a considerable voltage and current overshoot compared to the steady-state bias point. In contrast, a power switching device always switches from an off-state bias point (labeled as "(I)" in Figure 1.2) to an on-state bias point (labeled as "(II)"). The transition time from state (I) to state (II) is short compared to the switching period. Losses at off state and at on state should be as low as possible, which means that device leakage currents at off state and residual drain voltage at on state should be minimized. The exact time-dependent position of the load line strongly relates to the mode of device operation, extrinsic impedances of the device, external circuitry, and trapping effects in the device itself. For power switching devices, the transition time between the states should be as short as possible to ensure energy-efficient operation. If the aforementioned conditions are given, very large power levels can be switched without generating high levels of dissipated power (heat). This is different for microwave amplifier operation in an analogue amplification mode. Due to the position of the dynamic load point in the I/V characteristics, thermal losses are always generated, leading to very stringent conditions on heat dissipation from the device. This means, related to thermal design criteria, a microwave amplifying and a power switching GaN device are different, with more relaxed conditions for the power switching device.

Standard microwave AlGaN HEMTs are characterized by an inherent normally-on behavior due to the nature of the lateral two-dimensional electron gas (2DEG), acting as transistor channel [18]. For applications in power electronics,

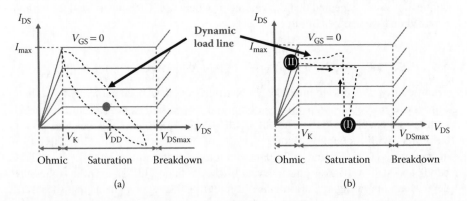

**FIGURE 1.2** Comparison of typical device operation conditions for (a) a microwave class AB power amplifier and (b) a power switching device. The dynamic load line of the power switching device is an example only. It strongly depends on external circuitry, parasitic device impedances, and mode of operation.

the normally-on switching characteristic is not a desired feature. For security reasons, normally-off devices are requested. This provides a safe margin of operation in system applications. Currently, considerable efforts are conducted worldwide to develop suitable normally-off AlGaN/GaN HEMTs. Generally, a high threshold voltage (>1 V, preferably >5 V) is essential for gaining acceptance for power electronic system applications.

Of course, dynamic switching properties are of paramount importance too. Together with the on- and off-state losses, they are contributing to the total loss of energy converters and thus strongly determining the conversion efficiency. As dynamic effects are also compromising the static advantages of GaN switching devices, their influence should be as low as possible to make GaN competitive to other technologies. For example, a design rule for 600-V switching GaN devices suggests that the dynamic increase of on-state resistance should be less than a factor of 2. Although considerable improvements in GaN technology have been presented so far, dynamic effects are still a major issue especially for high-voltage operations of 600 V and above [19,20]. Details are discussed in Section 1.3.

## 1.2 TECHNOLOGICAL DEVELOPMENTS TOWARD HIGH-VOLTAGE DEVICES

### 1.2.1 BREAKDOWN OF LATERAL GaN HIGH-ELECTRON-MOBILITY TRANSISTORS

#### 1.2.1.1 Breakdown Mechanisms

Many different physical effects are limiting the high-voltage performance of GaN devices. They all have to be taken into account when developing high-voltage GaN devices. The breakdown voltage of GaN devices is often defined as the voltage level at which the drain current of pinched-off transistors exceeds a normalized value of 1mA per millimeter of device width. Figure 1.3, taken from the study by Bahat-Treidel [21], provides a selection of breakdown events that have been observed. Device breakdown mainly depends on the specific epitaxial design of the buffer layer, its material quality, the lateral geometrical design of the devices, and their passivation technique. The latter issues are mainly dependent on processing technology, whereas the former parameters are due to the GaN/AlGaN epitaxial design, material properties, and interfacing between GaN/AlGaN material technologies and processing. Some mechanisms such as, for example, surface arcing between adjacent conductive structures with a large potential difference are known from standard power electronic device technologies. These mechanisms can be harnessed by applying adapted passivation and encapsulation schemes (see also Figure 1.3).

Breakdown mechanisms, being very specific for GaN power devices, need to be understood to successfully engineer high-voltage GaN devices. Table 1.3 summarizes the most important GaN-related breakdown mechanisms. Not all of them are related to classical breakdown events, characterized by a rapid increase of drain current at a given drain voltage (such as avalanche multiplication in Si devices). In many cases, the maximum allowable operation voltage is limited by excessive gate or drain leakage currents, which may have different technological root causes, as listed in Table 1.3.

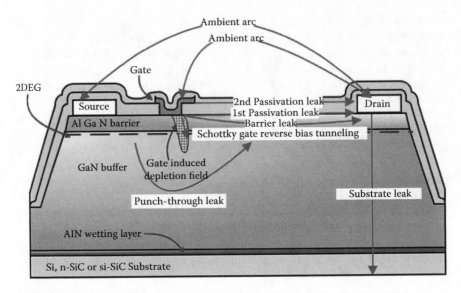

**FIGURE 1.3** Prominent leakage paths in GaN high-electron-mobility transistor devices for high-voltage applications. (From Bahat-Treidel, E., "GaN-based HEMTs for high voltage operation; design, technology and characterization," PhD thesis, Technical University, Berlin, Germany, 2012 [21].)

For example, the so-called punch-through mechanism [22] can substantially increase subthreshold leakage current (Table 1.3, column 2). In this case, if the channel is closed by a sufficiently negative gate bias electrons are bypassing the gate control region via the buffer. This is possible if the Fermi level in the buffer is close to the conduction band so that electrons can leave the channel region and find their way through the buffer layer [22,23]. As leakage and punch-through currents significantly increase with drain voltage, the devices increasingly dissipate thermal energy and are therefore not capable to operate efficiently at high bias levels. Often, classical breakdown cannot be observed as the devices approach their thermal limitations before the conditions for avalanche breakdown are reached. Moreover, it is known that excessive gate leakage currents are significantly compromising device reliability [24].

If leakage currents are suppressed by suitable technological means, the breakdown voltage of GaN HEMTs scales with increasing gate to drain separation, $d_{GD}$. Thus, $d_{GD}$ determines the maximum allowable operation voltage. At breakdown condition, the drain current rises rapidly and is mainly taken up from the gate terminal. This suggests electron flow from the gate to the drain, probably due to electron injection across the gate Schottky barrier (Table 1.3, column 3). Furthermore, in some cases the drain current splits into a source and a gate current component. As these effects are triggered by high electric field strength, particularly at the drain side edge of the gate or close to the drain electrode, field plates can suppress this effect and thus shift breakdown voltage to higher values [25–29].

If GaN power transistors are fabricated on conductive substrates such as Si or n-type SiC, device breakdown can occur vertically across the epitaxial layer stack. As the bias voltage further increases, the electric field between the active transistor

**TABLE 1.3**

**Collection of Mechanisms Limiting the High-Voltage Capability of GaN Field Effect Transistors Fabricated on Conductive Substrates such as Si or n- or p-Doped SiC**

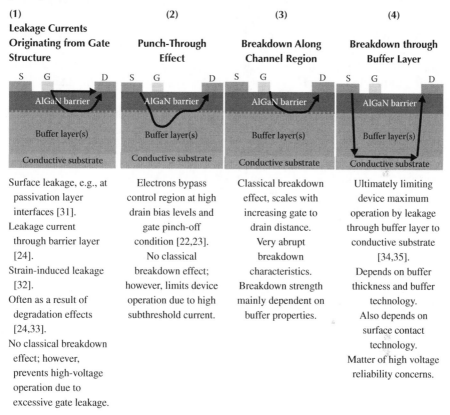

| (1) Leakage Currents Originating from Gate Structure | (2) Punch-Through Effect | (3) Breakdown Along Channel Region | (4) Breakdown through Buffer Layer |
|---|---|---|---|
| Surface leakage, e.g., at passivation layer interfaces [31]. Leakage current through barrier layer [24]. Strain-induced leakage [32]. Often as a result of degradation effects [24,33]. No classical breakdown effect; however, prevents high-voltage operation due to excessive gate leakage. | Electrons bypass control region at high drain bias levels and gate pinch-off condition [22,23]. No classical breakdown effect; however, limits device operation due to high subthreshold current. | Classical breakdown effect, scales with increasing gate to drain distance. Very abrupt breakdown characteristics. Breakdown strength mainly dependent on buffer properties. | Ultimately limiting device maximum operation by leakage through buffer layer to conductive substrate [34,35]. Depends on buffer thickness and buffer technology. Also depends on surface contact technology. Matter of high voltage reliability concerns. |

*Note:* The schematics show the principal paths of electron current through the device.

regions (source, gate, drain, and channel area) and the conductive substrate increases more and more until the breakdown strength of the buffer layer is reached. This opens a new leakage path bypassing the whole device, as indicated in Table 1.3, column 4.

### 1.2.1.2   Examples of Breakdown Effects

Figure 1.4 highlights the aforementioned mechanisms by comparing the scaling of gate to drain distance of devices grown on different epitaxial buffer structures. As shown by Würfl et al. [35], devices have been fabricated on n-SiC substrates with two different designs of the buffer layers, a carbon-compensated buffer layer doped to $1 \times 10^{18}$/cm$^3$ carbon followed by a GaN channel layer and an Al$_{0.05}$Ga$_{0.95}$N back barrier followed by a GaN channel. The AlGaN barrier layer of both designs is the same with an Al concentration of 25% and a thickness of 25 nm. For small gate to drain distances, the breakdown voltage monotonically increases, generally indicating a breakdown mechanism that is triggered by properties directly influenced by $d_{GD}$ scaling. In this case, breakdown occurs between the drain side edge of the gate and the channel region with a simultaneous increase of gate and drain current [23]. This is equivalent to the mechanism shown in Table 1.3, column 3. As $d_{GD}$ increases, the critical field for breakdown is obtained at an increasingly higher drain voltage. The slope at which the drain voltage increases with $d_{GD}$ is later referred to as the "breakdown strength" of the device. It strongly depends on the selected buffer composition. According to Figure 1.4, it approaches a value of 160 V/μm for the C-doped buffer structure and 50 V/μm for the AlGaN back-barrier structure. As the gate to drain scaling further increases, the breakdown voltage saturates at a certain drain voltage (~860 V for C-doped buffer and ~300 V for buffer with AlGaN back barrier). This means that the breakdown voltage is now independent on the gate to drain distance—another physical mechanism is dominating. Detailed DC measurements show that the gate current is no longer scaling with drain current at breakdown condition; instead of this, with further $d_{GD}$ scaling an ever-increasing part of the drain current is directly flowing to the substrate. For the $d_{GD}$ in the saturation region, this means that the breakdown current is now bypassing the gate control region. Detailed investigations have shown that the dominating current flow at breakdown condition is now originating from the source and flowing vertically through the buffer to the conductive substrate and back to the drain terminal, as indicated in Table 1.3, column 4. Figure 1.4 shows that the vertical breakdown strength of the C-doped buffer layer is considerably higher compared to the buffer structure with the AlGaN back barrier.

For further increasing device breakdown voltage, a sound understanding of the dependencies is necessary. Therefore, a simple test structure has been developed to probe vertical leakage current and breakdown effects. According to Figure 1.5a, the test pad consists of an ohmic contact on top of a HEMT epitaxial structure. It is surrounded by an isolation implantation region. Accordingly, only a vertical current flow from the metallic pad through the buffer to the substrate is possible. Voltage ramping between

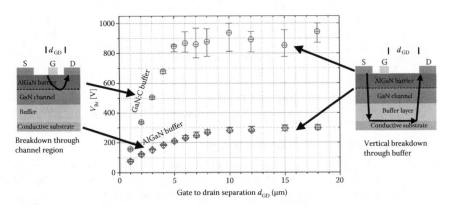

**FIGURE 1.4** Scaling of breakdown voltage, $V_{Br}$, with increasing gate to drain separation for GaN high-electron-mobility transistors grown on different epitaxial buffer structures (C-doped buffer with a C concentration of $1 \times 10^{18}/cm^3$ and AlGaN buffer with an Al concentration of 5%).

the test pad and the substrate provides the positive and negative branches of the vertical leakage characteristics. Figure 1.5 shows a selection of vertical breakdown profiles taken from HEMT structures grown on buffer layers with different epitaxial designs. The epitaxial layers were grown on n-type SiC wafers. Different buffer realizations feature very unique vertical current flow properties, as represented by the respective I/V characteristics. They differ in terms of symmetry between the positive branch and the negative branch and the absolute value of maximum allowable vertical voltage in the respective branch. For all three buffer types investigated, the negative branch always shows a higher vertical breakdown voltage. The most symmetric behavior along with the highest total breakdown strength has been observed on C-doped buffer layers. Vertical leakage across a Fe-doped buffer layer is very asymmetric and practically shows diode-like I/V characteristics with a very low vertical breakdown in the positive branch. The analyzed AlGaN buffer structure also shows comparably low total breakdown strength along with a rather low breakdown voltage in the positive branch. The vertical leakage

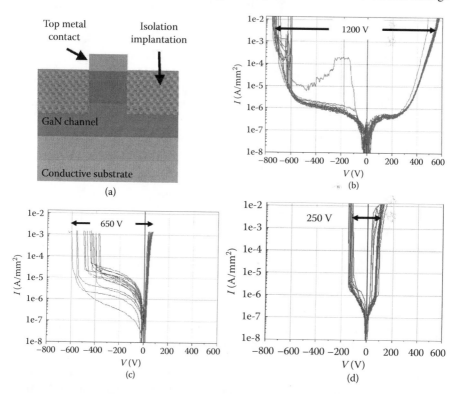

**FIGURE 1.5** Investigation of vertical I/V characteristics: The diagrams show the superposition of a multitude of measurements taken from different positions of the respective wafers. I/V characteristics plotted in the positive branches of the I/V characteristics represent the situation where the wafer top contact is positively biased, whereas the blue curves represent the opposite biasing direction. All measurements relate to an ohmic contact metal on the wafer front side. The voltage levels indicate the maximum possible breakdown voltage of a floating device on the respective wafer. (a) Schematic cross section of test structure for vertical breakdown evaluation, (b) carbon-doped buffer structure, (c) Fe-doped buffer structure, and (d) AlGaN back-barrier structure with an Al concentration of 5%.

current distribution over the negative and positive branches of the I/V characteristic also depends on the doping concentration of the substrate and on the nature of the metallic contacts on the wafer surface, for example, whether they have a Schottky or an ohmic behavior or whether they are placed on isolation implanted material [35]. Decreasing n-SiC substrate conductivity especially increases breakdown strength in the negative polarity. The breakdown strength in both polarities is compromised if the bias is applied to ohmic contacts at the wafer front side.

The physical background of the observations is not fully explained yet. It is believed that the energy level of the Fe- or C-acceptor states, material quality of the buffer, and band offsets between the GaN-based epitaxial layers and the substrate all play a significant role. Phenomenologically, the positive branch of the I/V characteristic may be explained as a vertical p–n diode with a very high series resistance biased in the on state; however, the proper physical interpretation requires further studies on this topic.

The maximum breakdown voltage that can be obtained from a device with an electrically floating substrate is finally limited to approximately the sum of the breakdown voltages of the negative and the positive branches of the vertical I/V characteristic. In many applications, a floating substrate cannot be accepted. The asymmetric behavior is particularly critical for half- or full-bridge arrangements of power switching transistors where two serially connected transistors are switching to a load in antiphase [34,35]. This mode of operation requires full isolation across the buffer for both switching polarities. For example, in case of Fe-doped buffer structures this would mean that the breakdown voltage of the transistor switching the positive polarity of the bias against a load connected to backside potential is significantly lower than that of the device switching the negative polarity (60 V in the positive branch compared to an average of 500 V in the negative branch).

### 1.2.2 Concepts of Increase in Breakdown Voltage

The findings discussed in the previous Section 1.2.1 lead to design rules for high-voltage GaN devices. Design optimizations address the topics discussed in Sections 1.2.2.1 through 1.2.2.3.

#### 1.2.2.1 Reduction of Electrical Field in Critical Device Regions

In GaN heterojunction field-effect transistors (HFETs), the highest electric fields appear in the vicinity of the gate, especially at the gate edge toward the source. The goal is to reduce the probability of electron injection from the gate into the AlGaN layer or into the passivation and thus prevent premature breakdown, as described in column 1 of Table 1.3. Field plates in close vicinity to the gate or drain field plates solve these issues. Appropriate configurations have been discussed in the literature to a large extent [25–30]. Figure 1.6 shows a selection of possible field plate configurations that are useful to spread out the electric fields in the vicinity of the gate and thus to reduce gate-assisted breakdown effects. As high electric fields are also triggering trap-related charging effects, properly designed field plates can modify the field distribution such that high field regions can be shifted from critical to less critical device regions and therefore reduce dispersion or current slump effects. This is discussed in more detail in Section 1.3.4.2.

Field plates are also implemented in GaN microwave devices, especially for S-band high power transistors. They usually consist of an asymmetric gate with an extension of the gate wing to the drain side in combination with a single source connected field plate overlapping the drain side gate wing by a certain dimension. This is proved to be a very successful approach for drain bias levels of around 50 V. Higher drain voltages require a more sophisticated technique to spread the electric fields efficiently without getting field crowding in dedicated device regions. Figure 1.6a shows a double source connected field plate, as often used for GaN power switching devices [25]. The dimensions of the field plates, that is, the overlap of the drain extensions and the thickness of passivation and metallization layers, have to be simulated and tested to effectively smoothen the electric field in gate vicinity. Principally, gate-connected field plates are also possible as the gate potential relative to a high voltage drain bias at off state is very close to the source potential. However, gate-connected field plates increase the drain to gate feedback capacitance, $C_{rss}$, and may therefore lead to instability problems. Source-connected field plates add to the drain source capacitance, $C_{oss}$, which is less harmful in terms of device stability. Of course, source-connected

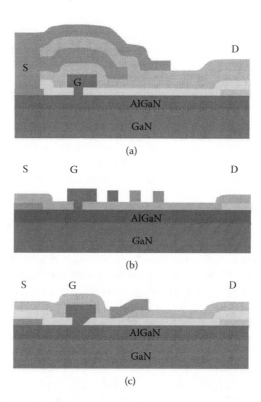

**FIGURE 1.6** Different field plate approaches for breakdown voltage enhancement and for reducing dispersive effects are shown. (a) Stacks of source-connected field plates in conjunction with an asymmetric T-gate structure. (b) Multiple grading field plates that can be deliberately biased at different potentials. (c) Slanted field plate structures.

field plates designed according to Figure 1.6a add gate source capacitance. This can be significantly reduced by applying air bridge-type field plate designs [29].

Multiple grating field plates (MGFPs) offer another very efficient possibility to spread out the electric field. As the drain access region of high-voltage GaN transistors has a structural dimension of typically 15–20 μm, it is comparably easy to place small metal lines in between and to connect them to either the gate or the source potential. This approach is shown in Figure 1.6, where, for example, the MGFP close to the gate is connected to gate potential and the other MGFPs are connected to source potential. A detailed analysis of MGFPs has been provided by Bahat-Treidel et al. [27].

Slanted field plates avoid high field peeking at the termination of the field plate as the geometrical distance between the field plate and the channel region increases gradually. Very different structural modifications are possible. For example, as shown in Figure 1.6, the gate structure can be slanted at the drain side edge [8]; additionally, other field plate structures placed further toward the drain electrode may be slanted toward the top of an additional passivation layer to ensure a gradual fading out of the electric field.

### 1.2.2.2 Confinement of Electrons in the Channel

This concept avoids the punch-through effect discussed previously and therefore increases the high-voltage operation capability. Punch-through is not a real breakdown effect as electrons are just bypassing the gate control region at high-voltage off-state biasing. Therefore, subthreshold leakage is considerably increased by this effect. High-voltage blocking behavior is obtained by designing potential barriers into the epitaxial buffer layer structures, thus preventing electrons from punching through the gate control region as shown in [22,23,36]. Figure 1.7a depicts the principal epitaxial design toward high-voltage GaN devices. AlGaN barrier layers, carbon- or iron-doped buffer layers, and combinations of these layers build up a repelling potential barrier for the channel electrons and thus avoid punch-through or shift it to higher voltage levels. For example, carefully designed epitaxial buffer structures on n-SiC substrates enable high-voltage devices above 1000 V [23,37]. As shown in Figure 1.4, carbon-doped buffer structures have a clear advantage over AlGaN buffer layers since high breakdown voltages can be obtained by comparably low gate to drain spacing. However, since C doping adds acceptor-type trap states into the buffer, high C concentrations in the vicinity of the channel lead to severe dispersion effects and thus significantly compromise switching properties. Details on dispersion effects are discussed in Section 1.3. In any case, breakdown voltage at a given gate to drain separation and thus on-state resistance has to be traded off against dispersion effects.

### 1.2.2.3 Increase of Vertical Breakdown

High vertical breakdown strength across the epitaxial structures is indispensable for GaN/AlGaN epitaxial layers grown on conductive SiC or Si substrates. For structures grown on insulating substrates such as si-SiC, these conditions are relaxed as the insulating substrate itself can take up the vertical voltage drop. As shown in the Section 1.2.2.2, the selection of an optimized buffer design plays a key role. The vertical breakdown strength, that is, the capability of the buffer to block a given vertical bias per unit buffer thickness, depends on the type and epitaxial growth parameters of buffer compensation that is used to avoid lateral punch-through at high drain bias. The blocking

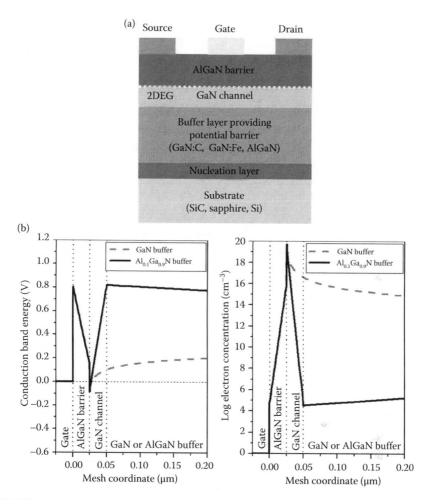

**FIGURE 1.7** Typical epitaxial layer structures used for high-voltage GaN devices. All of these buffer structures provide a potential barrier for channel electrons and thus confine electrons to the channel region as far as possible. (a) Schematic cross section. (b) Comparison of epitaxial transistor structure with and without potential barrier in the buffer in terms of energy band diagram (left) and electron concentration (right), according to Bahat-Treidel. (From Bahat-Treidel et al., *IEEE T. Electron Dev.*, 55(12), 3354–3359, 2010 [23].)

voltage of course scales with buffer layer thickness; however, as GaN layers are always strained on foreign substrates thick buffer layers lead to severe mechanical strain and wafer warping [38]. Therefore, buffer structures with as high as possible vertical blocking strength are envisaged. For example, C-doped buffer layers on Si or n-type SiC substrates are always characterized by low vertical leakage currents. On the other hand, as shown in Section 1.3, a quite complex trade-off between vertical blocking capability, on-state resistance, and dynamic switching properties has to be considered when selecting the suitable buffer design dependent on targeted device specs. Furthermore, due to the asymmetry of vertical leakage (Figure 1.5), the selection also depends on the biasing

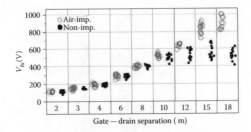

**FIGURE 1.8**  Effect of pre-epitaxy implantation of Ar in n-SiC substrates: scaling of device breakdown voltage with increasing gate to drain distance, $d_{GD}$, for a wafer comprising a GaN:Fe buffer. (From Hilt et al., *IEEE T. Electron Dev.*, 60(10), 3084–3090, 2013 [40]; Hilt et al., *ECS Trans.*, 58(4), 145–154, 2013 [50].)

conditions of the transistors operating in system environment especially whether or not the application allows for floating substrates. In the latter case, the asymmetry of the vertical buffer leakage is of minor concern only [35].

As demonstrated recently, the vertical buffer leakage can also be decreased by providing an isolation implantation into the conductive SiC substrate before epitaxial growth of the GaN/AlGaN structures [39]. It has been shown that pre-epitaxy implantation is possible without compromising the quality of epitaxial layers grown on top of these pretreated surfaces [40]. Depending on the actual implantation depth as well as the selection of ion species and implantation dose, the vertical breakdown can be significantly increased. For example, a 2.5 μm deep Ar isolation implantation into n-type SiC substrate considerably increases vertical breakdown voltage. For example, Figure 1.8 compares breakdown voltage scaling of devices fabricated on Ar-implanted n-SiC substrates with similar devices on non-implanted n-SiC substrates. As the gate-to-drain distance increases non-implanted devices depict an earlier voltage saturation compared to their pre-implanted counterparts. In case of isolation implantation of the n-SiC surface the critical field for vertical buffer breakdown will be reached at much higher drain voltage levels as the isolation implanted region can take up an additional portion of the total vertical voltage drop.

## 1.3  OPTIMIZATION OF SWITCHING EFFICIENCY

Some of the technological approaches toward high-voltage performance are increasing device breakdown voltage at the expense of switching properties. In GaN power devices, switching performance is often limited by an increase in on-state resistance as soon as the device is switched dynamically. These dependencies are discussed in Sections 1.3.1 through 1.3.4.

### 1.3.1  DYNAMIC ON-STATE RESISTANCE, $R_{on\_dyn}$

#### 1.3.1.1  Definition of $R_{on\_dyn}$

GaN-based power transistors can gain market acceptance only if they demonstrate switching properties superior to competing technologies. Due to trapping effects, the static on-state resistance $R_{on\_stat}$ may increase on dynamic switching of the device, leading to an elevated dynamic on-state resistance, $R_{on\_dyn}$. This effect is also known from GaN microwave transistors. Depending on the specific bias conditions during measurement, it is referred to as drain lag or gate lag, describing the dynamic decay of the drain current in the saturation region of the device. As power switching devices are never driven into saturation, for most of the switching sequence (only dynamically during a switching event) this definition is misleading. Therefore, the dynamic change of on-state resistance is taken as a measure of dynamic trapping effects. According to Figure 1.9, the on-state resistance is defined as the turn-on resistance of the device at 0-V drain bias. The dynamic on-state resistance often exponentially scales with device switching voltage and thus significantly challenges high-voltage GaN switching applications.

#### 1.3.1.2 Measurement of $R_{on\_dyn}$

The dynamic on-state resistance can be extracted from pulsed measurements of I/V characteristics starting from different bias conditions. Figure 1.9 explains this procedure in a practical measurement setup. For pulsed measurements, the device remains at the drain bias level given for most of the time (0.5 ms), only jumping to the measurement points and back for a duration of 200 ns. This ensures that the measured pulsed I/V characteristic represents the trapping properties at the given drain bias level. The example depicted in Figure 1.9 clearly shows a strong decrease in saturation current after pulsing at elevated drain voltages. This directly translates into an increase of $R_{on\_dyn}$ with increasing drain voltage.

To characterize switching transistors for power electronic applications being switched between two distinct bias points, as shown in Figure 1.2b, it is not necessary to measure the whole output characteristic of the transistor. Alternatively, the transistors are switched from an off-state bias point to an on-state condition. Figure 1.10a shows a suitable measurement setup. First, a capacitor is charged to the targeted switching voltage; the test transistor (DUT) is disconnected at this stage. Then, the capacitor is switched to the DUT biased in off state. Afterward, the transistor is turned on at the gate for 10 μs; the resulting current is set by the resistor $R_{Load}$. The on-state resistance is then calculated from the voltage drop at the transistor and the drain current.

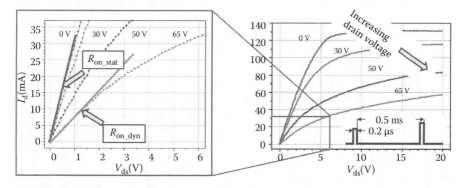

**FIGURE 1.9** Definition and determination of dynamic on-state resistance using pulsed measurements of direct current output characteristics where pulsing (index) has been started from different drain bias values, as indicated in the diagram.

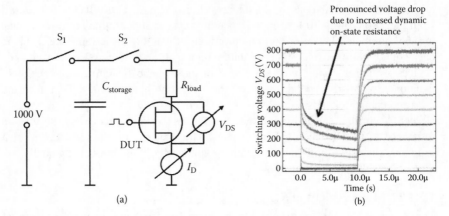

**FIGURE 1.10** Measurement of dynamic on-state resistance. (a) Measurement setup for determination of switching properties dependent on drain voltage level and drain current. The drain current is set by the load resistor $R_{Load}$. The device is switched on and off by controlling the gate voltage according to [41]. (b) Example of dynamic switching properties of a non-field-plated GaN heterojunction field-effect transistor showing considerable increase in dynamic on-state resistance after switching at drain bias voltages higher than 300 V. The device was fabricated on a heavily C-doped buffer structure and thus shows a very pronounced $R_{on\_dyn}$.

As an example of a device having a very pronounced increase in dynamic on-state resistance, Figure 1.10b highlights the off–on–off switching performance of a completely packaged 0.8-Ω GaN HFET fabricated on carbon-compensated GaN buffer doped to $2 \times 10^{19}$/cm³ at a distance of 35 nm from the 2DEG channel. The devices do not show a strong increase in dynamic on-state resistance up to 200-V drain bias; at higher voltages, the increasing dynamic on-state resistance causes a voltage drop across the transistor, which is clearly visible if the DUT is biased above 300 V. After switching at 800 V, a very strong dispersion effect is observed, which directly relates to a dynamic on-state resistance, $R_{on\_dyn}$, increase by a factor of more than 1000 after 800 V switching.

For the characterization of devices depicting a low $R_{on\_dyn}$, this method of measurement may become inaccurate as the drain voltage of the DUT at on state (Figure 1.10) has to be measured in the high-voltage setting of a fast oscilloscope. For this purpose, a very efficient clamping circuit has been proposed by Lu et al. [42], which effectively solves this problem. An example of nearly perfect switching properties is shown in Figure 1.25a.

Usually, $R_{on\_dyn}$ also depends on the bias level defined as the starting point for pulsed measurements and on the timing of the measurement. For example, this means that the off-state time at a given drain bias before starting the measurement pulse and the time interval of the measurement itself after pulsing is decisive. During long off-state time intervals, the influence of slowly responding traps will be visible. On the other hand, if the time interval between the switching event and the time window of the measurement increases short time responding traps may be blanked out and cannot be measured, which would result in a too optimistic determination of $R_{on\_dyn}$. These dependencies on the specific measurement conditions have to be taken into account when trying to compare $R_{on\_dyn}$ values from different manufacturers.

## 1.3.2 Switching Losses

As already pointed out in Section 1.1, the switching efficiency depends on the internal losses generated in the switching device. Suppose that two transistors arranged in a half-bridge configuration, as depicted in Figure 1.11a, are switched to a load Z in an antiphase drive mode. Then, if switching losses can be neglected, drain current and drain voltage of the transistor show a time dependency according to Figure 1.11b. An ideal switch is characterized by a fully conductive on state ($R_{on} = 0\ \Omega$); an infinitely resistive off state ($R_{on} = \infty\ \Omega$); and an infinitely small rise time, $\tau$, of current or voltage between the switching states. The goal of all switching transistor developments is to approach this ideal situation as close as possible.

In reality, devices suffer from numerous parasitic effects, leading to a time dependency of the switching cycle according to Figure 1.11c. The on-state resistance is limited by specific semiconductor parameters such as mobility and saturation velocity of the electrons and the device design itself (e.g., device width). Furthermore, $R_{on}$ can increase dynamically, meaning that $R_{on}$ increases with switching frequency and voltage. The higher the $R_{on}$, the higher the voltage drop across the device at on state. This creates thermal losses and thus compromises the maximum efficiency that can be obtained. At off-state condition, thermal loss is created by device leakage meaning that a residual drain current is flowing, although the device is turned off by the gate. Any compromise in switching speed, for example, when the device switches from off state to on state or vice versa, leads to a transient situation where the device

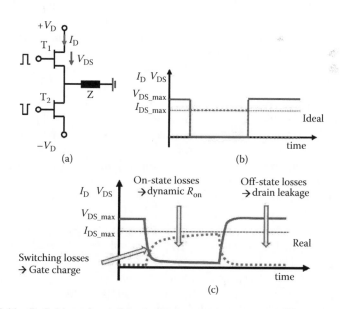

**FIGURE 1.11** Definition of switching losses created in power transistors: (a) Half-bridge arrangement. (b) Drain current and drain voltage cycle for an ideal switching situation. (c) Drain current and drain voltage cycle for a real switching situation with visualization of on-state, off-state, and dynamic losses.

is turned on while the drain voltage is still high. The dynamic load line of a power switch depicted in Figure 1.2 explains this situation. This may lead to temporary extremely high power dissipation in the device. Equation 1.1 describes the whole energy, $P_{diss}$, that is dissipated through a switch cycle of duration $T$:

$$P_{diss} = \int_0^T \frac{1}{T} V_{DS}(t) I_D(t)\, dt$$

(1.1)

### 1.3.3 MECHANISMS LIMITING SWITCHING EFFICIENCY

The dynamic on-state resistance $R_{on\_dyn}$ is caused by temporary charge trapping after switching, for example, from off-state to on-state conditions. If charge trapping happens close to the channel, charge neutrality requires a corresponding change in the 2DEG concentration in the affected device regions as soon as the device is switched on. Accordingly, negative charges close to the channel are depleting the 2DEG region. In contrast, positive charges would even result in 2DEG accumulation. Negative charge trapping usually occurs close to the channel, preferably in the vicinity of the drain side edge of the gate or in the drain access region. Trapping can take place at various places—in epitaxial layers, at passivation semiconductor interfaces, or even at interfaces between different passivation layers. As a consequence, electron transport in the channel region immediately after switching from on condition to another condition is impeded by negatively charged traps until they are emptied. This leads to increased switching losses, results in reduced system efficiency, and explains the time dependence of $R_{on\_dyn}$. $R_{on\_dyn}$ usually increases with operation voltage and can exceed the static value by more than two orders of magnitude [43,44].

Figure 1.12 visualizes this effect. At off-state condition, high electric field peaks appear at the drain side edge of the gate, close to the gate side edge of the drain, and close to the drain side edge of the metallic $\gamma$ gate in the passivation layer. The field is penetrating deeply into the buffer. If compensating acceptors are assumed in the buffer (e.g., due to Fe- or C-compensation doping or due to other imperfections), the high electric field may lead to a depletion of this region; correspondingly, the released holes would slowly move to the source. Additional electron trapping can also take place in this region if trap states are present and can be occupied. If off-state biasing persists for a longer time (longer than the trapping time constants), a steady-state negative charge distribution as indicated in Figure 1.12 builds up. On switching to on state, the charges cannot respond that fast as they are located in the highly insulating buffer layer and thus leave back a time-dependent negative space charge region. Charge neutrality rule would then force a depletion of the 2DEG region until the charges are reemitted to the valence band and recombine with the holes there.

Dispersion effects not only are related to buffer traps but can also be caused by trapping at the interface between semiconductor and passivation and/or between different passivation layers, as reported by Moens et al. [45]. Due to the changing field distribution at the drain side edge of the gate and in the drain access region with increasing drain bias, a complex interaction of electron trapping and de-trapping has been observed, leading to a maximum trapping at a certain drain bias (400 V in this case).

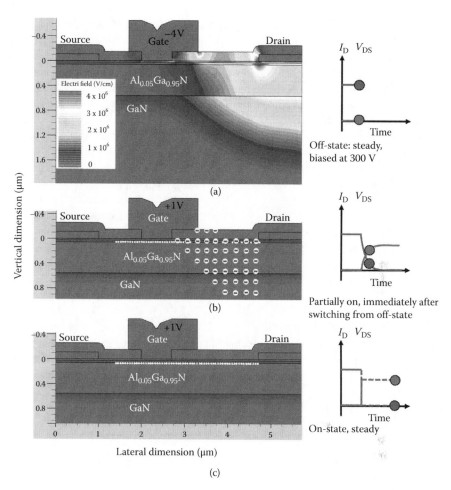

**FIGURE 1.12** Visualization of possible charge trapping effects after switching a GaN power transistor from off-state to on-state condition: (a) Simulation of electric field distribution of a GaN heterojunction field-effect transistor at reverse bias condition and at a drain voltage of 300 V. The two-dimensional electron gas (2DEG) is fully depleted. (b) Graphical representation of a possible trapping situation after the switching event to on state (e-fields are assumed to be similar compared to on-state condition). The 2DEG is partly depleted. (c) Simulation of electric field distribution at steady-state on-state condition (drain voltage 0 V). The 2DEG is fully conductive.

### 1.3.4 COUNTERMEASURES TOWARD OPTIMIZED POWER SWITCHING

#### 1.3.4.1 Optimized Epitaxial Buffer Design

As discussed earlier, design and composition of the buffer structure significantly influence dynamic properties. High values of breakdown strength, for example, 170 V/μm, have already been obtained in the past [47]. However, due to the high $R_{on\_dyn}$ these devices were not useful for switching applications. Therefore, different epitaxial buffer concepts were systematically compared to each other to find the best compromise between breakdown strength and increase in dynamic on-state resistance.

Figure 1.13 depicts this trade-off [20]: The presence of carbon acceptors in the buffer clearly increases device (gate/drain) breakdown strength with increasing C concentration, however at the expense of an increasing $R_{on\_dyn}$. Fe-doped buffer and AlGaN back-barrier structures are a better choice for reducing $R_{on\_dyn}$; but, on the other hand, the breakdown strength is reduced to only 40 or 50 V/μm [48,49]. Both findings are in agreement with the simulations performed by Uren et al. [51,52]. A good compromise can be obtained by combining AlGaN barrier and C doping in such a way that the AlGaN barrier is placed close to the channel, whereas the C-doped buffer is placed in regions underneath, which are not prone to large field changes during switching (denoted as "composite GaN:C" in Figure 1.13). A breakdown strength of 80 V/μm and an increase in $R_{on\_dyn}$ of only 10% at 65-V switching are obtained with this combination. For 600-V operation, the increase in $R_{on\_dyn}$ has been reduced to 25%.

In general, charge trapping can occur in all epitaxial device layers if trap states are incorporated either intentionally by doping or compensation or unintentionally in the form of defect states. Therefore, both design and quality of the epitaxial layers are decisive.

### 1.3.4.2  Harnessing Electric Fields

Field plate structures are decisive technological components for both breakdown voltage enhancement and harnessing dispersion effects. Numerous field plate approaches are discussed in literature. The goal of all approaches is to cut down the electric field peak present at the drain side corner of the gate to an acceptable level, which limits electron injection from the gate into adjacent semiconductor or passivation regions. The potential of field plates in conjunction with a reduction of dynamic

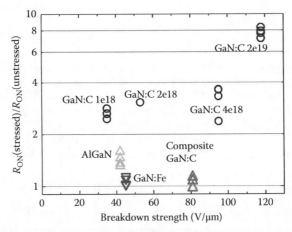

**FIGURE 1.13** Trade-off between dynamic on-state resistance increase and device breakdown strength depending on different technology tests for engineering an optimum high-voltage and practically dispersion-free buffer layer. The measurements have been taken by pulsing from an off-state bias point at 65 V to on state for 200 nanoseconds and then back to the original static bias point at a duty cycle of 1:2500.

on-state resistance has been discussed in literature [8,28,53,54]. Figure 1.6 provides an overview of different field plate concepts reported in literature.

## 1.4 NORMALLY-OFF GaN HIGH-ELECTRON-MOBILITY TRANSISTORS

Many approaches have been demonstrated toward normally-off GaN transistors. The general approach is to shift the conduction band of the channel region controlled by the gate above the Fermi level at zero gate bias and thus obtain a positive threshold voltage. For power applications, threshold voltage ranges above +1 V are desirable to ensure safe operation conditions. At the same time, on-state resistance should be kept as low as possible. This demand favors designs comprising low resistive device access regions (the areas between source and gate as well as between gate and drain) by making full use of the excellent transport properties of the two-dimensional electron channel (2DEG) at the GaN/AlGaN heterojunction interface. A number of normally-off approaches are reported in literature; an overview of applicable technologies having the potential toward threshold voltages above 2 V is given by Hahn et al. [55].

### 1.4.1 RECESSING OF AlGaN BARRIER LAYER

Realizing normally-off behavior by recessing according to Figure 1.14 means that the barrier layer underneath the gate is thinned until the conduction band in the channel region shifts above the Fermi level at 0-V gate bias. Depending on the actual Al concentration of the AlGaN barrier layer, the remaining barrier underneath the gate is reduced to a thickness of a few nanometers only (typically 3–5 nm) [56]. Gate recessing is often combined with gate dielectric isolator techniques to avoid increase in gate leakage [57]. Since the 2DEG is fully available outside the gate region, a low-resistivity access region is ensured. However, there is an insufficient trade-off between positive threshold voltage and maximum obtainable on-state current. Especially for devices having a high threshold voltage (which is desirable), the channel region remains extremely thin. This means that even if the gate is fully biased to on-state condition, the remaining conductivity of the channel underneath the gate is compromised.

Selective epitaxial techniques are able to mitigate this problem. This approach principally opens the possibility to independently design normally-off behavior and source/drain access region as epitaxial concepts being optimum for the respective device regions can be selected. Then, for example, the low-resistance access regions

Recessed schottky gate

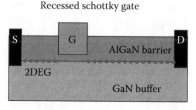

**FIGURE 1.14**   GaN normally-off transistor based on gate recess technology. In many cases, this technology is combined with a gate oxide underneath the gate itself.

are realized by selective overgrowth of all wafer areas except that of the gate region using a suitable material combination to enhance access region conductivity [58].

## 1.4.2 CHARGE MANAGEMENT IN GATE REGION

The placement of localized negative charges underneath the gate metal shifts depletes DEG as charge neutrality has to be always ensured. Positively charged counterparts are those charges that are created by differences of spontaneous and piezoelectric polarizations at the AlGaN/GaN heterojunction. Normally, they are compensated by electrons building the 2DEG. A sufficient amount of negative charges may completely deplete the channel in this region, which renders the transistor normally off. In view of the energy diagram, the conduction band of the channel region shifts above the Fermi level, which makes it unlikely for electrons to be present in this region at 0-V gate bias. On a positive gate bias, charge neutrality demands for a compensation by negative charges underneath the gate, and the 2DEG channel region starts to be populated with electrons.

There are different possibilities to realize fixed negative charges in the gate region at zero gate bias. The most popular method relies on a p-type doped GaN or AlGaN layer between the gate metal and the channel (see Figure 1.14). This technique was proposed by Uemoto et al. [59] from the Japanese company Panasonic in 2007 and since then further developed and brought to market [1,60]. The principal idea has also been taken up by several institutions, leading to very successful GaN normally-off devices [37,61,62]. At zero gate bias, the p-GaN layer is depleted, leaving back a fixed negative charge, which in turn depletes the channel. As the gate bias turns to positive polarity, the space charge region gets smaller, which means that less counter charges underneath the gate are needed to provide charge neutrality. As the number of positive charges is fixed (as defined by the difference of spontaneous and piezoelectric polarizations at the AlGaN/GaN heterojunction), channel electrons enter the region underneath the gate and the transistor starts to get conductive. As soon as the space charge region is fully neutralized, maximum 2DEG electron population can be achieved. The p-GaN (or p-AlGaN)/AlGaN/GaN junction can also be considered as a p–n heterojunction diode with the AlGaN barrier layer having only an unintentional and therefore low n-type doping concentration (basically i-AlGaN). If the positive gate bias increases further, holes are injected into the p–n heterojunction. As the hole mobility is much less than the electron mobility, the holes slowly drift toward the source. Some holes recombine with channel electrons; however, a steady-state hole density builds up underneath the gate and in the source side of the source gate access region. In this case, maintaining charge neutrality requires increased electron density in the adjacent channel region, which then results in a higher drain current of the device. This mechanism is called the "gate injection effect"; therefore, p-GaN-gated devices are also referred to as gate injection transistors (GITs). Ishida et al. [2] provide a more detailed description of the physical dependencies. Hole injection from the gate is of course associated with a hole current flowing from the metallic gate contact through the p-GaN/AlGaN gate. Practically, for system applications this means that driving the devices into the gate injection regime that forwards gate leakage increases considerably.

In principle, any fixed negative charges in the vicinity of the gate can deplete the channel at zero gate bias. In 2009, the Japanese company Sanken successfully investigated the properties of p-doped nickel oxide ($NiO_x$) on top of a AlGaN/GaN heterojunction [63]. Further, it was demonstrated by Chen et al. [64] that the incorporation of negatively charged fluorine in the semiconductor regions underneath the gate depletes the channel and renders the devices normally off (see Figure 1.14). The charged fluorine ions are introduced by a sophisticated plasma etching process. These activities have now been further developed, leading to normally-off devices with a high threshold voltage (Figure 1.15b). At the beginning of these developments, there was a debate on whether or not the incorporated fluorine is stable; however, publications of the time are available that demonstrate quite a few stable normally-off behaviors without drift effects of threshold voltage [64].

Generally, compared to recessed Schottky gate designs for normally-off applications the gate to channel distance of the aforementioned versions can be larger

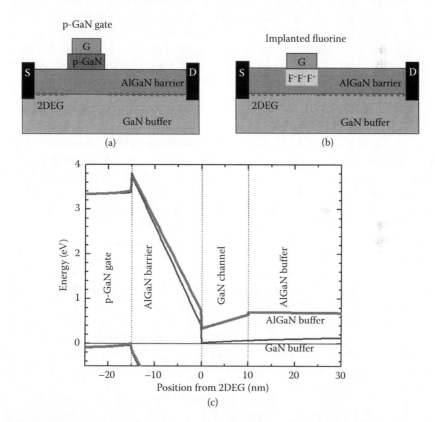

**FIGURE 1.15** Normally-off GaN transistors utilizing charge management in gate region: (a) GaN normally-off transistors using p-type doped GaN or AlGaN layer underneath the gate metal. (b) Incorporation of negative charges by F$^-$ implantation into the gate trench prior to gate metal deposition. (c) Energy band diagram of a p-GaN-type normally-off transistor comparing AlGaN buffer versus an unintentionally doped (UID) GaN buffer. (From Hilt et al., "Normally-off AlGaN/GaN HFET with p-type GaN gate and AlGaN buffer," Proceedings of the 22nd International Symposium on Power Semiconductor Devices (ISPSD), Hiroshima, Japan, 347–350, 2010 [65].)

(10–30 nm compared to a few nanometers only). This means that the intrinsic fields in the vicinity of the gate are reduced, which could be beneficial for reliability [36]. However, according to the author's knowledge, statistical reliability data on such devices are not available to date.

### 1.4.3 INVERSION TYPE OF NORMALLY OFF TRANSISTORS

This approach makes use of carrier inversion as the gate is biased in the positive direction. The AlGaN barrier layer is completely removed in the gate region. The gate is placed directly on top of the GaN channel layer and separated by a suitable gate insulator ($SiO_2$, $Si_3N_4$, and $Al_2O_3$ in most cases); see Figure 1.14d and the work by Kambayashi et al. [66]. Therefore, if the gate is forward biased an inversion layer is built under the isolator leading to a current flow underneath (Figure 1.16). This current is determined by the bulk properties of the channel; no 2DEG is formed in this region. Reliability issues are potentially related to the stability of the gate oxide and to the high-field regions in the channel region. However, as the AlGaN barrier layer is completely removed in gate vicinity, all degradation effects associated with a defect creation in the strained AlGaN barrier layer may be reduced in this case.

### 1.4.4 CASCODE TECHNOLOGY

Normally-off behavior can also be obtained by connecting a standard normally-off Si metal–oxide–semiconductor field-effect transistor (MOSFET) optimized for low-voltage operation and a high-voltage GaN normally-on transistor in cascode configuration. As depicted in Figure 1.14e, at off state the normally-on GaN transistor takes over the high-voltage part. Consequently, the standard normally-off Si transistor needs to be designed such that its on-state voltage drop is negligible. In an ideal case, both on-state and off-state properties should be determined by the GaN device, whereas the normally-off property is due to the Si device. If the Si-MOSFET is fully switched on, the source of the normally-on GaN transistor is basically grounded, which means that it is turned on at a gate voltage of about 0 V. This gate voltage level turns it on to a certain degree only; however, it is never biased fully on state (which would be the case at a gate bias of ~+2 V). This is certainly a disadvantage of the cascode approach, which could of course be compensated by the selection of a larger area normally-on GaN device. From an outside view, the packaged cascode configuration looks like a high-voltage normally-off

Inversion-type GaN MOSFET

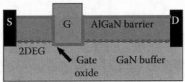

**FIGURE 1.16** Schematic cross section of an inversion-type lateral normally-off transistor. MOSFET, metal–oxide–semiconductor field-effect transistor.

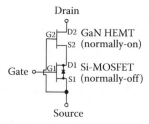

**FIGURE 1.17** Cascode-type hybrid integration of GaN normally-on and Si normally-off devices.

transistor. It combines the advantages of a large gate threshold voltage and gate voltage swing (feature of the Si normally-off transistor), a high breakdown voltage, and a comparably low on-state resistance as a consequence of the normally-on GaN HFET (Figure 1.17). The disadvantage is that the GaN HFET cannot be driven to full on-state condition, meaning that it has to be designed larger as necessary for a given current load. Additionally, the input capacitance is determined by the Si-MOSFET and therefore comparably high. If GaN transistors and Si-MOSFETs nominally having the same on-state resistance are compared to each other, the gate charge $Q_g$ of the Si-MOSFET device exceeds the $Q_g$ of the GaN device by roughly a factor of five at least [67]. However, as the on-state resistance of the Si-MOSFET needs to be small compared to the GaN transistor the situation may be even worse. This places more demands on the driver circuit and may slow down switching speed considerably. Nevertheless, the cascode design is successfully pursued by several GaN device manufacturers. As the dynamic switching properties of the cascode approach is quite complex and may affect conversion efficiency, this feature has been investigated thoroughly by several authors taking into account different circuit architectures for energy conversion [3,78,79,80].

## 1.5 VERTICAL GaN DEVICES

To date, most of the fabricated GaN transistors rely on lateral construction principles. Thus, they are making use of the 2DEG that is generated at the GaN/AlGaN heterostructure interface. The mobility and saturation velocity in this region are very high so that high current densities can be obtained. As many electronic applications demand ever-increasing operation voltages, the gate to drain distance of lateral devices has to expand accordingly. This creates additional losses and increases chip area. The benefit of 2DEG conductivity becomes increasingly less important as the operation voltage exceeds certain limits. Further, it will become more and more difficult to technologically handle large voltage differences on the chip surface. True vertical devices can overcome this problem, even if their device operation principle does not rely on 2DEG conductivity. In a similar manner as for many Si-based power devices, the operation voltage drops across a well-designed epitaxial layer at off state. At on state, the current directly flows through the device from the top (source contact) to the bottom (drain electrode). Current flow is controlled by a gate electrode on the wafer top. In the past, true vertical devices were hardly possible using GaN technology, since they require a low ohmic current path vertically through the wafer at on-state conditions. This is only possible if the devices can be realized on substrates that do not jeopardize vertical current flow by potential barriers, as is always the case for GaN devices grown on foreign substrates such as Si or SiC. Therefore, freestanding GaN substrates are a prerequisite for the efficient realization of vertical device concepts. As the technology of these substrates has matured in the last few years, true GaN vertical transistors are becoming feasible now.

Vertical topology GaN-based devices outperform lateral devices especially if high voltage operation and low inductance are targeted. First, the whole chip is taking care of high voltage potential separation; thus, destructive surface arcing is avoided. Second, due to the vertical architecture packaging becomes easier. Less bond wires may be required which would considerable reduce parasitic inductances of the packaged device. In many cases, chip area may be reduced too [3]. An additional benefit of any vertical device is the mitigation of DC-RF dispersion related to surface states [3,7], which would enable fast switching properties.

Several approaches for vertical GaN-based transistors, for example, current aperture vertical electron transistors (CAVETs) [68–71], vertical heterojunction field-effect transistors (VHFETs) [72], and gate trench MOSFETs [74–77], were presented during the past few years.

In most of the vertical GaN field-effect transistor (FET) concepts, the breakdown voltage is mainly determined by the thickness of the $n^-$-drift region and the quality of the material. It is therefore obvious that significant performance improvement will only be possible if the breakdown voltage versus on-state resistance trade-off can be handled. This requires high-quality substrate material.

### 1.5.1  GATE TRENCH METAL–OXIDE–SEMICONDUCTOR FIELD-EFFECT TRANSISTOR

The advantage of the gate trench MOSFET approach is that enhancement-mode operation with high threshold voltage levels can be obtained, which is an important issue for fail-safe operation of power electronic devices. A representative gate trench MOSFET structure grown on n-GaN substrate as proposed by Otake et al. [74] is shown in Figure 1.18. It consists of an $n^+$-GaN layer for drain contact, $n^-$GaN layer as drift region, Mg-doped p-GaN for the channel, and Si-doped $n^+$-GaN for source contact. The realized device shows a maximum drain current of 33.2 mA for $V_{DS} = 1$ V and $V_{GS} = 12$ V, and a corresponding on-state resistance of 9.3 m$\Omega$·cm$^2$. A threshold voltage of +3.7 V was measured, and a channel mobility of 131 cm$^2$/V·s was estimated. Off-leakage currents at $V_{GS} = 0$ V were on the order of 1 nA at $V_{DS}$ <5 V. The device breakdown voltage is determined by the p-GaN layer and was estimated to be 242 V. Kodama et al. [75] reported a GaN-based vertical transistor with a threshold voltage of +10 V and a device breakdown voltage of 180 V with a comparable epitaxial design but using an improved trench etching technique to reduce surface roughness in the trench. Also, a high voltage capability of 1600 V for GaN-trench MOSFETs was demonstrated by Oka et al. [77], reaching a threshold voltage of +7 V and an estimated specific on-state resistance of 12.1 m$\Omega$·cm$^2$. Ways to improve the on-resistance are decreasing the cell pitch and shrinking the cross section of the source electrodes, as discussed by Chowdhury et al. [70] and Oka

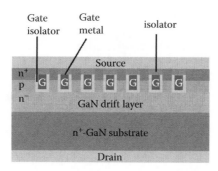

**FIGURE 1.18**   Trench gate inversion type of transistors similar to the version presented by Otake et al. [74]. On a positive bias to the gate, an electron inversion layer is formed at the tilted interface between the gate oxide and the p-GaN epitaxial layer, thus providing a conductive path through the device from top to bottom.

et al. [77]. This reduces chip area and results in improved current densities. Also, improvement of the substrate resistance including the drain contact resistance gives a possibility to reduce the on-resistance further. Published results show the feasibility of GaN trench gate MOSFETs for power electronic applications. Especially, the enhancement-mode operation is a big advantage compared to the normally-on operation of lateral AlGaN/GaN HEMTs. Therefore, further design and epitaxy improvement could improve the devices, making them competitive to vertical silicon and SiC devices and providing an interesting alternative to later AlGaN/GaN HEMTs.

### 1.5.2   Current Aperture Vertical Electron Transistor

An alternative approach to vertical gate trench GaN MOSFETs is the CAVET [68–71]. This concept combines the high mobility and saturation velocity of 2DEG with high-voltage operation ensured through a vertical drift region.

As shown in Figure 1.19, this device consists of an n+-GaN drain layer, a 3-μm-thick n-GaN drift layer, a current blocking layer (CBL) defining the aperture for the drain current, and the GaN/AlGaN heterostructure similar to a standard lateral GaN HEMT. To limit the vertical current flow to the area below the gate, a p-GaN current blocking layer is introduced. To realize aperture, it is necessary to etch back the p-GaN blocking current layer below and to regrow the GaN/AlGaN heterostructure afterward. The regrowth of the structured CBL leads to planarization problems and may impede further processing. Alternatively, Mg-implanted GaN can be used [69], having the advantage that regrowth of the AlGaN/GaN heterostructure can be continued on a planar surface and further standard processing can be applied.

Several challenges are to be overcome for this concept. The growth of the active HEMT layer without parasitic incorporation of Mg is of utmost importance. One method to avoid this effect is to perform the regrowth by molecular beam epitaxy (MBE) at reduced temperatures [70]. Also, insulated apertures of p-type may provide sufficient isolation of the active device from the drift region.

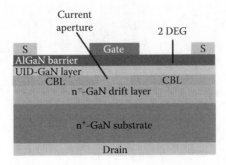

**FIGURE 1.19**   Schematic cross section of a current aperture vertical electron transistor similar to the version presented by Chowdhury et al. [70]. In this case, the current blocking layer consists of an Mg-doped p-GaN epitaxial layer. UID, unintentionally doped.

Depletion- and enhancement-mode devices with a threshold voltage of 0.6 V were demonstrated [69]. Up to now, CAVETs with on-resistance of 2.2 m$\Omega\cdot$cm$^2$ and a breakdown voltage of $-200$ V over a 3-$\mu$m n$^-$-GaN drift region have been shown. The maximum drain current obtained has been on the order of 3.8 kA/cm [70].

### 1.5.3  Vertical Heterojunction Field-Effect Transistors

Figure 1.20 shows a schematic structure of a VHFET relying on the presence of an inclined 2DEG. The VHFET has a V-grooved structure with an inclined 2DEG layer as channel, leading to a low on-resistance. In the on-state regime, electrons flow from the source on the top through the n$^+$-layer and the inclined 2DEG channel under the gate into the n$^-$-GaN drift layer. In off state, the p GaN layer is used as a blocking layer for the electrons. Additionally, the p$^-$-GaN layer serves as a back potential helping to realize normally-off operation. In addition, the reverse-biased p$^-$-GaN/n$^-$-GaN drift layer junction ensures high breakdown voltages.

To realize the VHFET device, an epitaxial structure (n$^-$-GaN, p$^-$-GaN, and n$^+$-GaN) is grown on c-plane freestanding GaN substrate by metalorganic chemical vapor deposition (MOCVD). Subsequently, a mesa is defined down to the n$^-$-GaN layer by dry etching techniques, allowing for graded walls with a glancing angle of 16°. To form the 2DEG in the desired concentration, undoped GaN and AlGaN are regrown in the appropriate thickness onto the sloped GaN layers. Accordingly, both normally-on and normally-off devices can be achieved.

Okada et al. [72] fabricated a VHFET with threshold voltages varying between $-3.2$ V and 0.3 V, depending on the AlGaN barrier thickness. Normally-off operation has been accomplished by using a 10-nm-thick Al$_{0.2}$Ga$_{0.8}$N barrier. A specific on-resistance of 7.6 m$\Omega\cdot$cm$^2$ and a maximum breakdown voltage of 672 V were achieved for a VHFET with a 28-nm-thick Al$_{0.2}$Ga$_{0.8}$N barrier. A drain current of about 270 A/cm$^2$ has been obtained. Further improvement is needed to achieve a more positive threshold voltage and safe enhancement-mode operation. Nevertheless, VHFETs lead to the highest ratio of breakdown voltage versus normalized on-state resistance ever reported for GaN-based vertical transistors.

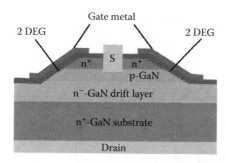

**FIGURE 1.20** Schematic cross section of a vertical heterojunction field-effect transistor with a tilted two-dimensional electron gas region, according to Okada et al. (From Okada et al., *Appl. Phys. Exp.*, 3, 054201, 2010 [72].)

## 1.6 STATE-OF-THE-ART GaN POWER DEVICES

### 1.6.1 GaN EPITAXY ON SI AND SIC SUBSTRATES

The epitaxial structure of GaN power switching transistors has to take care of the requirements toward high voltage capability and switching performance. Principally, this results in an epitaxial structure according to Figure 1.7, comprising a suitable back-barrier buffer design that focusses electrons to the channel region even at high drain bias. Additionally, the design has to guarantee suitable vertical breakdown and low parasitic dynamic effects.

The envisaged performance of GaN power switching devices will trigger a high volume market in future, which can, for example, be addressed by GaN-on-Si devices fabricated on standard Si process lines. Scaling of wafer diameter to 200 mm or more then leads to highly productive device processes. Therefore, considerable efforts have been undertaken to develop suitable GaN-on-Si epitaxial growth techniques on large-diameter Si wafers with (111) orientation. As already discussed in Section 1.1, GaN epitaxy on Si substrates faces the challenges of a comparably large lattice mismatch of 17% and a different thermal expansion coefficient between GaN and AlGaN epitaxial layers on the Si substrates (Table 1.2). Therefore, GaN-on-Si epitaxy development toward high power switching devices has to address the following technological issues:

- Provide suitable start of epitaxial growth on fairly lattice-mismatched material.
- Harness the mechanical strain issue caused by lattice mismatch and different thermal expansion coefficient.
- Provide mechanically flat wafers suitable for subsequent device processing.
- Optimize trade-off between technological concepts to optimize the aforementioned topics and vertical thermal heat conductivity.

Figure 1.21 shows an epitaxial concept that has been successfully applied for GaN-on-Si epitaxial structures [1]. Starting from the Si (111) substrate, it consists of

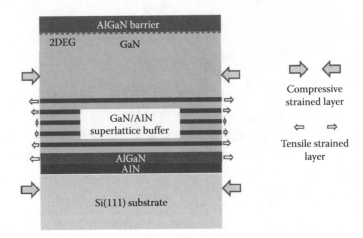

**FIGURE 1.21**　Principal design of a GaN-on-Si epitaxial structure for GaN power electronic devices on Si (111) substrates according to Ueda [1]. Due to the lattice mismatch between Si and GaN (17%) and the different coefficient of thermal expansion, mechanical strain builds up between the GaN layer and the Si substrate. The strain is compensated by a GaN/AlN superlattice buffer with many periods, with the AlN layers providing the tensile counterpart.

an AlN/AlGaN barrier layer providing a nucleation for subsequent epitaxial growth and preventing Ga diffusion into the Si crystal during growth. On top of these initial layers, a so-called superlattice buffer layer consisting of several alternating thin GaN and AlN layers follows. The superlattice buffer provides an adaptation of mechanical strain in such a way that after the epitaxial growth of the complete structure the wafer warp stabilizes within the specifications of subsequent processing steps. In this connection, projection lithography using wafer steppers requires pretty flat wafers, with a maximum flatness of a few tens of micrometers. Due to the thin AlN layers, the superlattice buffer is tensile strained and compensates the compressive strain of the GaN structures on top and the compressively strained initial portions of the Si substrate. Depending on design and geometrical placement of the respective AlN layers, the final mechanical properties of the wafer can be tuned. On top of the superlattice structure, a buffer layer ensuring vertical and lateral device breakdown properties as well as dispersion issues is grown followed by the active device layers. The thicker the GaN buffer layer (required for high vertical breakdown voltage levels), the higher the compressive strain that is provided by this layer. This would finally result in a convex wafer shape after epitaxy. The strain can of course be compensated to a certain extent by introducing the GaN/AlN superlattice; however, if a very high vertical breakdown voltage is required (thick buffer layer), then thicker Si (111) substrates than standard have to be selected and thinned to the required thickness after device epitaxy.

The thermal conductivity of the additional layers to adjust GaN to Si (111) substrate is still in discussion. It is known from Raman measurements that each of these layers can provide an additional thermal barrier, the so-called thermal boundary resistance, which impedes vertical heat flow from the heat source on top of the device to the Si substrate [73]. For devices operating in the saturation range of the I–V characteristic (e.g., most microwave amplifiers) (Figure 1.2), this may be a real issue. For power switching devices, the problem is not that severe as they are operated in the linear part of the I–V characteristic at on state where low power is dissipated. However, any self-heating leads to a reduction of switching efficiency and compromises device reliability.

A direct comparison of nominally identical GaN power devices fabricated on Si and GaN-on-si-SiC wafers shows very large differences in channel temperature if pulsed at a certain power density. Figure 1.22 compares the thermal properties at power levels of 3 and 5 W/mm. At 5 W/mm and after applying 200-µs pulses, excessive self-heating of the device fabricated on Si wafer has been observed. The temperature increase of the GaN-on-Si device has been five times compared to GaN-on-SiC device. For real power switching applications, the absolute difference might not be that drastic as the particular test device has been biased close to the knee region of the output I–V characteristic to highlight the differences.

### 1.6.2 Processing

Figure 1.23 provides a schematic cross section of a GaN high-voltage switching device processed at Ferdinand-Braun-Institut, Berlin, Germany [20]. Most of the concepts described in the Sections 1.2 through 1.4 have been implemented in the principal process design. The normally-off technology relies on the p-GaN gate approach [37]. Besides the gate and the drain metallic contacts, the process consists of two Au-based interconnect metallizations. Finally, benzocyclobutene (BCB) layers passivate the whole device. Source and drain fingers are additionally isolation implanted in their center to avoid spurious vertical leakage and thus to increase device yield. The process is applicable on Si and SiC substrates.

For GaN-on-Si devices processed on a Si process line, additional restrictions are applicable. To avoid cross-contamination with Si processes, Au-based metallizations are not applicable [57,83]. Similarly, Fe-containing epitaxial buffer layers are not possible.

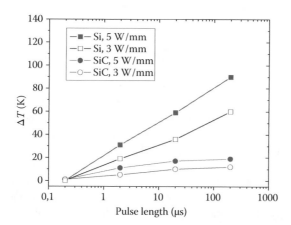

**FIGURE 1.22** Comparison of pulsed self-heating of GaN normally-off devices fabricated on Si and semi-insulating SiC substrate, having dissipated power densities of 3 and 5 W/mm, respectively [84]. The device channel has been heated up by pulses with a well-defined pulse length up to 200 µs. Device gate width is 250 µm, and the channel temperature has been determined by electrical measurements.

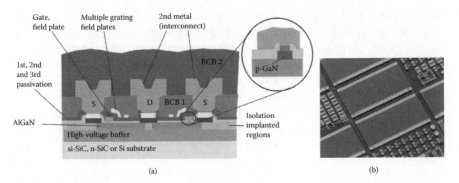

(a)                                                      (b)

**FIGURE 1.23** High-voltage normally-off switching GaN technology developed at Ferdinand-Braun-Institut [20]: (a) Schematic cross section; the inset shows details related to the p-type GaN gate technology. (b) Photograph of wafer section showing 75-mΩ transistors with 214-mm gate width. (From Würfl et al., "Techniques towards GaN power transistors with improved high voltage dynamic switching properties," IEEE International Electron Devices Meeting (IEDM), 6.1.1–6.1.4, 2013 [20].)

This means that special alternative technologies had to be developed. An overview on an Au-free GaN-on-Si technology based on GaN MISFETs developed by IMEC, Belgium, is provided by Van Hove et al. [57]; similarly, Kikkawa et al. [83] summarize details on the Fujitsu (United States, Japan) technology (Japan) also working on GaN MISFETs. A detailed report on Panasonic's technology based on p-AlGaN normally-off power devices has been published in the works by Ishida et al. [2] and Ueda [1]. More information on Au-free ohmic contacts to GaN is available in the literature [85,86].

The layout of GaN power transistors differs quite considerably. The most prominent design relies on multiple finger structures; however, especially for very-large-area devices fractal designs providing a better utilization of chip technology have also been investigated recently [87].

### 1.6.3 Actual Device Results

Very impressive device results have been published worldwide. They all show that GaN devices can meet the expectations and former predictions in terms of performance. Also, dynamic deficiencies—an initial problem of high-voltage devices—have been solved sufficiently [8,20]. Figure 1.24 summarizes the results on p-GaN gate normally-off power transistors fabricated at Ferdinand-Braun-Institut. The devices have been designed for 75-mΩ on-state resistance and 600-V operation. The devices have a total gate width of 214 mm and are based on a multi-finger transistor layout (Figure 1.23b). They demonstrated a gate threshold voltage between 1 and 1.5 V and rather low leakage below 60 μA at a drain bias of 600 V.

Similar to the properties demonstrated by Ishida et al. [2] on Panasonic devices, the Ferdinand-Braun-Institut devices also showed remarkable reverse conductivity (Figure 1.24b). This is quite an advantage, as at a gate voltage of 0 V the device can be basically operated in diode-like reverse characteristics. In this case, the threshold voltage for reverse conductivity is the same as the threshold voltage for

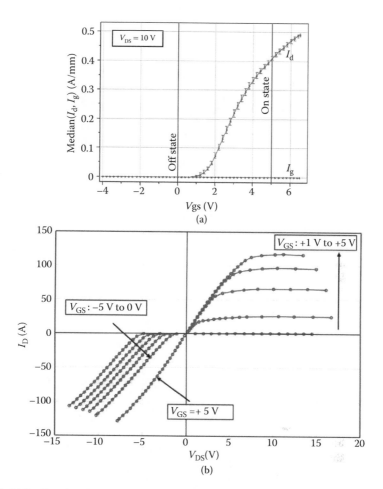

**FIGURE 1.24** Results obtained from a normally-off p-GaN-gated GaN power transistor optimized for 65-mΩ on-state resistance. The total gate width of the device is 214 mm. (a) Transfer characteristic demonstrating a turn-on voltage of +1 V. (b) Reverse conductivity: Output characteristic is measured in the first quadrant and the third quadrant.

turning on the device in forward mode. At a gate bias of +5 V, the linear part of the I–V characteristics can be practically mirrored. Thus the device provides a good reverse conductivity at this gate drive condition. This would further reduce inverter losses; however, on the expense of much more complicated gate drive regimes as also indicated by Ishida et al. [2].

Figure 1.25 depicts the dynamic properties of these devices. They show nearly ideal switching properties at 600 V with an increase of $R_{on\_dyn}$ of only 25% (compare with Figure 1.10b to evaluate the difference). According to Figure 1.25b, the tested device had a static $R_{on}$ of 56 mΩ, which increase to only 70 mΩ at 600-V operation. These devices already demonstrate the possibilities of GaN power switching devices. For example, Panasonic has systematically tested their normally-off devices in various inverter concepts where significant improvements of conversion losses could be obtained. For a

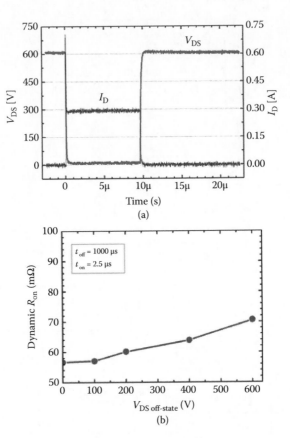

**FIGURE 1.25**   Dynamic properties: (a) 600-V switching transients of a test device design for 5-Ω on-state resistance (switches 300 mA) drain-source voltage $U_{DS}$ and drain current $I_D$ versus Time. (b) Dynamic on-state resistance of the 214-mm device according to Figure 1.24b versus drain voltage. The values have been taken after an off-state time of 1000 μs and measured 2.5 μs after the switching event.

GaN-powered 1.5-kW inverter, they have demonstrated an efficiency of 99.3% being practically constant with the power level. This is a significant improvement against conventional Insulated Gate Bipolar Transistor (IGBT), which have more losses especially at low power levels (about 60% more at 500-W operation compared to GaN HEMTs) [1,2].

Figure 1.26 benchmarks the recent results on GaN power devices against Si and SiC devices by means of the product of on-state resistance, $R_{on}$, and gate charge, $Q_g$. As already described in Section 1.1, $R_{on} \times Q_g$ can be interpreted as a measure of potential switching efficiency and provides a fair comparison of different transistor sizes (e.g., with different gate widths) as this quantity cancels out for first-order considerations. Remark: A larger area transistor of course has a smaller $R_{on}$; however, $Q_g$ also scales with increasing gate width, and the product should be independent of gate width. The collection of data shows that the newest generations of GaN switching devices definitely outperform their Si competitors. It is now the time to convert prototype devices to highly mature industrial devices and to assess their reliability.

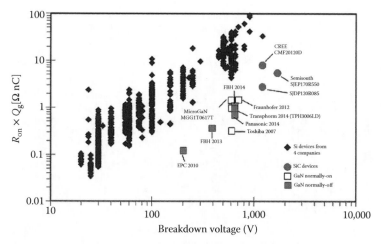

**FIGURE 1.26** Comparison of the product of specific on-state resistance, $R_{on}$, and gate charge, $Q_g$, versus breakdown voltage for different device families. As $R_{on}$ relates to the device losses during on state and the value of the gate charge $Q_g$ limits the maximum obtainable switching speed, the product can be considered as a measure of switching efficiency. Basic data were taken from the study by Reiner et al. [87], and additional recent data were added. (Data from Reiner et al., "Fractal structures for low-resistance large area AlGaN/GaN power transistors," Proceedings of ISPSD, 341–345, 2012 [87].)

### 1.6.4 PRELIMINARY RELIABILITY DATA

As GaN high power switching devices are maturing considerably, their degradation mechanisms need to be analyzed in terms of statistical evaluations and physical material analyses. This will then provide an understanding of the physical mechanisms leading to degradation and will allow for a lifetime extrapolation based on certain test conditions. In principle, as GaN power switching devices rely on similar materials and similar principles of operation compared to GaN microwave transistors, some findings that are valid for microwave devices can be transferred to power switching GaN devices. Figure 1.27 provides an overview of the basic degradation mechanisms taking place in GaN HEMTs. Classically, in microwave devices GaN HEMT degradation is often associated with an increase of gate leakage and a reduction of drain current. Defect creation at the drain side edge of the gate causes ever-increasing gate leakage, finally leading to percolation paths to the channel region. Zanoni et al. [88] provide an overview of general GaN device degradation aspects in their work.

However, in the case of GaN power switching devices there are many technological aspects that differ from microwave devices. They need to be considered in detail as they may be responsible for new degradation mechanisms that are not known from microwave devices such as the following:

- Vertical leakage effects that may create vertical percolation paths between the top surface of the device and the conductive substrate [89] or may be associated to noise fluctuations prior to actual breakdown effects [92].

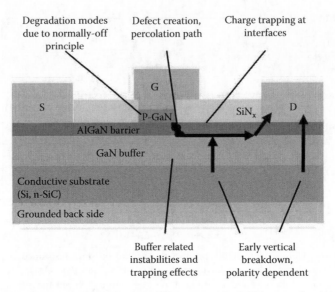

**FIGURE 1.27**  Schematic visualization of important degradation mechanisms reported from high-voltage GaN devices.

- Degradation mechanisms related to new normally-off gate modules. For example, Meneghini et al. [90] have identified trapping effects in p-AlGaN-gated normally-off GaN devices leading to considerable knee walkout.
- Buffer technology: Many papers have been published on the influence of buffer technology on trapping and memory effects in GaN HFETs [22,51,52,81,91]. For high-voltage switching devices, these mechanisms gain importance and have to be avoided by specially adapted technological countermeasures.

An overview of the current state of understanding GaN high-voltage degradation effects has been presented elsewhere [93]. In general, understanding on the relevant physical mechanisms leading to device degradation is ever increasing thanks to the concentrated research done in many universities, research institutes, and companies worldwide. The first very promising statistical data on GaN high-voltage performance have been published by Whitman [94].

## ACKNOWLEDGMENTS

The author is very grateful to his colleagues at the Business Area GaN Electronics at Ferdinand-Braun-Institut, Leibniz Institut für Höchstfrequenztechnik, Berlin, Germany, for supporting this work. Special thanks are devoted to Dr. Oliver Hilt, Dr. Bahat-Treidel, Przemyslaw Kotara, and Rimma Zhytnytska for sharing the results of their work and for fruitful scientific discussions and to Prof. Sibylle Dieckerhoff and Jan Böcker from the Technical University, Berlin, for special measurements on GaN power devices.

# REFERENCES

1. T. Ueda, "Recent advances and future prospects on GaN-based power devices," International Power Electronics Conference (IPEC-Hiroshima 2014–ECCE-ASIA), 2014, pp. 18–21, 2075–2078.
2. M. Ishida, T. Ueda, T. Tanaka, D. Ueda, "GaN on Si technologies for power switching devices," *IEEE T Electron Dev*, vol. 60, no. 10, 2013, pp. 3053–3059.
3. U. Mishra, "AlGaN/GaN transistors for power electronics," 2010 IEEE International Electron Devices Meeting (IEDM), 2010, pp. 312–315.
4. E.A. Jones, F. Wang, B. Ozpineci, "Application-based review of GaN HFETs," IEEE Workshop on Wide Bandgap Power Devices and Applications (WiPDA), 2014, pp. 24–29.
5. C. Abi Abboud, M. Chahine, C. Moussa, H.Y. Kanaan, E.A. Rachid, "Modern power switches: the gallium nitride (GaN) technology," IEEE 9th Conference on Industrial Electronics and Applications (ICIEA), 2014, pp. 2203–2208.
6. J. Millan, P. Godignon, X. Perpina, A. Perez-Tomas, J. Rebollo, "A survey of wide bandgap power semiconductor devices," *IEEE T Power Electr*, vol. 29, no. 5, 2014, pp. 2155–2163.
7. M. Kanechika, T. Uesugi, T. Kachi, "Advanced SiC and GaN power electronics for automotive systems," 2010 IEEE International Electron Devices Meeting (IEDM), 2010, pp. 13.5.1–13.5.4.
8. M. Kuzuhara, H. Tokuda, "Low-loss and high-voltage iii-nitride transistors for power switching applications," *IEEE T Electron Dev*, vol. PP, no. 99, 2015, pp. 1,1.
9. K. Nishikawa, "GaN for automotive applications," IEEE Bipolar/BiCMOS Circuits and Technology Meeting (BCTM), 2013, pp. 143–150.
10. W. Saito, I. Omura, T. Ogura, H. Ohashi, "Theoretical limit estimation of lateral wide band-gap semiconductor power-switching device," *Solid State Electronics*, vol. 48, no. 9, 2004, p. 1555.
11. M.A. Brierre, "The power electronics market and the status of gan based power devices," Compound Semiconductor Integrated Circuit Symposium (CSICS), Proceedings, 1, 2011.
12. N. Ikeda, R. Tamura, T. Kikkawa, H. Kambayashi, Y. Sato, T. Nomura et al., "Over 1.7 kV normally-off GaN hybrid MOS-HFETs with a lower on-resistance on a Si substrate," Proceedings of the 23rd International Symposium on Power Semiconductor Devices & IC's, 2011, pp. 284–287.
13. B.J. Baliga, "Semiconductors for High-Voltage, Vertical Channel Field-Effect Transistors," *J Appl Phys*, vol. 53, no. 3, 1982, pp. 1759–1764.
14. B.J. Baliga, "Power semiconductor device figure of merit for high-frequency applications," *IEEE Electr Device L*, vol. 10, 1989, pp. 455–457.
15. A. Johnson, "Physical limitations on frequency and power parameters of transistors," *RCA Rev*, vol. 26, 1965, pp. 163–177.
16. N. Zhang, "High voltage GaN HEMTs with low on-resistance for switching applications," PhD diss., University of California, Santa Barbara (UCSB), CA, 2002.
17. N. Kaminski, O. Hilt, "GaN device physics for electrical engineers," ECPE GaN and SiC User Forum, Proceedings, Birmingham, United Kingdom, 2011.
18. R.S. Pengelly, S.M. Wood, J.W. Milligan, S.T. Sheppard, W.L. Pribble, "A review of GaN on SiC high electron-mobility power transistors and MMICs," *IEEE T Microw Theory*, vol. 60, issue 2, part 2, 2012, pp. 1764–1783.
19. R. Chu, A. Corrion, M. Chen, R. Li, D. Wong, D. Zehnder et al., "1200 V normally-off GaN-on-Si field effect transistors with low dynamic on resistance," *IEEE Electr Device L*, vol. 32, no. 5, 2011, pp. 632–634.
20. J. Würfl, O. Hilt, E. Bahat-Treidel, R. Zhytnytska, P. Kotara, F. Brunner et al., "Techniques towards GaN power transistors with improved high voltage dynamic switching properties," IEEE International Electron Devices Meeting (IEDM), 2013, pp. 6.1.1–6.1.4.

21. E. Bahat-Treidel, "GaN based HEMTs for high voltage operation; design, technology and characterization," PhD, Technical University, Berlin, Germany, 2012.

22. M.J. Uren, K.J. Nash, R.S. Balmer, T. Martin, E. Morvan, N. Caillas et al., "Punch-through in short-channel AlGaN/GaN HFETs," *IEEE T Electron Dev*, vol. 53, no. 2, 2006, pp. 395–398.

23. E. Bahat-Treidel, O. Hilt, F. Brunner, J. Würfl, G. Tränkle, "Punch-through-voltage enhancement of AlGaN/GaN HEMTs using AlGaN double-heterojunction confinement," *IEEE T Electron Dev*, vol. 55, no. 12, 2010, pp. 3354–3359.

24. E. Zanoni, M. Meneghini, A. Chini, D. Marcon, G. Meneghesso, "AlGaN/GaN-based HEMTs failure physics and reliability: mechanisms affecting gate edge and schottky junction," *IEEE T Electron Dev*, vol. 60, no. 10, 2013, pp. 3119–3131.

25. X. Huili, Y. Dora, A. Chini, S. Heikman, S. Keller, U. Mishra, "High breakdown voltage AlGaN–GaN HEMTs achieved by multiple field plates", *IEEE Electr Device L*, vol. 25, no. 4, 2004, pp. 161–163.

26. Y. Dora, A. Chakraborty, L. McCarthy, S. Keller, S.P. DenBaars, U. Mishra, "High breakdown voltage achieved on AlGaN/GaN HEMTs with integrated slant field plates," *IEEE Electr Device L*, vol. 27, no. 9, 2006, p. 713.

27. E. Bahat-Treidel, O. Hilt, F. Brunner, V. Sidorov, J. Würfl, G. Tränkle, "AlGaN/GaN/AlGaN DH-HEMTs breakdown voltage enhancement using multiple grating field plates," *IEEE T Electron Dev*, vol. 57, no. 6, 2010, pp. 1208–1216.

28. Z. Li, R. Chu, D. Zehnder, S. Khalil, M. Chen, X Chen et al., "Improvement of the dynamic on-resistance characteristics of GaN-on-Si power transistors with a sloped field-plate," 72nd Annual Device Research Conference (DRC), 2014, pp. 257–258.

29. G. Xie, E. Xu, J. Lee, N. Hashemi, W.T. Ng, B. Zhang et al., "Breakdown voltage enhancement for power AlGaN/GaN HEMTs with air-bridge field plate," 24th International Symposium on Power Semiconductor Devices and ICs (ISPSD), 2012, pp. 337–340.

30. A. Petru, "Breakdown voltage enhancement in lateral AlGaN/GaN heterojunction FETs with multiple field plates," 10th IEEE International Conference on Solid-State and Integrated Circuit Technology (ICSICT), 2010, pp. 1344, 1346.

31. G.I. Ng, H. Zhou, S. Arulkumaran, Y.K.T. Maung, "Reduced surface leakage current and trapping effects in AlGaN/GaN high electron mobility transistors on silicon with $SiN/Al_2O_3$ passivation," *Appl Phys Lett*, vol. 98, no. 11, 2011, p. 113506.

32. S.A. Chevtchenko, P. Kurpas, N. Chaturvedi, R. Lossy, J. Würfl, "Investigation and reduction of leakage current associated with gate encapsulation by $SiN_x$ in AlGaN/GaN HFETs," International Conference on Compound Semiconductor Manufacturing Technology (CS ManTech 2011), Proceedings, 2011, pp. 237–240.

33. P. Ivo, A. Glowacki, E. Bahat-Treidel, R. Lossy, J. Würfl, C. Boit et al., "Degradation mechanism of GaN HEMTs in dependence on buffer quality and gate technology," *Microelectron Reliab*, vol. 51, no. 2, 2011, pp. 217–223.

34. D. Visalli, M. Van Hove, J. Derluyn, P. Srivastava, D. Marcon, J. Das et al., "Limitations of field plate effect due to the silicon substrate in AlGaN/GaN/AlGaN DHFETs," *IEEE T Electron Dev*, vol. 57, no. 2, 2010, pp. 3333–3339.

35. J. Würfl, E. Bahat-Treidel, F. Brunner, M. Cho, O. Hilt, A. Knauer et al., "Device breakdown and dynamic effects in GaN power switching devices: dependencies on material properties and device design," *ECS Trans*, vol. 41, no. 8, 2012, pp. 127–138.

36. J. Würfl, E. Bahat-Treidel, F. Brunner, E. Cho, O. Hilt, P. Ivo et al., "Reliability issues of GaN based high voltage power devices," *Microelectron Reliab*, vol. 51, no. 9–11, 2011, pp. 1710–1716.

37. O. Hilt, F. Brunner, E. Cho, A. Knauer, E. Bahat-Treidel, J. Würfl, "Normally-off high-voltage p-GaN gate GaN HFET with carbon-doped buffer," Proceedings of Symposium on Power Semiconductor Devices and ICs (ISPSD), 2011, pp. 239–243.

38. I.B. Rowena, S.L. Selvaraj, T. Egawa, "Buffer thickness contribution to suppress vertical leakage current with high breakdown field (2.3 MV/cm) for GaN on Si," *IEEE Electr Device L*, vol. 32, no. 11, 2011, pp. 1534–1536.

39. P. Kotara, R. Zhytnytska, O. Hilt, E. Cho, F. Brunner, A. Thies et al., "Vertical blocking voltage improvement of GaN HEMT structures on n-SiC by pre-epitaxial substrate implantation," *ECS J Solid State Sci Technol*, vol. 2, no. 8, 2013, pp. N3064–N3067.

40. O. Hilt, P. Kotara, F. Brunner, A. Knauer, R. Zhytnytska, J. Würfl, "Improved vertical isolation for normally-off high voltage GaN-HFETs on n-SiC substrates," *IEEE T Electron Dev*, vol. 60, no. 10, 2013, pp. 3084–3090.

41. J. Würfl, O. Hilt, E. Bahat-Treidel, R. Zhytnytska, P. Kotara, O. Krüger et al., "Breakdown and dynamic effects in GaN power switching devices," *Phys Stat Sol (c)*, vol. 10, no. 11, 2013, pp. 1393–1396.

42. B. Lu, T. Palacios, D. Risbud, S. Bahl, D.I. Anderson, "Extraction of dynamic on-resistance in GaN transistors: under soft- and hard-switching conditions," IEEE Compound Semiconductor Integrated Circuit Symposium (CSICS) 2011, conference proceedings, 2011, pp. 1–4.

43. W. Saito, T. Nitta, Y. Kakiuchi, Y. Saito, K. Tsuda, I. Omura et al., "Suppression of dynamic on-resistance increase and gate charge measurements in high-voltage GaN-HEMTs with optimized field-plate structure," *IEEE T Electron Dev*, vol. 54, no. 8, 2007, pp. 1825–1830.

44. W. Saito, "Reliability of GaN-HEMTs for high-voltage switching applications," Proceedings IEEE International Reliability Physics Symposium, 2011, pp. 417–421.

45. P. Moens, C. Liu, A. Banerjee, P. Vanmeerbeek, P. Coppens, H. Ziad et al., "An industrial process for 650 V rated GaN-on-Si power devices using in-situ SiN as a gate dielectric," Proceedings of the 26th International Symposium on Power Semiconductor Devices & IC's (ISPSD), 2014, pp. 374–377.

46. R. Chu, A. Corrion, M. Chen, R. Li, D. Wong, D. Zehnder et al., "1200 V normally-off GaN-on-Si field-effect transistor with low dynamic on-resistance," *IEEE Electr Device L*, vol. 32, no. 5, 2011, pp. 632–634.

47. O. Hilt, E. Bahat-Treidel, E. Cho, S. Singwald, J. Würfl, "Impact of buffer composition on the dynamic on-state resistance of high-voltage AlGaN/GaN HFETs," Proceedings of Symposium on Power Semiconductor Devices and ICs (ISPSD), 2012, pp. 345–348.

48. E. Bahat-Treidel, F. Brunner, O. Hilt, M. Cho, J. Würfl, G. Tränkle, "AlGaN/GaN/GaN:C back-barrier HFETs with breakdown voltage of over 1 kV and low RON × A," *IEEE T Electron Dev*, vol. 57, no. 11, 2010, pp. 3050–3057.

49. J. Wuerfl, E. Bahat-Treidel, F. Brunner, M. Cho, O. Hilt, A. Knauer et al., "Device breakdown and dynamic effects in GaN power switching devices: dependencies on material properties and device design," *ECS Trans*, vol. 50, no. 3, 2012, pp. 211–222.

50. O. Hilt, E. Bahat-Treidel, F. Brunner, A. Knauer, R. Zhytnytska, P. Kotara et al., "Normally-off GaN transistors for power switching applications," *ECS Trans*, vol. 58, no. 4, 2013, pp. 145–154.

51. M.J. Uren, J. Möreke, M. Kuball, "Buffer design to minimize current collapse in GaN/AlGaN HFETs," *IEEE T Electron Dev*, vol. 59, no. 12, 2013, pp. 3327–3333.

52. M.J. Uren, M. Silvestri, M. Casar, G.A.M. Hurkx, J.A. Croon, J. Sonsky et al., "Intentionally carbon-doped AlGaN/GaN HEMTs: necessity for vertical leakage paths," *IEEE Electr Device L*, vol. 35, no. 3, 2014, pp. 327–329.

53. W. Saito, "Reliability of GaN-HEMTs for high-voltage switching applications," Proceedings of the IEEE International Reliability Physics Symposium, 2011, pp. 417–421.

54. W. Saito, Y. Kakiuchi, T. Nitta, Y. Saito, T. Noda, H. Fujimoto, A. Yoshioka, T. Ohno, M. Yamaguchi: "Field-Plate Structure Dependence of Current Collapse Phenomena in High-Voltage GaN-HEMTs", *IEEE Electron Device Lett*, vol. 31, no. 7, 2010, pp. 659–661.

55. H. Hahn, F. Benkhelifa, O. Ambacher, F. Brunner, A. Noculak, H. Kalisch et al., "Threshold voltage engineering in GaN-based HFETs: a systematic study with the threshold voltage reaching more than 2 V," *IEEE T Electron Dev*, vol. PP, no. 99, 2015, pp. 1,1.

56. W. Saito, Y. Takada, M. Kuraguchi, K. Tsuda, I. Omura, "Recessed-gate structure approach towards normally off high-voltage AlGaN/GaN HEMT for power electronics applications," *IEEE T Electron Dev*, vol. 53, no. 2, 2006, pp. 356–362.

57. M. Van Hove, K. Xuanwu, S. Stoffels, D. Wellekens, N. Ronchi, R. Venegas et al., "Fabrication and performance of Au-free AlGaN/GaN-on-silicon power devices with $Al_2O_3$ and $Si_3N_4/Al_2O_3$ gate dielectrics," *IEEE T Electron Dev*, vol. 60, no. 10, 2013, pp. 3071–3078.

58. U. Singisetti, M.H. Wong, S. Dasgupta, Nidhi, B. Swenson, B.J. Thibeault et al., "Enhancement-mode n-polar GaN MISFETs with self-aligned source/drain regrowth," *IEEE Electr Device L*, vol. 32, no. 2, 2011, 137.

59. Y. Uemoto, M. Hikita, H. Ueno, H. Matsuo, H. Ishida, M. Yanagihara et al., "Gate injection transistor (GIT)—a normally-off AlGaN/GaN power transistor using conductivity modulation," *IEEE T Electron Dev*, vol. 54, no. 12, 2007, 3393–3399.

60. S. Nagai, Y. Kawai, O. Tabata, H. Fujiwara, Y. Yamada, N. Otsuka et al., "A drive-by-microwave isolated gate driver with a high-speed voltage monitoring," IEEE 26th International Symposium on Power Semiconductor Devices & IC's (ISPSD), 2014, pp. 434, 437.

61. J. Kim, S.-K. Hwang, H. Hwang, H. Choi, S. Chong, H.-S. Choi et al., "High threshold voltage p-GaN gate power devices on 200 mm Si," 25th International Symposium on Power Semiconductor Devices and ICs (ISPSD), 2013, pp. 315–318.

62. A. Lidow, "GaN as a displacement technology for silicon in power management," IEEE Energy Conversion Congress and Exposition (ECCE), 2011, pp. 1–6.

63. N. Kaneko, N. Kaneko, O. Machida, M. Yanagihara, S. Iwakami, R. Baba, H. Goto, A. Iwabuchi et al., "Normally-off AlGaN/GaN HFETs using $NiO_x$ gate with recess," Proceedings of Symposium on Power Semiconductor Devices and ICs (ISPSD), 2009, pp. 25–28.

64. K.J. Chen, L. Yuan, M.J. Wang, H. Chen, S. Huang, Q. Zhou et al., "Physics of fluorine plasma ion implantation for GaN normally-off HEMT technology," IEEE International Electron Devices Meeting (IEDM), 2011, pp. 19.4.1–19.4.4.

65. O. Hilt, A. Knauer, F. Brunner, E. Bahat-Treidel, J. Würfl, "Normally-off AlGaN/GaN HFET with p-type GaN gate and AlGaN buffer," Proceedings of the 22nd International Symposium on Power Semiconductor Devices (ISPSD), Hiroshima, Japan, 2010, pp. 347–350.

66. H. Kambayashi, Y. Satoh, Y. Niiyama, T. Kokawa, M. Iwami, T. Nomura et al., "Enhancement mode GaN hybrid MOS-HEMT on Si substrates with over 70 A operation," ISPSD, 2009, pp. 21–24.

67. R. Reiner, P. Waltereit, F. Benkhelifa, S. Muller, M. Wespel, R. Quay et al., "Benchmarking of large-area GaN-on-Si HFET power devices for highly-efficient, fast-switching converter applications," IEEE Compound Semiconductor Integrated Circuit Symposium (CSICS), 2013, pp. 1–4.

68. M. Kanechika, M. Sugimoto, N. Soejima, H. Ueda, O. Ishiguro, M. Kodama et al., "A vertical insulated gate AlGaN/GaN heterojunction field-effect transistor," *Jpn J Appl Phys*, vol. 46, 2007, p. L503.

69. S. Chowdhury, B.L. Swenson, U.K. Mishra, "Enhancement and depletion mode AlGaN/GaN CAVET with Mg-ion-implanted GaN as current blocking layer," *IEEE Electr Device L*, vol. 29, 2008, pp. 543–545.

70. S. Chowdhury, M.H. Wong, B.L. Swenson, U.K. Mishra, "CAVET on bulk GaN substrates achieved with MBE-regrown AlGaN/GaN layers to suppress dispersion," *IEEE Electr Device L*, vol. 33, 2012, pp. 41–43.

71. S. Chowdhury, U.K. Mishra, "Lateral and vertical transistors using the AlGaN/GaN heterostructure," *IEEE T Electron Dev*, vol. 60, no. 10, 2013, pp. 3060–3066.

72. M. Okada, Y. Saitoh, M. Yokoyama, K. Nakata, S. Yaegassi, K. Katayama et al., "Novel vertical heterojunction field-effect transistors with re-grown AlGaN/GaN two-dimensional electron gas channels on GaN substrates," *Appl Phys Exp*, vol. 3, 2010, p. 054201.

73. A. Sarua, H. Ji, K.P. Hilton, D.J. Wallis, M.J. Uren, T. Martin et al., "Thermal boundary resistance between GaN and substrate in AlGaN/GaN electronic devices," *IEEE T Electron Dev*, vol. 54, no. 12, 2007, pp. 3152, 3158.

74. H. Otake, S. Egami, H. Ohta, Y. Nanishi, H. Takasu, "GaN-based trench gate metal oxide semiconductor field effect transistors with over 100 cm$^2$/(V·s) channel mobility," *Jpn J Appl Phys*, vol. 46, 2007, p. L599.

75. M. Kodama, M. Sugimoto, E. Hayashi, N. Soejima, O. Ishiguro, M. Kanechika et al., "GaN-based trench metal oxide semiconductor field-effect transistor fabricated with novel wet etching," *Appl Phys Exp*, vol. 1, 2008, p. 021104.

76. H. Otake, K. Chikamatsu, A. Yamaguchi, T. Fujishima, H. Ohta, "Vertical GaN-based trench gate metal oxide semiconductor field-effect transistors on GaN bulk substrates," *Appl Phys Exp*, vol. 1, 2008, p. 011105.

77. T. Oka, Y. Ueno, T. Ina, K. Hasegawa, "Vertical GaN-based trench metal oxide semiconductor field-effect transistors on a free-standing GaN substrate with blocking voltage of 1.6 kV," *Appl Phys Exp*, vol. 7, 2014, p. 021002.

78. X. Huang, Q. Li, Z. Liu, F.C. Lee, "Analytical loss model of high voltage GaN HEMT in cascode configuration," *IEEE T Power Electr*, vol. 29, no. 5, 2014, pp. 2208–2219.

79. T. Hirose, M. Imai, K. Joshin, K. Watanabe, T. Ogino, Y. Miyazaki et al., "Dynamic performances of GaN-HEMT on Si in cascode configuration," Twenty-Ninth Annual IEEE Applied Power Electronics Conference and Exposition (APEC), 2014, pp. 174–181.

80. Y.-F. Wu, J. Gritters, L. Shen, R.P. Smith, B. Swenson, "kV-class GaN-on-Si HEMTs enabling 99% efficiency converter at 800 V and 100 kHz," *IEEE T Power Electr*, vol. 29, no. 6, 2014, pp. 2634–2637.

81. M.J. Uren, M. Kuball, "GaN transistor reliability and instabilities," 10th International Conference on Advanced Semiconductor Devices & Microsystems (ASDAM), 2014, pp. 1–8.

82. L. Zhongda, T.P. Chow, "Design and simulation of 5–20-kV GaN enhancement-mode vertical superjunction HEMT," *IEEE T Electron Dev*, vol. 60, no. 10, 2013, pp. 3230–3237.

83. T. Kikkawa, T. Hosoda, S. Akiyama, Y. Kotani, T. Wakabayashi, T. Ogino et al., "600 V GaN HEMT on 6-inch Si substrate using Au-free Si-LSI process for power applications," IEEE Workshop on Wide Bandgap Power Devices and Applications (WiPDA), 2013, pp. 11–14.

84. O. Hilt, R. Zhytnytska, J. Boecker, E. Bahat-Treidel, F. Brunner, A. Knauer et al., "70 mΩ/600 V normally-off GaN transistors on SiC and Si substrates," Proceedings of the 27th International Symposium on Power Semiconductor Devices & IC's (ISPSD), 2015, pp. 237–240.

85. A. Malmros, H. Blanck, N. Rorsman, "Electrical properties, microstructure, and thermal stability of Ta-based ohmic contacts annealed at low temperature for GaN HEMTs," *Semicond Sci Technol*, vol. 26, no. 7, 2011, pp. 075006-1–075006.

86. H.S. Lee, D.S. Lee, T. Palacios, "AlGaN/GaN high-electron mobility transistors fabricated through a Au-free technology," *IEEE Electr Device L*, vol. 32, no. 5, 2011, pp. 623–625.

87. R. Reiner, P. Waltereit, F. Benkhelifa, S. Müller, H. Walcher, S. Wagner et al., "Fractal structures for low-resistance large area AlGaN/GaN power transistors," Proceedings of ISPSD, 2012, pp. 341–345.

88. E. Zanoni, M. Meneghini, A. Chini, D. Marcon, G. Meneghesso, "AlGaN/GaN-based HEMTs failure physics and reliability: mechanisms affecting gate edge and Schottky junction," *IEEE T Electron Dev*, vol. 60, no. 10, 2013, pp. 3119–3131.

89. C. Fleury, R. Zhytnytska, S. Bychikhin, M. Cappriotti, O. Hilt, D. Visalli et al., "Statistics and localisation of vertical breakdown in AlGaN/GaN HEMTs on SiC and Si substrates for power applications," *Microelectron Reliab*, vol. 53, no. 9–11, 2013, pp. 1444–1449.

90. M. Meneghini, C. de Santi, T. Ueda, T. Tanaka, D. Ueda, E. Zanoni et al., "Time- and field-dependent trapping in GaN-based enhancement-mode transistors with p-gate," *IEEE Electr Device L*, vol. 33, no. 3, 2012, pp. 375–377.

91. M. Meneghini, I. Rossetto, D. Bisi, A. Stocco, A. Cester, G. Meneghesso et al., "Role of buffer doping and pre-existing trap states in the current collapse and degradation of AlGaN/GaN HEMTs," IEEE International Reliability Physics Symposium, 2014, 6C.6.1–6C.6.7.

92. P. Marko, A. Alexewicz, O. Hilt, G. Meneghesso, E. Zanoni, J. Würfl et al., "Random telegraph signal noise in gate current of unstressed and reverse-bias-stressed AlGaN/GaN high electron mobility transistors," *Appl Phys Lett*, vol. 100, 2012, pp. 143507-1–143507-3.

93. J. Würfl, "Drift and reliability mechanisms in GaN based power devices for high voltage switching applications: the current understanding," Tutorial at the 25th European Symposium on Reliability of Electron Devices, Failure Physics and Analysis (ESREF 2014), 2014.

94. C.S. Whitman, "Methodology for predicting off-state reliability in GaN power transistors," *Microelectron Reliab*, vol. 54, 2014, pp. 354–359.

# 2 AlGaN/GaN High-Electron-Mobility Transistors Grown by Ammonia Source Molecular Beam Epitaxy

*Yvon Cordier*

## CONTENTS

## 2.1 INTRODUCTION

Since it has been shown that an AlGaN/GaN heterostructure can generate a two-dimensional electron gas (2DEG) [1], GaN-based high-electron-mobility transistors (HEMTs) have been developed and are now established as the most interesting III-nitride electron devices for high-frequency power amplification as well as power switching. The reason for this is a combination of many factors [2]: the possibility of achieving high sheet carrier concentration in the 2DEG ($\sim 1 \times 10^{13}$/cm$^2$) with a high saturated velocity ($>1.5 \times 10^7$ cm/s), quite a high electron mobility (up to about 2000 cm$^2$/V·s at RT), and a breakdown electron field exceeding 3 MV/cm. Moreover, the chemical inertness and the wide energy bandgap of GaN guarantee the thermal stability of the devices.

The most commonly used growth techniques for III-nitride heterostructures are metalorganic vapor phase epitaxy (MOVPE) and molecular beam epitaxy (MBE), each having its own advantages and drawbacks. Whereas MOVPE is widely used due to larger throughput and larger wafer size handling capability, MBE operates at lower temperatures under high vacuum with much less consumption of source products and is equipped with useful in situ inspection tools like reflection high-energy electron diffraction (RHEED). MBE production tools for III-nitrides are rare, but a lot of research reactors are used worldwide and are at the origin of demonstrations of optoelectronic and electronic devices. In this chapter, we first discuss the advantages of using ammonia as the nitrogen source for MBE growth. This is followed by a description of the growth of AlGaN/GaN HEMT heterostructures on GaN-on-sapphire templates, freestanding GaN, and foreign substrates like silicon and silicon carbide. Finally, the behavior of transistor devices is described and high-frequency power density results are presented.

## 2.2 CHARACTERISTICS OF AMMONIA SOURCE MOLECULAR BEAM EPITAXY

The main features of ammonia-MBE (NH3–MBE) are the following. Ammonia thermally decomposes at the surface of the films at temperatures beyond 450°C [3]. The typical growth temperature for GaN is 800°C, but thick AlN films necessitate higher temperatures. Usually, AlGaN films can be grown in the 800°C–875°C temperature range, the optimum temperature depending on the Al content and the thickness of the film to be grown. Growth rates in the range of 0.5–1.5 μm/h are quite easy to achieve. However, compared to MOVPE (GaN grown typically at 1000°C) lateral growth is very limited for MBE and whenever a significant roughening has developed it is difficult to recover a flat surface, which is detrimental for electron transport in the channel of lateral devices such as HEMTs. The study described here was done in a Riber Compact 21 MBE reactor equipped with an NH3 gas injector [4,5] and a plasma source (ADDON RFN50/63) connected to an N2 gas line [6,7]. The reactor configuration was optimized for uniform films on 2-in.-diameter substrates [4]. Nevertheless, even with this configuration the GaN thickness uniformity deviation is below 3% on 3 in. diameter.

As described by Vézian [8], when the growth proceeds ammonia-MBE grown GaN exhibits a transition from a spiral growth mode to a mixed growth mode where two-dimensional nucleation is sufficiently active to give rise to kinetic roughening. According to this study, the spiral growth mode occurs at the beginning of the growth, thanks to the step flow growth mode in the presence of screw dislocations. So, it results in a coarsening of growth mounds correlated with the decrease in dislocation density. Thick films (>2 μm) exhibit root mean square (RMS) roughness on the order of 4 to 5 nm (Figure 2.1a) and an increase in the correlation length saturating around 1 μm.

We discuss here the influence of the nitrogen source flow rate on the growth of GaN. The case of ammonia is quite simple. The optimum growth conditions are N-rich, which means that a large ammonia flow rate is required (200 sccm in our MBE reactor). At the optimum growth temperature of 800°C and for a fixed Ga flux, the growth rate and the surface morphology slightly change while reducing the ammonia flow rate. Then, a first regime is reached with pits corresponding to the facet development of threading dislocations (TDs) opening at the surface. Moreover, further reduction of the ammonia flow rate leads to a decrease in the growth rate, as well as further development of the surface roughness due to the insufficient amount of available active nitrogen species compared to incoming Ga species [3,9]. The nitrogen plasma growth (PA-MBE) is very different. First, the most frequently reported growth condition for smooth GaN films is growth at 720°C–730°C with a fine-tuning of nitrogen flow rate, radio-frequency (RF) cell power, and Ga flow rate to keep a thin metallic Ga film (two to three monolayers) floating on the growing surface and a resulting surface roughness of typically 1 nm (Figure 2.1b). Out of this equilibrium, either a rough film is obtained or an excess of gallium generates droplets at the surface [6]. To mitigate these effects, growth at a higher temperature (780°C–790°C) under a high nitrogen flow rate has been proposed [10], but the sensitivity to threading defects makes the growth

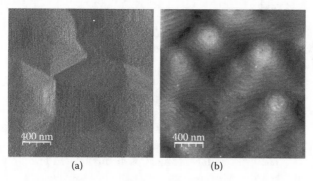

(a)                                             (b)

**FIGURE 2.1**  Tapping mode atomic force microscopy view of the surface of GaN layers grown (a) with ammonia and (b) with nitrogen plasma. The picture on the left is a derivative mode image to highlight the monolayer height steps present at the surface.

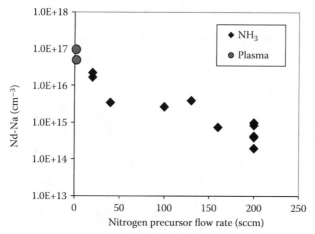

**FIGURE 2.2**  Donor concentration estimated by C–V as a function of the flow rate of the nitrogen precursor ($NH_3$ or nitrogen for plasma).

sometimes more difficult to monitor. A crucial parameter for electron devices such as field effect transistors is the residual doping level in the channel and the buffer resistivity. As seen in Figure 2.2, the ammonia flow rate has a crucial influence on the residual doping level [9]. Secondary ion mass spectroscopy indicates that silicon and oxygen are the main donors, even in the case of nitrogen plasma growth. Moreover, it seems that the flow rate itself has more influence than the nitrogen species, ammonia, or nitrogen molecules. One last point concerns the growth of AlGaN alloys. The desorption of Ga and Al species is negligible when the HEMT/AlGaN barrier is grown at 800°C with a large ammonia flow rate, so that composition and growth rate calibrations are facilitated.

## 2.3  HOMOEPITAXY OF GaN HIGH-ELECTRON-MOBILITY TRANSISTORS

The availability of highly resistive GaN-on-sapphire templates and freestanding GaN substrates is very helpful for the development of HEMT structures on high-crystal-quality GaN. Unless significantly thicker GaN is regrown, the regrowth of GaN just replicates the substrate threading dislocation density (TDD) of roughly $1 \times 10^8$ to $5 \times 10^8/cm^2$ in the GaN-on-sapphire templates and $1 \times 10^7$ to $5 \times 10^7/cm^2$ in the freestanding substrates. Recently, bulk GaN substrates grown by ammonothermal method with an ultralow dislocation density of $10^4/cm^2$ and wafers up to 2 in. in diameter have been shown to be compatible with HEMT structures' regrowth by MOVPE [11]. The growers have to face problems related to regrowth interface pollution. Even when the substrate/template is highly resistive/semi-insulating, the regrowth interface is contaminated with shallow donors such as silicon. The doping of GaN with elements such as carbon [12,13], beryllium [14], magnesium [15], and iron [16] is efficient in increasing the resistivity of GaN. So, the introduction of such elements in a regrown GaN layer is a solution to compensate the source of leakage in transistors. The incorporation rate can depend on the growth technique as well as the growth conditions. For instance, the incorporation of carbon from $CBr_4$ is highly dependent on temperature and not very efficient in the case of high-temperature MBE [17]. Then, ionization of methane is preferred for ammonia-MBE [13] since a high growth temperature is suitable.

However, such doping sources are not always available in growth reactors dedicated to HEMTs. Some authors propose the growth of a thin AlN layer to upraise the conduction band at the regrown interface [18]. Another alternative is to develop GaN templates or substrates ready for epitaxial regrowth with a reduced electrical leakage. In our case, we have developed MOVPE GaN with iron doping [19,20]. In these templates, the amount of iron available at the regrowth interface is low enough to avoid unrecoverable surface roughening but sufficient to compensate the effect of silicon or oxygen contaminants. The transistors fabricated on such epi-ready insulating GaN-on-sapphire templates exhibit drain leakage currents as low as 10 μA/mm at $V_{ds} = 10$ V when regrown by ammonia-MBE. As a high temperature favors the diffusion of iron, the leakage can be further reduced by about three orders of magnitude when regrown by MOVPE [20].

## 2.4 HETEROEPITAXY OF GaN HIGH-ELECTRON-MOBILITY TRANSISTORS

The growth of GaN HEMTs on foreign substrates requires both insulating and stress-mitigating buffer layers. On sapphire, the difference in thermal expansion coefficients (TECs) induces a residual compressive strain in GaN, which is responsible for a noticeable convex bowing of the wafers. The large lattice parameter mismatch is responsible for a large number of dislocations, which helps in trapping the carriers related to the residual doping at the initial stages of growth. The further thickening of the GaN buffer drastically reduces the number of TDs. The growth of AlN is an alternative to increasing the buffer layer resistivity because AlN is a wider bandgap material and the polarization electric field depletes the GaN/AlN regrowth interface from eventual free carriers. In the present study, we chose to grow with $NH_3$–MBE a HEMT structure on a 1-μm-thick MOVPE AlN-on-sapphire template. Contrary to sapphire, GaN grown on SiC or Si suffers a tensile strain induced by the TEC mismatch. To compensate for this effect and the associated risk of layer cracking, GaN is grown on an AlN nucleation layer to benefit from an initial compressive strain related to the 2.5% lattice parameters mismatch strain between the two materials. Moreover, AlN appears to be a useful solution to avoid reactions between gallium and the silicon substrate [21]. But, if the high temperature (800°C) of $NH_3$–MBE favors dislocation bending, interactions, and elimination it also promotes the relaxation of this strain [22] so that more complex structures with additional intercalated AlN layers have been grown to obtain 2-μm-thick crack-free HEMT structures on Si (111) [23] and SiC.

## 2.5 ELECTRICAL PROPERTIES

Figure 2.3 shows the main kinds of structures we have grown. Table 2.1 summarizes the best results that we obtained on these kinds of structures: 2DEG carrier density, the low field electron mobility assessed by Hall effect, as well as the buffer residual donor density obtained by exploiting the C–V measurements beyond pinch-off. The TDD assessed by atomic force microscopy (AFM) or x-ray diffraction (XRD) is also reported. At room temperature, the electron mobility is mainly limited by optical phonon scattering, interface roughness, and alloy scattering. Due to the high carrier density, the screening of TD fields is quite efficient and reduces the influence of TDD with respect to other scattering mechanisms. Moreover, it is obvious that the insertion of a thin AlN spacer at the AlGaN/GaN interface enhances room temperature mobility by more than 200 cm²/V·s. The mobility reaches 2000 cm²/V·s when the TDD is sufficiently low (below $5 \times 10^9$ cm⁻²). A consequence of the good crystal quality is that the carrier density in the 2DEG is mainly determined by barrier thickness and Al molar content, as well as GaN cap thickness. At low temperatures, the TDD reduction has a noticeable impact on electron mobility enhancement.

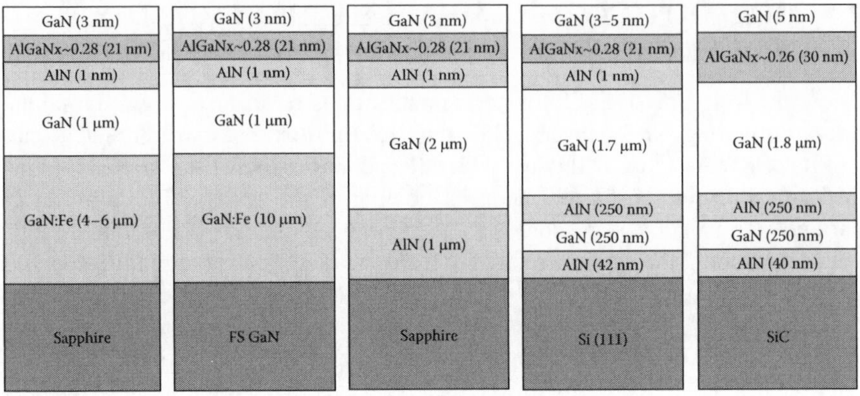

Structure 1 (Sapphire):
GaN (3 nm) / AlGaNx~0.28 (21 nm) / AlN (1 nm) / GaN (1 µm) / GaN:Fe (4–6 µm) / Sapphire

Structure 2 (FS GaN):
GaN (3 nm) / AlGaNx~0.28 (21 nm) / AlN (1 nm) / GaN (1 µm) / GaN:Fe (10 µm) / FS GaN

Structure 3 (Sapphire):
GaN (3 nm) / AlGaNx~0.28 (21 nm) / AlN (1 nm) / GaN (2 µm) / AlN (1 µm) / Sapphire

Structure 4 (Si (111)):
GaN (3–5 nm) / AlGaNx~0.28 (21 nm) / AlN (1 nm) / GaN (1.7 µm) / AlN (250 nm) / GaN (250 nm) / AlN (42 nm) / Si (111)

Structure 5 (SiC):
GaN (5 nm) / AlGaNx~0.26 (30 nm) / GaN (1.8 µm) / AlN (250 nm) / GaN (250 nm) / AlN (40 nm) / SiC

**FIGURE 2.3** Schematic cross section of the typical high-electron-mobility transistor structures grown by NH3–MBE.

## TABLE 2.1
## 2DEG Carrier Density and Mobility, Buffer Residual Donor Density, and TDD in the Studied HEMTs

|  | GaN/Al$_2$O$_3$ | GaN FS | AlN/Al$_2$O$_3$ | H-SiC | Si (111) |
|---|---|---|---|---|---|
| Al | 28% | 28% | 28% | 26% | 28% |
| Ns @ 300 K (cm-2) | $10 \times 10^{12}$ | $10 \times 10^{12}$ | $9 \times 10^{12}$ | [a]$8 \times 10^{12}$ | $9 \times 10^{12}$ |
| at 300 K (cm$^2$/V·s) | 2,080 | 2,140 | 2,085 | [a]1,769 | 2,000 ([a]1,780) |
| at <10 K (cm$^2$/V·s) | 30,000 ([a]12,500) |  |  | [a]8,740 | 12,700 ([a]7,880) |
| Nd-Na (cm$^{-3}$) | $3 \times 10^{13}$ | $3 \times 10^{13}$ | $<1 \times 10^{13}$ | $3 \times 10^{13}$ | $1 \times 10^{14}$ to $3 \times 10^{14}$ |
| TDD (cm$^{-2}$) | $0.4 \times 10^9$ to $1 \times 10^9$ | $1 \times 10^7$ to $2 \times 10^7$ | $2.5 \times 10^9$ | $3 \times 10^9$ | $3 \times 10^9$ to $4 \times 10^9$ |

*Note:* GaN FS: freestanding GaN.
[a]No AlN spacer between the AlGaN barrier and the GaN channel.

## 2.6 EVALUATION OF TRANSISTORS

To evaluate the potentialities of these structures, test devices including transmission line model (TLM) and isolation patterns, diodes, and transistors have been fabricated by photolithography. The device process starts with mesa definition by reactive ion etching (RIE) step in a Cl$_2$/Ar/CH$_4$ mixture. After a short RIE etching, TiAlNiAu stacks are deposited by e-beam evaporation. The ohmic contacts are achieved after rapid thermal annealing at 750°C for 30 seconds. NiAu films are then evaporated for gate Schottky contact as well as for access pads. These devices are not passivated.

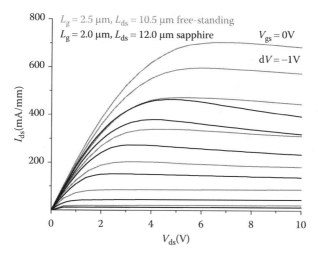

**FIGURE 2.4** DC output characteristics of AlGaN/GaN high-electron-mobility transistors regrown on GaN/sapphire template and on freestanding GaN substrate. Device dimensions are $W = 150$ μm, $L_g = 2.0–2.5$ μm, and $L_{sd} = {\sim}10.5–12.0$ μm.

Figure 2.4 shows the DC output characteristics of a transistor on a freestanding GaN substrate compared to one fabricated on a GaN-on-sapphire template. The 2.5 μm × 150 μm gates are deposited in nominal 11-μm source drain spacings. Despite the contact resistance and the sheet resistance being not very different (0.9 Ω·mm and 280 Ω/sq on sapphire, 0.7 Ω·mm and 270 Ω/sq on freestanding GaN), one notices that the I–V curves diverge while $V_{gs}$ increases. We suspect here a large influence of the device's self-heating on the electrical behavior. To confirm this, we first analyze the output characteristics of transistors fabricated with similar dimensions on different kinds of substrates. We then define a relative DC drain current collapse, $R$, as follows:

$$R = (\mathrm{Id}_{knee} - \mathrm{Id}_{20\,V})/\mathrm{Id}_{knee}$$

where $\mathrm{Id}_{knee}$ is the maximum drain current and $\mathrm{Id}_{20\,V}$ is the drain current obtained at $V_{ds} = 20$ V for a given gate bias $V_{gs}$. This is illustrated in the insert of Figure 2.5 on a transistor fabricated on Si (111).

When plotting the relative drain current collapse as a function of the drain current, one not only clearly sees the increase of the collapse with the drain current but also notices the similarity of collapse for transistors on sapphire with AlN or GaN buffer. Moreover, the transistors on SiC present a smaller collapse, whereas an intermediate one is obtained on silicon and freestanding GaN. Such a collapse seems to be related not to the defect density reported in Table 2.1 but to the thermal conductivity of the substrate. However, one type of transistor developed on silicon presents a relative collapse as low as that on SiC (dashed line in Figure 2.5). The peculiarity of this transistor is the unusual buffer layer thickness (0.5-μm GaN on a 0.5-μm AlN/GaN stress-mitigating stack).

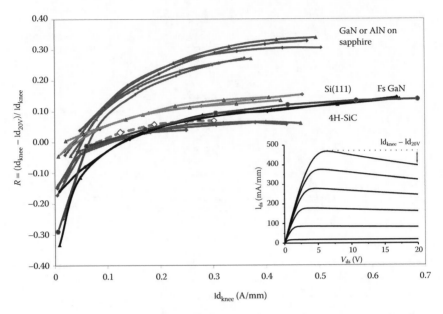

**FIGURE 2.5**  Relative drain current collapse measured in DC conditions. Device dimensions are $W = 150$ μm, $L_g = 3$ to 4 μm, and $L_{sd} = 9$–12 μm.

To go further, pulsed measurements have been performed on transistors fabricated on sapphire and Si (111) [24]. Pulsed current–voltage measurements (600 ns, 100 Hz) performed on devices on sapphire and on silicon substrates confirm the primary importance of thermal effects (Figure 2.6). As expected, the pulsed drain currents obtained after biasing at $V_{gs} = 0$ V and $V_{ds} = 0$ V quiescent point are systematically higher than those recorded under DC conditions. The difference dramatically increases on sapphire (Figure 2.6c). However, as shown in Figure 2.6b, it appears that the device on silicon with the thicker GaN buffer layer (here 1.7-μm GaN on a 0.5-μm AlN/GaN stress-mitigating stack) and then with a resulting reduced TDD shows an increase of the pulsed drain current collapse with quiescent points at $V_{gs} = -8$ V, $V_{ds} = 0$ V (gate lag) and $V_{gs} = -8$ V, $V_{ds} = 20$ V (drain lag). This trend is similar to the one previously observed for the DC current collapse (Figure 2.5). Then, it seems that in these bias conditions the presence of dislocations helps in achieving more stable device operation. However, GaN transistors generally operate under more severe bias conditions. The development of an efficient surface passivation drastically mitigates the current collapse. For instance, the combination of an N2O plasma surface treatment with the deposition of SiN/SiO2 passivation layers leads to small dispersions, as recently demonstrated on 0.25-μm gate transistors developed on similar structures with a resulting state of the art $L_g \times F_t = 15$ GHz·μm product and a power density of 1.5 W/mm at 40 GHz [25].

The critical parameters for transistors, especially power transistors, are buffer resistivity and breakdown voltage. Often, buffer resistivity directly impacts the drain leakage current evidenced when the transistor is in off state. Sometimes, it appears that the gate leakage itself is the source of leakage current in the off-state

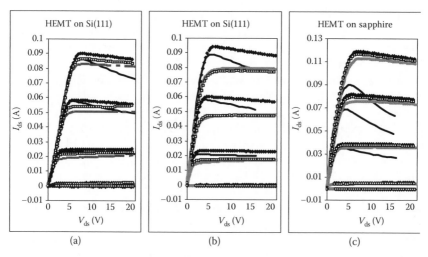

**FIGURE 2.6** DC and pulsed measurements performed on high-electron-mobility transistors on silicon with (a) thin buffer and (b) thick buffer and on (c) sapphire substrate. On silicon, bias conditions are as follows: line (DC,DC), diamonds [0 V,0 V], empty squares [−8 V,0 V], and filled squares [−8 V,20 V]. On sapphire, bias conditions are as follows: line (DC,DC), diamonds [0 V,0 V], empty squares [−4 V,0 V], and filled squares [−4 V,10 V].

**TABLE 2.2**
**Buffer and Transistor Leakage Currents Measured at $V_{ds}$ = 100 V**

| Substrate | GaN/Al O 2 3 | AlN/Al O 2 3 | SiC | Si (111) | Si (111) |
|---|---|---|---|---|---|
| Buffer layer | GaN:Fe | GaN/AlN | GaN/AlN | GaN/AlN | AlGaN/AlN |
| Buffer leakage at 100 V | <10 μA/mm | <40 μA/mm | <200 μA/mm | 20–100 μA/mm | 20 μA/mm |
| Transistor $W$ = 150-μm leakage at 100 V | <70 μA | 1 mA/mm @ 85 V | <30 μA | 4–120 μA | <30 μA |

configuration. The buffer leakage currents measured between ohmic contacts fabricated on HEMT active layers and separated by mesa etching spacings superior to 10 μm are reported in Table 2.2. With such spacings, electric fields propagate down to the substrate so that all regions of the buffer layer can contribute. The drain leakage in transistors with gate lengths of 2 to 3 μm and source to drain spacings of 12 to 13 μm is also reported for a drain bias of 100 V and a gate bias, $V_{gs}$, of approximately 2 V beyond the threshold voltage. The transistor development, $W$, is 150 μm. It is clear from Table 2.2 that both the buffer leakage current and the drain leakage current are on the same order of magnitude, which confirms the primordial influence of buffer resistivity. Nevertheless, a closer look shows that in some cases the leakage currents of transistors at pinch-off systematically overpass the buffer leakage currents. This is probably due to the increase in electric field crowding in the vicinity of the gate and/or to the presence of significant self-heating. The device on Fe-doped GaN-on-sapphire is very satisfying in terms of trade-off between crystal quality and drain leakage, contrary to the device on AlN-on-sapphire whose breakdown voltage is below 100 V, as illustrated in Figure 2.7a.

The transistor leakage on hexagonal SiC better scales with buffer leakage, indicating a more stable behavior with respect to the electric fields and the self-heating. However, the buffer on SiC is not the most resistive one, as shown in Figure 2.7a. This can be due to the possible diffusion of silicon from the substrate, as evocated by Hoke [26] in the case of PA-MBE grown AlN. The present buffer layer contains a thin AlN layer and a 0.5-μm-thick GaN/AlN stack (Figure 2.3), and an increase by two orders of magnitude of the buffer leakage current is obtained in absence of the AlN interlayer. Thickening of AlN layers as well as reduction in the growth temperature of these layers are possible ways to enhance the electrical resistance of buffer layers on SiC.

Our most studied devices are grown on Si (111). Figure 2.7a shows that depending on growth conditions buffer resistivity can vary significantly. Moreover, the GaN thickness in structures like the one described in Figure 2.3 can influence the buffer resistivity in an expected way, with an increase of the leakage current when the GaN thickness is reduced. As this occurs while the dislocation density increases, we suspect that the particular arrangement of dislocations (dislocation loops, bended dislocations, and number of screw-type dislocations) in the GaN grown compressively strained on the AlN interlayer is responsible for this. Indeed, for 0.5 μm of GaN thickness, the buffers grown without interlayers on Si (111) present much lower leakage currents (Figure 2.7b), but with clearly worse crystal quality (TDD > $2 \times 10^{10}/cm^2$) and resulting electron mobility on the order of 1300 $cm^2$/V·s. Coming back to thick buffer layers with interlayer stacks, a solution to stabilize the electrical

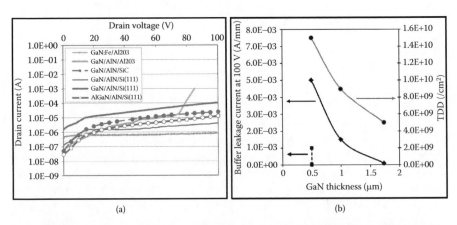

(a)                                        (b)

**FIGURE 2.7**  (a) Transistors' drain leakage current under pinch-off conditions. (b) Buffer leakage current at 100 V and threading dislocation density (TDD) in high-electron-mobility transistor structures grown on Si (111) with different GaN buffer thicknesses. Circles represent TDD, diamonds the leakage for buffers on silicon with interlayers, and squares the leakage for buffers on silicon without interlayers.

resistivity can be the replacement of GaN in the buffer with a larger bandgap material such as AlGaN. However, the growth of such a layer on silicon is more difficult due to the smaller lattice parameter of AlGaN and the resulting stress. Nevertheless, we succeeded in growing up to 1.5-μm crack-free AlGaN with 5%–10% of Al and with low crystal quality degradation [25,27]. This enables us to achieve more stable drain leakage currents below 30 μA at $V_{ds} = 100$ V in the studied devices.

These devices are not passivated, so they are not able to sustain large drain and gate biases. In spite of a drain leakage current that can be as low as a few microamperes, the breakdown voltage obtained in air is around 200 V (234 V at best) with destruction of the metal contacts. The possible leakage current paths are represented in Figure 2.8. Electrons can be injected from the source to the substrate through the buffer layer and the nucleation layer, so that a current can flow via the buried part of the buffer or via the substrate (when not insulating) and then reach the positively biased drain contact. When the leakage arising from the gate (path $\gamma_2$) is small enough, the drain current leakage at pinch-off can follow path $\gamma_1$, so that measuring the vertical leakage is helpful to identify the detailed mechanisms of the drain leakage.

The behaviors of HEMT devices on Si, GaN-on-sapphire template, and free-standing GaN substrates have been studied in this vertical configuration by Pérez-Tomás [28]. The drain is positively biased while the substrate is grounded. In this configuration, the leakage through the device on Si is a combination of Poole-Frenkel (trap-assisted) conduction and resistive conduction with an estimated resistance of 72 kΩ up to a soft breakdown at 420 V. We should remember here that the total buffer thickness is less than 2.3 μm, leading to an average breakdown field of more than 1.8 MV/cm. Due to the insulating sapphire, more than 350 V is

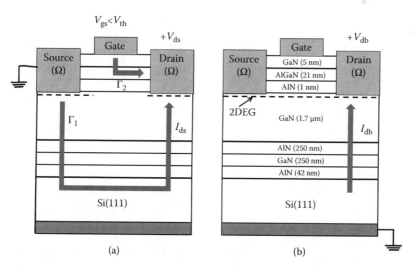

**FIGURE 2.8** (a) Leakage paths in the transistor on silicon under pinch-off conditions. (b) Vertical leakage path in the device under vertical current configuration. (Adapted from Pérez-Tomás et al., *J. Appl. Phys.*, 113, 174501, 2013 [28].)

necessary to notice a vertical leakage current with a resistive conduction (85 kΩ) up to 1 kV without any irreversible breakdown phenomena. This similar resistance is explained by the similar growth conditions of the thick GaN grown by MBE either on the silicon substrate or on the GaN-on-sapphire template. The difference in terms of TDD seems to be not sufficient to have a noticeable effect on the vertical conduction. On the other hand, the device on the freestanding GaN substrate follows a resistive behavior only, with a resistance of 7 MΩ up to the destructive breakdown at 840 V. The absence of AlN layers in the buffer and the low TDD are efficiently compensated by the presence of iron in the 10-μm-thick GaN layer. However, when the temperature is increased the vertical leakage current on Si and freestanding GaN rapidly increases with an activation energy of 0.35 eV. In reverse bias conditions (when the drain is negatively biased), the leakage on the silicon substrate is larger, with an activation energy of 0.1 eV and a resistance of 20 kΩ, due to the band lineup between the materials. A detailed analysis performed on silicon shows that for a broad range of drain biases the gate to source leakage current is more than one order of magnitude lower than the leakage current between the drain and the source, which itself is the same as the vertical leakage current flowing from the substrate to the drain. This suggests that the preferential leakage path is $G_l$ and then any enhancement of the resistivity of the nucleation layer region will benefit the transistor power DC operation.

## 2.7 HIGH-FREQUENCY DEVICES

Both the buffer and the substrate resistivities are crucial for RF performances. Even when it is possible to extract the losses in access regions of transistors (de-embedding) [29], performing functional devices require minimal losses especially at frequencies of several tens of gigahertz (GHz). For these reasons, substrates with high resistivity are used (typically superior to 3 kΩ·cm). However, the presence of Al atoms in the early stage of the growth and the high temperature usually employed to obtain a good-crystal-quality film can reduce overall substrate resistivity. To limit the capacitive coupling of the device with the substrate, a thick buffer layer and may be a lower temperature growth technique like MBE are preferable. For instance, Lecourt et al. [30] measured RF losses of 0.4 dB/mm at 50 GHz on a 2.5-μm-thick structure grown on resistive Si (111) (ρ > 1 kΩ·cm) by ammonia-MBE.

Figure 2.9 shows the state of the art in terms of power densities obtained at 40 GHz for devices with varying gate lengths. These results have been obtained with bias points $V_{ds}$ = 15–20 V, and the dependence of the performance is almost linear. The power density results of 1.5 W/mm with 250-nm gate [25] and 3.3 W/mm with 60-nm gate [33] have been obtained with AlGaN/GaN HEMTs grown by ammonia-MBE on silicon. Two intermediate results [31,32] have been obtained with MOVPE-grown structures. Of course, the achievable output power density increases with decreasing operation frequency (Table 2.3). Results on GaN substrates are still quite rare but at least as good as on silicon. Devices on SiC substrates clearly benefit from a better thermal conductivity, which is crucial for high-frequency power devices. It is often difficult to find direct comparisons between devices made with the same technology on structures grown with different techniques. Palacios [41] did this for

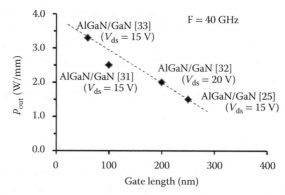

**FIGURE 2.9** Gate length dependence of the state-of-the-art output power densities obtained at 40 GHz with AlGaN/GaN high-electron-mobility transistors on silicon substrates.

**TABLE 2.3**

**High-Frequency Power Density Results Reported on Silicon, GaN, and Silicon Carbide Substrates with Different Growth Techniques**

| Substrate | Growth Technique | Gate Length (μm) | Frequency (GHz) | Drain Bias (V) | Output Power Density (W/mm) | Reference |
|---|---|---|---|---|---|---|
| Si | NH$_3$-MBE | 0.25 | 40 | 15 | 1.5 | 25 |
| Si | NH$_3$-MBE | 0.06 | 40 | 15 | 3.3 | 33 |
| Si | NH$_3$-MBE | 0.25 | 18 | 35 | 5.1 | 34 |
| Si | NH$_3$-MBE | 0.3 | 10 | 40 | 7 | 35 |
| GaN | MOVPE | 0.15 | 10 | 25 | 4.8 | 36 |
|  |  |  |  | 50 | 9.4 |  |
| GaN | PA-MBE | 0.5 | 10 | 25 | 4.8 | 37 |
| SiC | NH$_3$-MBE | 0.6 | 10 | 30 | 6.3 | 38 |
|  |  |  |  | 48 | 11.2 |  |
| SiC | PA-MBE | 0.25 | 10 | 31 | 6.3 | 39 |
| SiC | NH$_3$-MBE | 0.2 | 30 | 40 | 6.5 | 40 |
| SiC | PA-MBE | 0.16 | 40 | 30 | 8.6 | 41 |
| SiC | MOVPE | 0.16 | 40 | 30 | 10.5 | 41 |

HEMT structures grown by PA-MBE and MOVPE on SiC. Ammonia-MBE and PA-MBE produce similar results, which are not very far from the ones obtained using MOVPE. However, the long-term reliability of HEMTs is still an issue, in particular the degradation induced by electrical stress. An interesting study has been reported by Puzyrev et al. [42] on HEMTs grown either by PA-MBE under Ga-rich and N-rich conditions or by ammonia-MBE or MOVPE (both NH$_3$-rich conditions). According to the study, in samples grown by PA-MBE (Ga rich and N rich), initially hydrogenated Ga-vacancies can loose hydrogen atoms during the stress with hot electrons. This results in a positive threshold voltage shift and a degradation of the transconductance. On the other hand, in the case of ammonia-MBE the author proposes that the dehydrogenation of N-antisite defects is responsible for a negative threshold voltage shift and a reduced transconductance change due to the lower charge state of the defects. The degradation of metalorganic chemical vapor deposition (MOCVD)-grown devices is similar to that of ammonia-MBE-grown devices, showing that ammonia favors antisite-related defects themselves passivated by hydrogen.

## 2.8   CONCLUSION

MBE is able to produce device quality materials for field effect transistors such as HEMTs. Compared with ammonia-MBE and MOVPE, the low growth rate of PA-MBE (0.3–0.5 µm/h) has been considered for a long time as a drawback for production. A first attempt to solve this has been to use several nitrogen plasma cells simultaneously, allowing to reach more than 0.8 µm/h growth rate [43]. More recently, new developments on plasma cells have allowed to reach more than 2 µm/h growth rate for GaN with unchanged surface morphology and optical properties [44,45]. However, the uniformity has not been demonstrated yet. With a growth temperature intermediate between PA-MBE and MOVPE, ammonia-MBE presents the advantage of combining a simple N-rich growth regime with acceptable growth rate and uniformity [4,9]. Moreover, it seems that ammonia-MBE offers the possibility to obtain a better purity material with a lower residual donor concentration, making the incorporation of additional compensation species not always necessary. Compared with MOVPE, another advantage of MBE is the possibility of developing structures at temperatures compatible with the integration of already processed devices like silicon metal–oxide–semiconductor field effect transistors (MOSFETs), for instance [46,47]. All these points explain why MBE, and in our case why ammonia-source MBE, is so useful in developing electron device structures like HEMTs.

## ACKNOWLEDGMENTS

The author thanks all of his colleagues in France, Spain, and Canada who contributed to the results presented in this chapter and who will recognize themselves in the references. He is also grateful to the company RIBER SA and to the agencies that supported a part of these studies. Among these agencies, the French Délégation Générale de l'Armement and the National Research Agency are deeply acknowledged.

## REFERENCES

1. M. Asif Khan, J. N. Kuznia, J. M. Van Hove, N. Pan, and J. Carter, Observation of a Two-Dimensional Electron Gas in Low Pressure Metalorganic Chemical Vapor Deposited GaN-AlGaN Heterojunctions, *Appl. Phys. Lett.*, vol. 60, p. 3027 (1992).
2. Bernard Gil (Editor), *III-Nitride Semiconductors and Their Modern Devices*, Chapter 11, Oxford Science Publications, Oxford University Press Oxford, UK (2013).
3. M. Mesrine, N. Grandjean, and J. Massies, Efficiency of NH$_3$ as Nitrogen Source for GaN molecular Beam Epitaxy, *Appl. Phys. Lett.*, vol. 72, p. 350 (1998).

4. Y. Cordier, F. Pruvost, F. Semond, J. Massies, M. Leroux, P. Lorenzini et al., Quality and Uniformity Assessment of AlGaN/GaN Quantum Wells and HEMT Technostructure Grown by Molecular Beam Epitaxy With Ammonia Source, *Phys. Stat. Sol C*, vol. 3, p. 2325 (2006).
5. Y. Cordier, F. Semond, J. Massies, M. Leroux, P. Lorenzini, and C. Chaix, Developments for the Production of High Quality and High Uniformity AlGaN/GaN Heterostructures by Ammonia MBE, *J. Crystal Growth*, vol. 301–302, p. 434 (2007).
6. F. Natali, Y. Cordier, C. Chaix, and P. Bouchaib, Advances in Quality and Uniformity of (Al,Ga)N/GaN Quantum Wells Grown by Molecular Beam Epitaxy With Plasma Source, *J. Crystal Growth*, vol. 311, p. 2029 (2009).
7. F. Natali, Y. Cordier, J. Massies, S. Vezian, B. Damilano, and M. Leroux, Signature of Monolayer and Bilayer Fluctuations in the Width of (Al,Ga)N/GaN Quantum Well, *Phys. Rev. B*, vol. 79, p. 035328 (2009).
8. S. Vézian, F. Natali, F. Semond, and J. Massies, From Spiral Growth to Kinetic Roughening in Molecular-Beam Epitaxy of GaN(0001), *Phys. Rev. B*, vol. 69, p. 125329 (2004).
9. Y. Cordier, F. Natali, M. Chmielowska, M. Leroux, C. Chaix, and P. Bouchaib, Influence of Nitrogen Precursor and Its Flow Rate on the Quality and the Residual Doping in GaN Grown by Molecular Beam Epitaxy, *Phys. Status Solidi C*, vol. 9, no. 3–4, p. 523 (2012).
10. G. Koblmüller, F. Reurings, F. Tuomisto, and J. S. Speck, Influence of Ga/N Ratio on Morphology, Vacancies, and Electrical Transport in GaN Grown by Molecular Beam Epitaxy at High Temperature, *Appl. Phys. Lett.*, vol. 97, p. 191915 (2010).
11. P. Kruszewski, P. Prystawko, I. Kasalynas, A. Nowakowska-Siwinska, M. Krysko, J. Plesiewicz et al., AlGaN/GaN HEMT Structures on Ammono Bulk GaN Substrate, *Semicond. Sci. Technol.*, vol. 29, p. 075004 (2014).
12. C. Poblenz, P. Waltereit, S. Rajan, S. Heikman, U. K. Mishra, and J. S. Speck, Effect of Carbon Doping on Buffer Leakage in AlGaN/GaN High Electron Mobility Transistors, *J. Vac. Sci. Technol. B*, vol. 22, no. 3, p. 1145 (2004).
13. J. B. Webb, H. Tang, S. Rolfe, and J. A. Bardwell, Semi-insulating C-doped GaN and High-Mobility AlGaN/GaN Heterostructures Grown by Ammonia Molecular Beam Epitaxy, *Appl. Phys. Lett.*, vol. 75, p. 953 (1999).
14. D. F. Storm, D. S. Katzer, J. A. Mittereder, S. C. Binari, B. V. Shanabrook, X. Xu et al., Homoepitaxial Growth of GaN and AlGaN/GaN Heterostructures by Molecular Beam Epitaxy on Freestanding HVPE Gallium Nitride for Electronic Device Applications, *J. Crystal Growth*, vol. 281, p. 32 (2005).
15. T. M. Kuan, S. J. Chang, Y. K. Su, J. C. Lin, S. C. Wei et al., High-performance GaN/InGaN Heterostructure FETs on Mg-doped GaN Current Blocking Layers, *J. Crystal Growth*, vol. 272, p. 300 (2004).
16. A. Corrion, F. Wu, T. Mates, C. S. Gallinat, C. Poblenz, and J. S. Speck, Growth of Fe-doped GaN by RF Plasma-assisted Molecular Beam Epitaxy, *J. Crystal Growth*, vol. 289, p. 587 (2006).
17. S. W. Kaun, M. H. Wong, U. K. Mishra, and J. Speck, Molecular Beam Epitaxy for High-performance Ga-face GaN Electron Devices, *Semicond. Sci. Technol.*, vol. 28, p. 074001 (2013).
18. Y. Cao, T. Zimmermann, H. Xing, and D. Jena, Polarization-engineered Removal of Buffer Leakage for GaN Transistors, *Appl. Phys. Lett.*, vol. 96, p. 042102 (2010).
19. Y. Cordier, M. Azize, N. Baron, S. Chenot, O. Tottereau, and J. Massies, AlGaN/GaN HEMTs Regrown by MBE on Epi-Ready Semi-insulating GaN-on-Sapphire With Inhibited Interface Contamination, *J. Crystal Growth*, vol. 309, p. 1 (2007).
20. Y. Cordier, M. Azize, N. Baron, Z. Bougrioua, S. Chenot, O. Tottereau et al., Subsurface Fe Doped Semi-insulating GaN Templates for Inhibition of Regrowth Interface Pollution in AlGaN/GaN HEMT Structures, *J. Crystal Growth*, vol. 310, p. 948 (2008).

21. A. Watanabe, T. Takeuchi, K. Hirosawa, H. Amano, K. Hiramatsu, and I. Akasaki, The Growth of Single Crystalline GaN on a Si Substrate Using AlN as an Intermediate Layer, *J. Crystal Growth*, vol. 128, p. 391 (1993).

22. Y. Cordier, N. Baron, S. Chenot, P. Vennéguès, O. Tottereau, M. Leroux et al., Strain Engineering in GaN Layers Grown on Silicon by Molecular Beam Epitaxy: The Critical Role of Growth Temperature, *J. Crystal Growth*, vol. 311, p. 2002 (2009).

23. N. Baron, Y. Cordier, S. Chenot, P. Vennéguès, O. Tottereau, M. Leroux et al., The Critical Role of Growth Temperature on the Structural and Electrical Properties of AlGaN/GaN High Electron Mobility Transistor Heterostructures Grown on Si(111), *J. Appl. Phys.*, vol. 105, p. 033701 (2009).

24. Y. Cordier, N. Baron, F. Semond, M. Ramdani, M. Chmielowska, E. Frayssinet et al., Effects of Substrate and Buffer Layer Quality on the Behavior of AlGaN/GaN HEMTs: Thermal Effects Versus Electron Trapping, *Proceedings of the 35th Workshop on Compound Semiconductor Devices and Integrated Circuits*, WOCSDICE 2011, Catania, Italy, May 29th to June 1st, 2011. pp. 89–90.

25. S. Rennesson, F. Lecourt, N. Defrance, M. Chmielowska, S. Chenot, M. Lesecq et al., Optimization of $Al_{0.29}Ga_{0.71}N$//GaN High Electron Mobility Transistor Heterostructures for High Power/Frequency Performances, *IEEE T. Electron Dev.*, vol. 60, p. 3105 (2013).

26. W. E. Hoke, A. Torabi, J. J. Mosca, R. B. Hallock, and T. D. Kennedy, Rapid Silicon Outdiffusion from SiC Substrates During Molecular-Beam Epitaxial Growth of Algan/Gan/Aln Transistor Structures, *J. Appl. Phys.*, vol. 98, p. 084510 (2005).

27. Y. Cordier, F. Semond, M. Hugues, F. Natali, P. Lorenzini, H. Haas et al., Structural and Electrical Properties of AlGaN/GaN HEMTs Grown by MBE on SiC, Si(111) and GaN Templates, *J. Crystal Growth*, vol. 278/1–4, p. 383 (2005).

28. A. Pérez-Tomás, A. Fontserè, J. Llobet, M. Placidi, S. Rennesson, N. Baron et al., Analysis of the AlGaN/GaN Vertical Bulk Current on Si, Sapphire, and Free-Standing GaN Substrates, *J. Appl. Phys.*, vol. 113, p. 174501 (2013).

29. E. M. Chumbes, A. T. Schremek, J. A. Smart, Y. Wang, N. C. MacDonald, D. Hogue et al., AlGaN/GaN High Electron Mobility Transistors on Si(111) Substrates, *IEEE T. Electron Dev.*, vol. 48, p. 420 (2001).

30. F. Lecourt, Y. Douvry, N. Defrance, V. Hoel, Y. Cordier, and J. C. De Jaeger, *Proceedings of the 5th European Microwave Integrated Circuits Conference (EuMIC)*, Analysis of AlGaN/GaN Epi-material on Resistive Si(111) Substrate for MMIC Applications in Millimeter Wave Range, Paris, France, September 27–28, 2010, pp. 33–36.

31. F. Medjdoub, M. Zegaoui, B. Grimbert, D. Ducatteau, N. Rolland, and P. Rolland, First Demonstration of High-Power GaN-on-Silicon Transistors at 40 GHz, *IEEE Electr. Device L.*, vol. 33, p. 1168 (2012).

32. D. Marti, S. Tirelli, A. Alt, J. Roberts, and C. Bolognesi, 150-GHz Cutoff Frequencies and 2-W/mm Output Power at 40 GHz in a Millimeter-Wave AlGaN/GaN HEMT Technology on Silicon, *IEEE Electr. Device L.*, vol. 33, p. 1372 (2012).

33. A. Soltani, J. C. Gerbedoen, Y. Cordier, D. Ducatteau, M. Rousseau, M. Chmielowska et al., Power Performance of AlGaN/GaN High-Electron-Mobility Transistors on (110) Silicon Substrate at 40 GHz, *IEEE Electr. Device L.*, vol. 34, p. 490 (2013).

34. D. Ducatteau, A. Minko, V. Hoel, E. Morvan, E. Delos, B. Grimbert et al., Output Power Density of 5.1/mm at 18 GHz With an AlGaN/GaN HEMT on Si Substrate, *IEEE Electr. Device L.*, vol. 27, p. 7 (2006).

35. D. C. Dumka, C. Lee, H. Q. Tserng, P. Saunier, and R. Kumar, AlGaN/GaN HEMTs on Si Substrate With 7 W/mm Output Power Density at 10 GHz, *Electronics Lett.*, vol. 40, no. 16, p. 1023 (2004).

36. K. K. Chu, P. C. Chao, M. T. Pizzella, R. Actis, D. E. Meharry, K. B. Nichols et al., 9.4-W/mm Power Density AlGaN-GaN HEMTs on Free-Standing GaN Substrates, *IEEE Electr. Device L.*, vol. 25, no. 9, p. 596 (2004).

37. D. F. Storm, J. A. Roussos, D. S. Katzer, J. A. Mittereder, R. Bass, S. C. Binari et al., Microwave Power Performance of MBE-grown AlGaN/GaN HEMTs on HVPE GaN Substrates, *Electronics Lett.*, vol. 42, no. 11, p. 663 (2006).
38. C. Poblenz, A. L. Corrion, F. Recht, Chang Soo Suh, Rongming Chu, L. Shen et al., Power Performance of AlGaN/GaN HEMTs Grown on SiC by Ammonia-MBE at 4 and 10 GHz, *IEEE Electr. Device L.*, vol. 28, no. 11, p. 945 (2007).
39. N. X. Nguyen, M. Micovic, W. S. Wong, P. Hashimoto, L. M. McCray, and P. Janke, High Performance Microwave Power GaN/AlGaN MODFETs Grown by RF-Assisted MBE, *Electronics Lett.*, vol. 36, no. 5, p. 468 (2000).
40. Y. Pei, C. Poblenz, A. L. Corrion, R. Chu, L. Shen, J. S. Speck et al., X- and Ka-Band Power Performance of AlGaN/GaN HEMTs Grown by Ammonia-MBE, *Electronics Lett.*, vol. 44, no. 9, p. 598 (2008).
41. T. Palacios, A. Chakraborty, S. Rajan, C. Poblenz, S. Keller, S. P. DenBaars et al., High-Power AlGaN/GaN HEMTs for Ka-Band Applications, *IEEE Electr. Device L.*, vol. 26, no. 11, p. 781 (2005).
42. Y. S. Puzyrev, T. Roy, M. Beck, B. R. Tuttle, R. D. Schrimpf, D. M. Fleetwood et al., Dehydrogenation of Defects and Hot-Electron Degradation in GaN High-Electron-Mobility Transistors, *J. Appl. Phys.*, vol. 109, p. 034501 (2011).
43. R. Aidam, E. Diwo, N. Rollbühler, L. Kirste, and F. Benkhelifa, Strain Control of AlGaN/GaN High Electron Mobility Transistor Structures on Silicon (111) by Plasma Assisted Molecular Beam Epitaxy, *J. Appl. Phys.*, vol. 111, p. 114516 (2012).
44. B. M. McSkimming, F. Wu, T. Huault, C. Chaix, and J. S. Speck, Plasma Assisted Molecular Beamepitaxy of Gan With Growth Rates > 2.6 μm/h, *J. Crystal Growth*, vol. 386, p. 168 (2014).
45. Y. Kawai, S. Chen, Y. Honda, M. Yamaguchi, H. Amano, H. Kondo et al., Achieving High-Growth-Rate in GaN, Homoepitaxy Using High-Density, Nitrogen Radical Source, *Phys. Status Solidi C*, vol. 8, p. 2089 (2011).
46. W. E. Hoke, R. V. Chelakara, J. P. Bettencourt, T. E. Kazior, J. R. Laroche, T. D. Kennedy et al., Monolithic Integration of Silicon CMOS and GaN Transistors in a Current Mirror Circuit, *J. Vac. Sci. Technol. B*, vol. 30, no. 2, p. 02B101 (2012).
47 R. Comyn, Y. Cordier, V. Aimez, and H. Maher, "Reduction of the Thermal Budget of AlGaN/GaN Heterostructures Grown on Silicon: A Step Towards Monolithic Integration of GaN-HEMTs With CMOS," *Phys. Status Solidi* A 212, p. 1145 (2015).

# 3 Gallium Nitride Transistors on Large-Diameter Si(111) Substrate

*Subramaniam Arulkumaran and Geok Ing Ng*

## CONTENTS

## 3.1 OVERVIEW OF GaN-ON-Si TECHNOLOGY

Nitride semiconductors are quickly transforming our world by enabling new solid-state lighting, highly efficient amplifiers for wireless communications, advanced power electronics with ultra low losses, and a large array of new high-performance devices. Today, the need for high-power and high-frequency transistors are still increasing steadily, commensurate with the huge demand for wireless telecommunications. Higher power, higher frequency bandwidths, better linearity, and improved efficiency are still driving the current development of radio frequency (RF) semiconductor devices (i.e., applications such as defense, wireless telecom, very small aperture terminal, and community antenna television). The market needs devices capable of handling all these specifications at an affordable price. Over the last few years, silicon laterally diffused metal oxide semiconductor coverage of high-power RF amplification applications in the frequency range of 2–6 GHz decreased from 92% to 76% with the remaining 24% market share covers by GaAs pseudomorphic high-electron-mobility transistor (HEMT) technology, HEMTs GaN, and Si bipolar junction transistor. Military applications were the first to use GaN devices and better performing devices continue to emerge with the support by Defense Advanced Research Projects Agency and Department of Defense funded research programs in the United States and European Space Agency in Europe. Apart from RF devices, very high demand also exists for high-power switching device applications such as compact adaptors for laptops, DC–AC and AC–DC invertors, efficient energy conversion from solar panels, hybrid cars and so on. Currently, silicon-based device technologies dominate the inverters market. However, such devices are very big in size and lossy during the conversion.

For the invention of blue light-emitting diodes (LEDs) using direct bandgap InGaN/GaN quantum wells, the energy-saving white light sources started to change the world's lighting technology by replacing the conventional incandescent lamps. In 2014, the Nobel Prize in Physics was awarded jointly to the inventors: Isamu Akasaki (Nagoya University, Nagoya), Hiroshi Amano (Nagoya University, Nagoya), and Shuji Nakamura (University of California, Santa Barbara) "for the invention of efficient blue light-emitting diodes which has enabled bright and energy-saving white light sources." Globally, more than 40% (average) energy saving is possible if all adopt energy-saving light sources (e.g., white light emitting diodes) [1].

Currently most of the LEDs are produced by the growth and process technology of GaN-on-sapphire and GaN-on-SiC. The cost of a simple 40-W LED bulb is ~US$20. Due to the limitations of wafer size and substrate cost (sapphire ~0.8 $/cm$^2$ and SiC ~6.6 $/cm$^2$), it is difficult to reduce the cost of the LEDs further. Out of the different substrates available for the growth of GaN (i.e., bulk GaN, Si, sapphire, and SiC), the Si substrate has traditionally been considered as a low-cost and lower-performance option. The industries (e.g., Toshiba, Lattice Power, Plessey Semiconductors, etc.) are investing in R&D to reduce the cost by using a GaN-on-Si technology to manufacture the energy-saving lighting technology [2]. The material properties of different substrates are discussed in Section 3.2.1. Market researchers at M/s. IHS, Inc., forecasted that the share of GaN-on-Si wafers in the LED market (See Figure 3.1) will increase at a compound annual growth rate of 69% between 2013 and 2020, reaching 40% of all GaN LEDs manufactured [3]. In 2013, it was estimated that 95% of GaN LEDs were produced on sapphire, whereas only 1% (~0%) were made on silicon. Growth in GaN-on-Si LED manufacturing is also expected to grab market share from SiC substrates, which are often used for high-brightness devices because of its superior thermal conductivity. Although the thermal conductivity of silicon is less than that of silicon carbide, it is much better than that of sapphire. Figure 3.2 shows the history of GaN-on-Si technology development for material growth (small size sample to 200-mm-diameter substrate), power transistors, and RF transistors from 1991 to 2014.

According to Yole development RF market (April 2014), the cost of 200-mm (8-inch)-diameter silicon substrate (Si ~0.1 $/cm$^2$) is not enough to justify the transition to GaN- on-Si technology. The main driver is the ability to manufacture in

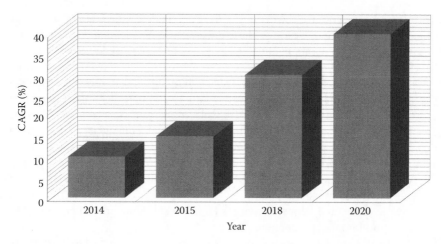

**FIGURE 3.1**    Forecasted share of GaN-on-Si wafers in the light-emitting diode (LED) market by market researchers at HIS, Inc.

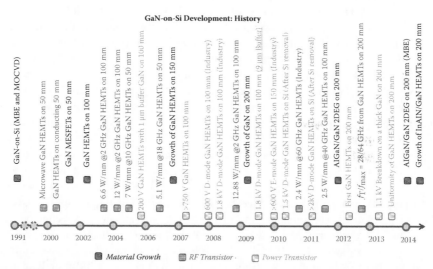

**FIGURE 3.2**    History of GaN-on-Si development from 1991 to 2014 for material growth, radio frequency transistors, and power transistors.

existing, depreciated complementary metal–oxide–semiconductor (CMOS) fabs in 150 mm or 200 mm. The manufacturing companies want to use the underutilized 150-mm or 200-mm-diameter silicon CMOS line for the mass production of LEDs and transistors. Through the usage of GaN-on-Si technology, it is possible to integrate the GaN technology with a mainstream silicon technology. So far, three different methods can be used to integrate GaN on silicon substrates. First approach is the direct epitaxial growth of GaN on large diameter wafer Si(111), which provides an important competitive advantage for the development of GaN-based high-performance power electronics [4–16]. This method is a way for large-scale device and circuit manufacturing. These devices, with an estimated potential market above $10 billion dollar, could be a key to energy savings equivalent to 10% of the global electricity consumption. Since the CMOS process technology is using Si(100) substrate, researchers have used hybrid integration by direct wafer bonding of a GaN-on-Si sample to a Si(100) wafer. This method has been used to demonstrate the first on-wafer integration of GaN electronics and Si(100) CMOS circuits [17]. The integration allows the combination of the high complexity and flexibility of Si circuits with the vast array of new devices enabled by GaN: LEDs, transistors, energy-harvesting devices, and filters. A third approach involves the fabrication of N-face GaN transistors by removing the Si substrate of a Ga-face-grown sample. By using this layer transfer method, the fabricated N-face GaN HEMTs had also exhibited good DC and RF performances [18]. This bonding process may not be a viable method for mass production.

GaN HEMTs have the benefits of high blocking voltage with faster switching speed, higher efficiency, and lower energy loss for large commercial markets such as mobile phone and DC power switching for automobile vehicles [10,19–21]. Based on the GaN-on-Si technology platform, HEMT can be produced in large volume using the existing growth technology of GaN on 150-mm-diameter Si(111) substrate. Many leading power device manufacturers such as International Rectifiers, Infineon, Fairchild, Samsung, and Panasonic are also focusing on the development of GaN-on-Si power switches using the growth technology of GaN on 100-mm to 150-mm-diameter Si(111) substrate [10,19–22]. According to Yole Development's market analysis, the GaN power device market can reach US$0.35 billion in 2015, then it will increase sharply in the following several years. In the next decade, GaN power device will become a strong competitor to the traditional Si power device and share the large discrete power device market valued at US$10 billion per year. It is projected that GaN-based devices could reach more than 7% of the overall power device market by 2020. GaN-on-Si wafers will be able to capture more than 1.5% of the overall power substrate volume, representing more than 50% of the overall GaN-on-Si wafer volume, subjecting to the hypothesis that the 600 V devices would take off in 2014–2015. Through the massive development in the material growth and process technology of GaN-on-Si, Transphorm, Inc., has started selling a qualified 600 V normally OFF-type GaN power switch for US$5.89 [23]. To further slash down the cost per chip at least 50% to 60%, large diameter GaN-on-Si wafer technology (>200-mm diameter) as well as the matured CMOS-compatible process line is necessary [24].

In this chapter, two of the most immediate challenges, namely, epitaxial growth and non-gold ohmic contact technologies, to realize low-cost GaN HEMTs on large diameter silicon substrates up to 200-mm diameter will be discussed. In Section 3.2, details of the epitaxial growth techniques for GaN on 200-mm-diameter silicon substrates are presented. The device fabrication and electrical results as well as the uniformity studies of the material and electrical parameters will also be given. In Section 3.3, we will summarize and report the non-gold process technologies for CMOS-compatible process. Finally, submicron gate transistor characteristics of both AlGaN/GaN and InAlN/GaN HEMTs using non-gold metal scheme will also be presented.

## 3.2    GaN ON 200-mm-DIAMETER Si (111) SUBSTRATE

After the first attempt by Manasevi et al. [25] in 1971 for the growth of AlN on Si using metal–organic chemical vapor deposition (MOCVD), Nagoya University had successfully grown GaN on Si using two different interlayers 3C-SiC [26] and AlN [27]. Nagoya Institute of Technology realized the difficulty of using GaN as a nucleation layer (NL) [28]. Figure 3.2 shows the history of GaN-on-Si growth technology and its advancement from small size sample to multiple wafers on 200-mm-diameter Si(111) substrates. Nowadays, universities and manufacturing industries have put in significant efforts to achieve reproducible large diameter GaN-on-Si wafers (up to 200-mm diameter). Though the GaN growth on 200-mm-diameter Si(111) is quite challenging, few research groups have demonstrated AlGaN/GaN HEMTs on 200-mm-diameter Si substrate [8–14]. After the first demonstration of GaN grown on 200-mm-diameter Si substrate [8], Chen et al. demonstrated AlGaN/GaN/AlGaN double heterostructures on a 200-mm-diameter Si(111) substrate by MOCVD with the two-dimensional-electron-gas (2DEG) mobility of 1766 cm$^2$/V·s [11]. The ultimate aim is to use the grown GaN on 200-mm-diameter Si substrate in a CMOS-compatible fab. Hence, the wafer quality should meet strict CMOS process standards. For example, the wafer bowing value needs to be well controlled below ±40 μm to reduce the failure of wafer handling by robotic arm during the lithography in the stepper.

De Jaeger et al. have successfully demonstrated the first non-gold CMOS-compatible metal–insulator–semiconductor HEMTs on a 200-mm-diameter Si(111) substrate using AlGaN/GaN/AlGaN double heterostructures [24]. Subsequently, Tripathy et al. have also successfully grown AlGaN/AlN/GaN single heterostructures on a 200-mm-diameter Si(111) substrate with 2DEG mobility of 1900 cm$^2$/V·s [13]. Recently, Arulkumaran et al. [12] have also successfully demonstrated the world's first microwave performance of AlGaN/GaN HEMTs grown on 200-mm-diameter Si(111) substrate using conventional III-V process. The Nagoya Institute of Technology team (Christy et al.) have also demonstrated 1.5-μm-gate AlGaN/GaN HEMTs on 200-mm-diameter Si(111) substrate with OFF-state breakdown voltage of 1.1 kV using conventional III-V process [14]. More recently, researchers have also reported the uniformity of AlGaN/AlN/GaN HEMTs on 200-mm-diameter Si(111) substrate [14]. All these encouraging results suggest that it is feasible to grow GaN on large diameter silicon substrates.

### 3.2.1    Growth of AlGaN/GaN Heterostructures

GaN is typically grown on foreign single crystalline substrates namely, sapphire, SiC, and Si with high crystalline quality. Nowadays, though native GaN substrates are available up to 2 inches, the cost is very expensive and hence it is mainly confined to R&D arena. There are several factors when considering the choice of substrate materials such as thermal conductivity, thermal expansion coefficient, cost, wafer diameter, and integration. Table 3.1 shows the material properties of different substrates for the growth of GaN. Due to the difficulty in the substrate making technology, small-size GaN substrates are now available in the market with very high cost. To increase the internal quantum efficiency, the miss cut or semipolar (20-21) free-standing GaN

**TABLE 3.1**

**Properties of Commonly Used Substrates for the Growth of GaN**

| Properties | GaN | Si | Sapphire | SiC | Diamond |
|---|---|---|---|---|---|
| Bandgap (eV) | 3.42 | 1.11 | 9.9 | 3.26 | 5.45 |
| Thermal conductivity at 300 K, $k$ (W/cm-K) | 2.0 | 1.5 | 0.35 | 4.9 | 180 |
| Lattice mismatch with GaN (%) | 0 | 17 | 14 | 3.5 | 89 |
| Thermal expansion coefficient ($\times 10^{-6}$ K$^{-1}$) | 5.5 | 2.6 | 7.5 | 4.46 | 1.0 |
| Thermal expansion coefficient mismatch GaN (%) | 0 | 52.7 | 36.4 | 18.9 | 81.8 |
| Substrate size (mm) | 30 | 300 | 150 | 150 | 10 |
| Substrate cost | Very high | Very low | Medium | High | Extremely high |

substrates have also been used for the demonstration of green lasers [29,30]. For high-power operations, the thermal conductivity of the substrate is expected to be an important design issue and the relatively poor thermal conductivity of sapphire ($k$Sapp = 0.35 W/cm-K) is expected to limit the device output power [7,31]. For the production of LEDs, industries are now using 150-mm-diameter sapphire as a substrate for the growth of GaN LED structure using laser lift-off technique [32]. SiC on the other hand, has good thermal conductivity ($k_{SiC}$ = 4.9 W/cm-K) and is an obvious choice for a substrate when thermal considerations are important. AlGaN/GaN high-power HEMTs are usually fabricated on Si–SiC substrates, which are commercially available but not cost-effective [33]. Although diamond has excellent thermal conductivity, the growth of GaN on diamond is still challenging due to a large lattice mismatch and a thermal expansion coefficient mismatch [34–36]. Hence, researchers have instead used bonding technology and demonstrated GaN devices on CVD diamond [37,38].

The large-size Si substrate can overcome these limitations of sapphire, SiC, and diamond to realize high-volume manufacturability. In addition, the thermal conductivity of Si ($k_{Si}$ = 1.5 W/cm-K) is also better than sapphire substrate. Normally, the device quality epitaxial growth of GaN on Si(111) substrates was carried out using both MOCVD [4–16,20,21,24–28] and molecular-beam epitaxy (MBE) [39–42] tools. So far, research groups have demonstrated GaN LEDs and HEMTs up to 200-mm-diameter Si(111) substrates [8–16,24]. Growth of GaN on 200-mm silicon is quite challenging because the larger wafer diameter leads to problems such as high wafer bow that can lead to nonuniform material properties and make further device processing difficult. The growth of GaN on 200-mm-diameter Si substrate has been mostly realized by MOCVD [8–16,24]. Initially, researchers have tried thick 200-mm-diameter Si substrate up to 1.5 mm, which helps to suppress the wafer bowing values [12,13]. De Jaeger et al. [24] have used 1.15-mm-thick 200-mm-diameter Si wafers (instead of the commonly used 730-μm thick) for the growth of ~3-μm-thick AlGaN/GaN/AlGaN epitaxial stack. The achieved bowing value was in the acceptable range of lower than ±50 μm, which is essential for device fabrication using conventional contact lithography or stepper. Subsequently, the researchers have also realized small wafer bowing values down to ~25 μm using 1.0-mm-thick silicon substrate [15].

For a high-quality HEMT epitaxial layer, it is very critical to have a good transition layer on the silicon substrates to circumvent the large lattice and thermal mismatched between GaN and Si. In general, the growth start with the heating of the Si(111) substrate to a high temperature (~1000°C) in $H_2$ atmosphere and annealed for about 3 minutes to remove native oxide from the surface. Then, a high-temperature AlN NL was normally grown to prevent the melt-back etching [28] between Ga and Si and also to improve the wetting properties of III–nitrides on Si. The AlN NL (thickness 100–140 nm) is optimized in terms of both structural quality and surface morphology. After the NL growth, specially designed stress mitigation layers were grown to achieve crack-free GaN layers. There are various proprietary approaches used by different groups [11–15]. For example, one approach uses a three-step-graded $Al_xGa_{1-x}N$ buffer layers with an Al content of 75%, 50%, and 25% that were grown by maintaining the chamber pressure of 100 Torr. Then, about 2.4- to 2.5-μm-thick GaN was overgrown on these graded buffers using a single-strain-compensating low-temperature AlN interlayer. Finally, the $Al_xGa_{1-x}N$ barrier (0.2> x ≤0.3) and GaN cap layer (3 nm) were grown to complete the growth of AlGaN/GaN HEMT structure [13]. Another approach uses $Al_{0.3}Ga_{0.7}N$ (40 nm)/AlN (100 nm) interlayers that were grown at 1100°C on AlN NLs. Then the GaN/AlN (20/5 nm) superlattice structures (SLS) were also grown with a thickness of 1.25–3 μm. Finally, the unintentionally doped GaN (UID-GaN) layer and the $Al_{0.26}Ga_{0.74}N$ barrier layer with GaN cap layer were grown to complete the HEMT structure growth [14].

Figure 3.3 shows (a) double heterostructure AlGaN/GaN/AlGaN HEMTs with graded AlGaN NL [11], (b) single heterostructure AlGaN/GaN HEMT with graded AlGaN and low-temperature grown AlN interlayer [12,13,15], and (c) single heterostructure AlGaN/GaN HEMTs with AlN NL and AlN/GaN superlattices [14]. Chen et al. [11] have reported crack-free AlGaN/GaN/AlGaN double heterostructures on 200-mm-diameter, 1.15-mm-thick Si substrate with a bowing value as low as 33 μm (See Figure 3.3a). Tripathy et al. [13] reported a crack-free epilayers with total stack thickness of ~3.5 μm with a bowing value of <25 μm using 1.0-mm-thick Si(111) substrates (See Figure 3.3b). Subsequently, Arulkumaran et al. [15] reported uniformity studies of electrical and device properties of grown AlGaN/GaN HEMTs on 200-mm-diameter Si substrate with AlN spacer layer, which can reduce the alloy scattering followed by the enhancement of 2DEG mobility. Christy et al. [14] used 80 pairs of GaN/AlN (20/5 nm) SLS to mitigate the stress between the silicon substrate and GaN (See Figure 3.3c). The bowing value of 50 μm was achieved for 3.8-μm-thick buffer GaN. More recently, Institute of Materials Research and Engineering (IMRE) research team demonstrated AlGaN/GaN HEMTs on 1.0-mm-thick, 200-mm-diameter Si(111) substrate using AIXTRON CCS MOCVD system [16,43]. The total thickness of the nitride stack is 4.3 μm, which consists of 400–nm-thick AlN NL, three-step-graded $Al_xGa_{1-x}N$ intermediate layers (AlGaN I, AlGaN II, and AlGaN III) with a composition tuning range of Al from 60% to 20%, followed by an uninterrupted growth of GaN with a thickness >2.3 μm without any interlayers. Subsequently, they have also demonstrated both AlGaN/GaN HEMTs [43] and InAlN/GaN HEMTs [16] by growing $Al_{0.24}Ga_{0.76}N$ barrier and $In_{0.16}Al_{0.84}N$ barrier, respectively, on 200-mm-diameter Si substrate.

Figure 3.4 shows the photograph of crack-free AlGaN/GaN HEMT structures grown on a full 200-mm-diameter Si(111) with a starting substrate resistivity of (a) ~100 Ω-cm [13] and (b) 0.01 Ω-cm using the MOCVD system [15]. The bowing value decreases down to 25 μm for 1.0-mm-thick, 200-mm-diameter substrate with a resistivity of 0.01 Ω-cm. Figure 3.5 shows atomic force micrograph (AFM) of AlGaN/GaN HEMTs on 200-mm-diameter Si substrate (scan area: 5 μm$_2$). The surface root-mean-square (RMS) roughness is 0.25 nm, which are similar to the reported data in the literature in the cases of AlGaN/GaN HEMT structures grown on 100- to 150-mm-diameter Si substrates [4,44–46]. Figure 3.6 shows the cross-sectional high angle annular dark field scanning transmission electron microscopy (STEM) image of the (a) full HEMT structure, (b) GaN cap + AlGaN barrier layer, (c) stress mitigation layers grown on 200-mm-diameter Si(111). This HEMT structure has two AlN interlayers grown at low temperature to suppress the dislocation density. The average full width at half maximum (FWHM) of X-ray rocking curves of (002) and (102) diffraction planes of GaN are 418–420 and 612–625 arcsecs, respectively. The observation of small FWHM of 504 arcsecs for (105) diffraction

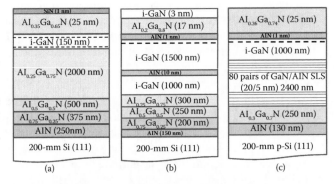

**FIGURE 3.3** (a) Double heterostructure AlGaN/GaN/AlGaN high-electron-mobility transistors (HEMTs) with graded AlGaN nucleation layer (NL), (b) single heterostructure AlGaN/GaN HEMTs with graded AlGaN and low-temperature grown AlN interlayer (IL), and (c) single heterostructure AlGaN/GaN HEMT with AlN NL and AlN/GaN superlattices.

**FIGURE 3.4** Photograph of crack-free AlGaN/GaN HEMT structures grown on a full 200-mm-diameter Si(111) with a starting resistivity of (a) ~100 Ω-cm and (b) 0.01 Ω-cm using the metal–organic chemical vapor deposition system.

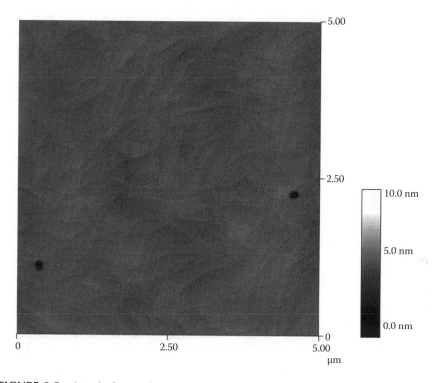

**FIGURE 3.5** Atomic force micrograph of AlGaN/GaN HEMTs on 200-mm-diameter Si substrate (scan area: 5 μm²).

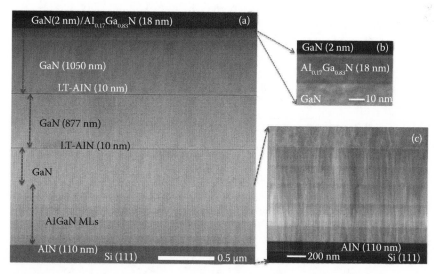

**FIGURE 3.6** Cross-sectional high angle annular dark field scanning transmission electron microscopy (STEM) image of the (a) full HEMT structure, (b) GaN cap + $Al_{0.17}$GaN barrier layer, and (c) stress mitigation layers grown on 200-mm-diameter Si(111).

plane rocking curve is an indicative of high crystalline device quality ~2.9 μm-thick GaN on 200-mm Si substrate [13]. Christy et al. [14] claims that the FWHM of (004), (100), and (102) are 478, 1257, and 854 arcsecs, respectively, is an indicative of high-crystalline device grade GaN epitaxy on 200-mm Si substrate. The average edge- and screw-dislocation densities of the AlGaN/GaN HEMT structure (See Figure 3.3c) on the samples were in the order of $1.5 \times 10^9$ and $5.3 \times 10^8$ cm$^{-2}$, respectively [12]. The HEMT structure with SLS buffer (See Figure 3.3c) exhibited the edge-dislocation densities of $8.0 \times 10^9$ cm$^{-2}$ and screw-dislocation densities of $4.4 \times 10^8$ cm$^{-2}$, respectively. The edge-dislocation density values are about seven times lower for the HEMT structure without SLS layers [5,14]. To make highly resistive GaN for electronic applications, carbon autodoping has been used (by changing the V/III [N/Ga] ratio). Some researchers have also intentionally incorporated carbon doping to increase the buffer breakdown voltage. Alternatively, thick GaN buffer layer can also be used to enhance the breakdown voltage ($BV_{gd}$) of the transistors [5,9].

To study the uniformity of the grown wafer, data from five samples were taken from the different locations of 200-mm-diameter Si substrate (Figure 3.3a). Figure 3.7 shows room temperature 2DEG mobility ($\mu_H$) and sheet resistance ($R_{sh}$) of AlGaN/GaN HEMT measured from van der Pauw Hall samples taken from center for edge of 200-mm-diameter Si substrate. Inset of Figure 3.7 shows the Hall samples from different locations of the 200-mm-diameter Si substrate. The HEMT structure exhibited the average $\mu_H$ of 1550 cm$^2$/V·s, sheet carrier concentration ($n_s$) of $0.84 \times 10^{13}$ cm$^{-2}$, and an average $R_{sh}$ of <400 Ω/sq. IMRE researchers have reported $\mu_H$ of 1370 cm$^2$/V·s, $n_s$ of $1.35 \times 10^{13}$ cm$^{-2}$, and an average $R_{sh}$ of <500 Ω/sq. Recently, M/s. Raytheon has also realized the growth of AlGaN/GaN HEMT structure on 200-mm-diameter Si substrate with 2DEG properties by MBE system [42]. The MBE grown AlGaN/GaN HEMT were also exhibited equally good $\mu_H$ and $n_s$ values. M/s. Veeco has also reported the highest $\mu_H$ of 2000 cm$^2$/V·s from AlGaN/GaN HEMTs with AlN spacer layer on 200-mm-diameter Si substrate [47]. The highest $n_s \times \mu_H$ of $2.21 \times 10^{16}$/V·s was reported by the research team at the

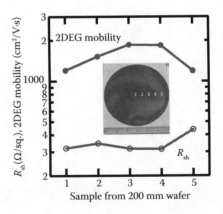

**FIGURE 3.7**  Room temperature two-dimensional-electron-gas (2DEG) mobility ($\mu_H$) and sheet resistance ($R_{sh}$) of AlGaN/GaN HEMTs measured from van der Pauw Hall samples. Inset photo shows the Hall sample locations of 200-mm-diameter Si(111) substrate.

Nanyang Technological University using AlGaN/GaN HEMT structure with AlN spacer layer [14,48]. Figure 3.8 shows the cross-sectional high-resolution transmission electron microscopic (HR-TEM) image of AlGaN/GaN HEMT structure on 200-mm-diameter Si(111) with AlN spacer layer. Table 3.2 shows the list of reported AlGaN/GaN HEMTs with 2DEG properties on 200-mm-diameter Si(111) substrate.

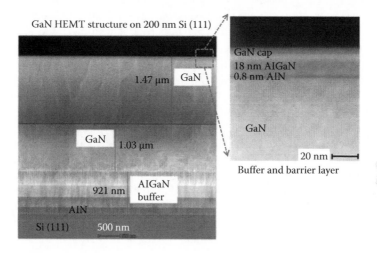

**FIGURE 3.8**    Cross-sectional high-resolution transmission electron microscopy (HR-TEM) image of AlGaN/GaN HEMT structure on 200-mm-diameter Si(111) with 0.8-nm-thick AlN spacer layer.

**TABLE 3.2**
**AlGaN/GaN HEMTs on 200-mm-Diameter Si(111) Substrate**

| Affiliation | Year | Growth Method | Substrate Thickness (μm) | Total Buffer Thickness (μm) | Bowing (μm) | Sheet Resistance, $R_{sh}$ (Ω/sq) | 2DEG Mobility, $\mu_H$ (cm²/V·s) | Sheet Carrier Density, $n_s$ (×10¹³ cm⁻²) | Product ($\mu_H \times n_s$) (×10¹⁶/V·s) |
|---|---|---|---|---|---|---|---|---|---|
| IMEC [11] | 2012 | MOCVD | 1300 | — | 20 | 306 | 1766 | 1.16 | 2.05 |
| NTU [12] | 2012 | MOCVD | 1500 | 2.9 | 90 | 484 | 1550 | 0.84 | 1.30 |
| IMRE [13] | 2012 | MOCVD | — | 3.3 | 20 | — | 1900 | — | — |
| NIT [14] | 2012 | MOCVD | — | 3.71 | 60 | — | 1660 | 0.85 | 1.41 |
| NTU [15] | 2013 | MOCVD | 1000 | 3 | 20 | 387 | 1990 | 1.11 | 2.21 |
| Raytheon [42] | 2014 | MBE | — | — | — | 451 | 1522 | 0.92 | 1.40 |
| Raytheon & IQE [42] | 2014 | MOCVD | — | — | 30 | 500 | 1600 | — | — |
| IMRE [43] | 2014 | MOCVD | 1000 | 2.3 | — | 390 | 1540 | 1.35 | 2.08 |
| Veeco [47] | 2013 | MOCVD | — | 3 | 50 | — | 2000 | 0.8 | 1.60 |
| NIT [48] | 2013 | MOCVD | — | 3.8 | 50 | — | 1500 | 1.37 | 2.06 |
| IMRE [49] | 2014 | MOCVD | 1000 | 4.4 | 50 | 484 | 1370 | 1.35 | 1.85 |

### 3.2.2 Device Fabrication

Figure 3.9 shows a typical process flow of submicron-gate AlGaN/GaN HEMTs based on conventional III-V gold-based process technology. The mesa isolation for HEMT device fabrication was accomplished by dry etching down to GaN buffer layer using $Cl_2$ /$BCl_3$ plasma-based inductive coupled plasma system. The ohmic contacts were formed using a conventional four-layer metallization scheme (Ti/Al/Ni/Au) (25/120/40/50 nm) followed by rapid thermal annealing (RTA) at 825°C for 30 seconds. The measured contact resistance ($R_c$) was about 1.8 Ω-mm. Due to the existence of AlN spacer layer in the AlGaN/GaN HEMT structure, the $R_c$ values are higher than the standard typical $R_c$ values (0.18 to 0.3 Ω-mm). To reduce the $R_c$ values, optimized ohmic-recess etching was also realized [50]. Figure 3.10 shows the typical channel current and buffer leakage current of AlGaN/GaN HEMTs. The measured $I_{ON}/I_{OFF}$ ratio is $3.74 \times 10^7$ and $1.36 \times 10^5$ at a bias voltage of 20 V and 100 V, respectively. The grown buffer GaN is having acceptable range of buffer leakage current, which is suitable for the fabrication of transistor. After the ohmic contact formation, the submicron mushroom gate (0.3-μm T-gate) and 2-μm gate was defined by electron beam lithography (EBL) and optical contact lithography, respectively. To realize the normally OFF devices, gate-recess etching (~18 nm) was also carried out on AlGaN/GaN HEMTs on 200-mm-diameter Si substrate using optimized dry etching process [50]. Then, the evaporation by electron beam evaporation and lift-off

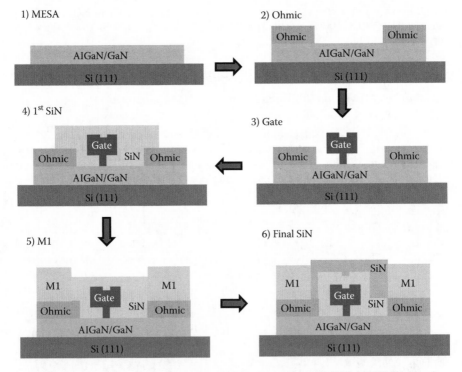

**FIGURE 3.9** Typical device process flow of submicron gate AlGaN/GaN HEMTs based on conventional III-IV gold-based process technology.

of Ni/Au (50/400 nm) was used to form the Schottky gates. Finally, the devices were passivated with 120-nm-thick SiN deposited by plasma-enhanced chemical vapor deposition (PECVD) [51]. The details of device processing can also be found elsewhere [12,15,52]. Figure 3.11 shows the (a) top-view scanning electron microscopic image of the formed T-gate with ~0.3-μm gate length by EBL.

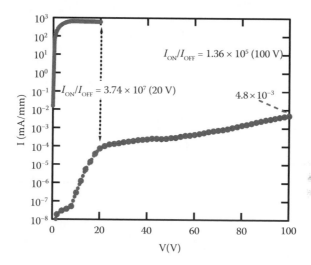

**FIGURE 3.10** Typical channel current and buffer leakage current of AlGaN/GaN HEMTs on silicon substrate.

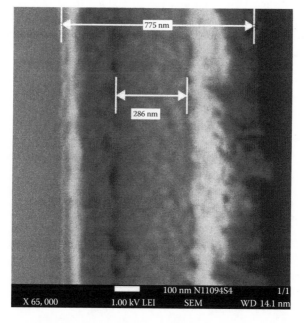

**FIGURE 3.11** (a) Top-view scanning electron microscopic image of the formed T-gate with ~0.3-μm gate-length on AlGaN/GaN HEMT structure.

### 3.2.3 DC CHARACTERISTICS OF AlGaN/GaN HEMTs

To measure the 2DEG density as a function of channel depth, capacitance–voltage ($C–V$) measurements were carried out at 1.0 MHz on the Schottky diodes [12,44]. Figure 3.12 shows the measured $C–V$ profile of the Schottky diode (diameter 50 μm). Similar $C–V$ profile was also measured using mercury probe measurements (See Section 3.2.5). The 2DEG density of around $1.8 \times 10^{20}$ cm$^{-3}$ peaks at a depth of 20.2 nm, which matches well with the AlGaN barrier thickness as seen from STEM imaging (see Figure 3.6). The background carrier density in the buffer GaN is about $4.98 \times 10^{14}$ cm$^{-3}$, and this value is comparable to the HEMT structures previously reported on 100-mm Si substrate [4,44]. The $(N_{D2DEG})$peak/$(N_{D2DEG})$min ratio is $3.6 \times 10^5$, which correlates well with the channel $I_{ON}/I_{OFF}$ ratio in such devices.

Figure 3.13 shows typical $I_{DS}–V_{DS}$ characteristics of (a) unrecessed- and (b) recessed-gate AlGaN/GaN HEMTs on 200-mm-diameter Si(111). Figure 3.14 shows the transfer characteristics of unrecessed- and recessed-gate AlGaN/GaN HEMTs. The dimensions of the devices are the following: source-gate distance, $L_{sg} = 0.8$ μm; gate width, $W_g = (2 \times 25)$ μm; gate length, $L_g \sim 0.3$ μm; gate-drain distance, $L_{gd} = 1.25$ μm; and gate-gate distance, $L_{gg} = 12$ μm. The device exhibited good pinch-off characteristics with the maximum drain current ($I_{Dmax}$) of 853 mA/mm and maximum extrinsic transconductance ($g_{mmax}$) of 178 mS/mm. The threshold voltage ($V_{th}$) is −3.4 V for unrecessed gate AlGaN/GaN HEMTs. However, the recessed-gate AlGaN/GaN HEMTs exhibited $V_{th}$ of +0.1 V by optimized gate-recess etching process. The recessed-gate GaN HEMTs exhibited the $I_{Dmax}$ of 225 mA/mm and the $g_{mmax}$ of 180 mS/mm with good pinch-off at $V_g = 0$ V.

The three-terminal OFF-state breakdown voltage ($BV_{gd}$) and the two-terminal lateral buffer breakdown ($BV_{Buff}$) characteristics of AlGaN/GaN HEMTs were carried out by a fixed current compliance of 0.5 mA/mm. The 0.3-μm-gate HEMTs with $W_g/L_{sg}/L_{gd}/L_{gg} = (2 \times 25)/0.8/2/12$ μm exhibited a $BV_{gd}$ of 60 V. However, the

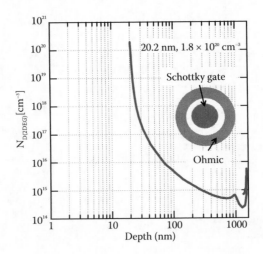

**FIGURE 3.12** $C–V$ profile of the Schottky diode on AlGaN/GaN HEMTs on 200-mm-diameter Si. Inset: Guard ring type Schottky diode with 50-μm diameter.

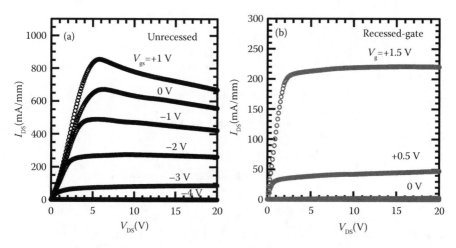

**FIGURE 3.13** Typical $I_{DS}$–$V_{DS}$ characteristics of (a) unrecessed- and (b) recessed-gate AlGaN/GaN HEMTs on 200-mm-diameter Si(111).

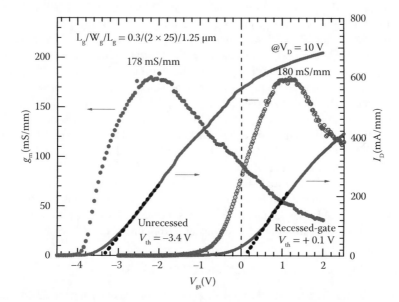

**FIGURE 3.14** Transfer characteristics of unrecessed- and recessed-gate AlGaN/GaN HEMTs on 200-mm-diameter Si(111).

2.0-μm-gate HEMTs with $W_g/L_g/L_{sg}/L_{gd}$ = (1 × 80)/2/2/6 μm exhibited a $BV_{gd}$ of 188 V, which is close to the $BV_{Buff}$ value (see Figure 3.15). The $BV_{Buff}$ of 192 V was measured between the two adjacent ohmic pads (50 × 50 μm²) with 10 μm gap. The breakdown strength ($BV_{gd}/L_{gd}$) of ~0.3 MV/cm has been occurred for both 0.3-μm-gate and 2.0-μm-gate AlGaN/GaN HEMTs. Normally, the $BV_{gd}$ was limited by the GaN buffer thickness and its crystalline quality [9]. Thick high crystalline quality GaN buffer layers are often desired to increase the $BV_{gd}$ of the devices [5,9].

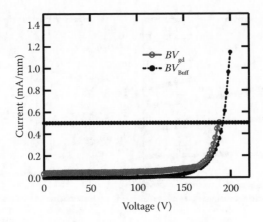

**FIGURE 3.15** The three-terminal OFF-state breakdown voltage ($BV_{gd}$) and the two-terminal lateral buffer breakdown ($BV_{Buff}$) characteristics of 2-μm-gate AlGaN/GaN HEMTs.

Christy et al. [14] reported 1100 V of $BV_{gd}$ for devices ($W_g/L_{gd}/L_g = 15/20/1.5$ μm) for 4-μm-thick buffer GaN with specific ON-resistance of 2.3 mΩ-cm$^2$.

This is the highest $BV_{gd}$ achieved so far in the conventional single heterostructure AlGaN/GaN HEMTs on 200-mm-diameter Si substrate. These values are almost closer to the values achieved using 7.0-μm-thick buffer GaN on 100-mm-diameter Si substrate by Selvaraj et al. [9]. However, Chen et al. [11] have reported only the two-terminal buffer breakdown voltage of 1380 V with a total buffer thickness of 4.6 μm for the double heterostructure AlGaN/GaN/AlGaN HEMTs on 200-mm-diameter Si substrate. The observation of enhanced buffer breakdown is normal for double heterostructure AlGaN/GaN/AlGaN HEMTs due to wider bandgap of AlGaN buffer layer. The power device figure-of-merit (Power-FoM = $BV_{gd}^2/R_{DS[ON]}$) for AlGaN/GaN HEMTs on 200-mm diameter [14] is $5.3 \times 10^8$ V$^2$·Ω$^{-1}$·cm$^{-2}$, which is better than the HEMTs with 2.4-μm buffer GaN ($4 \times 10^8$ V$^2$·Ω$^{-1}$·cm$^{-2}$) on 100-mm-diameter high-resistivity Si(111) substrate [21]. Table 3.3 shows the device DC parameters of AlGaN/GaN HEMTs on 200-mm-diameter Si substrate from different research groups.

### 3.2.4 RF Characteristics of AlGaN/GaN HEMTs

The small-signal characterization of the devices was carried out using HP 8510c vector network analyzer [12,52,54,55]. Figure 3.16 shows the small-signal microwave performance of $(2 \times 75)$ μm wide-gate (a) unrecessed (D-mode) and (b) recessed-gate (E mode) AlGaN/GaN HEMTs on a 200-mm Si(111). A unit current gain cutoff frequency ($f_{max}$) of 28 GHz and a maximum oscillation frequency ($f_{max}$) of 64 GHz was achieved in the D-mode HEMTs at $V_D = 10$ V and $V_g = -2.4$ V. The $f_T$ of 28 GHz and $f_{max}$ of 32 GHz was achieved in the E-mode HEMTs at $V_D = 10$ V and $V_g = +0.1$ V. The $f_T \times L_g$ is 8.4 GHz·μm, which is comparable to the AlGaN/GaN HEMTs fabricated on smaller diameter Si substrate [55]. The trend of $f_T$ and $f_{max}$ for different bias conditions ($V_g$ and $V_D$) for both types of devices (unrecessed and recessed gate) are shown in the Figure 3.17, which is consistent with the device transfer characteristics (See Figure 3.14). This is the

**TABLE 3.3**

**Device DC Parameters of AlGaN/GaN HEMTs on 200-mm-Diameter Si Substrate from Different Research Groups**

| Affiliation | $R_c$ (Ω-mm) | $I_{ON}/I_{OFF}$ Ratio (×10⁶) | Buffer Break-down (V) | $L_{sg}$ (μm) | $L_{gd}$ (μm) | $L_g$ (μm) | $W_g$ (μm) | $V_{th}$ (V) | $I_{Dmax}$ (mA/m m) | $g_{mmax}$ (mS/ mm) | Device Break-down (V) |
|---|---|---|---|---|---|---|---|---|---|---|---|
| IMEC [11] | 1.25 | ~1 | 1380 | — | — | — | — | — | 650 | — | — |
| NTU [12] | 1.80 | 0.14 | 192 | 0.8 | 1.25 | 0.30 | 50 | −3.8 | 853 | 180 | 188 |
| IMRE [13] | — | — | — | — | — | 1.50 | 100 | −3.0 | 660 | 210 | — |
| NIT [14] | — | ~0.12 | — | — | 20 | 1.50 | 15 | −4.2 | 850 | 150 | 1100 |
| NTU [15] | 0.31 | 9 | 200 | 0.80 | 1.25 | 0.30 | 50 | −3.0 | 685 | 202 | 170 |
| Raytheon [42] | — | — | — | — | — | 0.25 | 10 | — | 910 | — | — |
| IMRE [43] | 9.02 | 10 | — | 2.00 | 10 | 1.50 | 100 | — | 550 | 197 | — |
| NIT [48] | — | ~0.12 | — | — | 20 | 1.50 | 15 | −4.2 | 856 | 153 | 1110 |
| IMRE [53] | — | — | — | 2.00 | 9.5 | 2.50 | — | — | 550 | — | 414 |

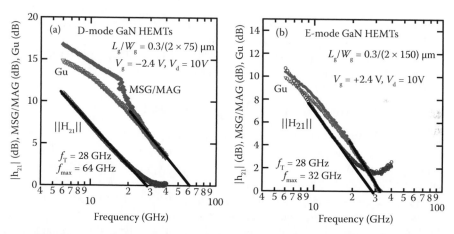

**FIGURE 3.16** Small-signal microwave performance of (2 × 75) μm wide-gate (a) unrecessed (D-mode) and (b) recessed-gate (E-mode) AlGaN/GaN HEMTs on a 200-mm Si(111).

first small-signal measurement data of AlGaN/GaN HEMTs on 200-mm-diameter Si substrate [12]. Due to the high output conductance ($g_{ds}$), E-mode HEMTs exhibited low $f_{max}$, which has been confirmed by intrinsic parameter extraction. A conductive buffer layer can introduce parasitic capacitances (extrinsic capacitances), which can lower the available power gains of the HEMT at high-microwave frequencies. However, the ratio of $f_{max}/f_T > 1$ (~2.28) supports the fact that the grown buffer GaN does not have additional charge coupling effects by parallel conduction in the buffer GaN [56]. The Johnson's figure-of-merit (JFoM = $f_T × BV_{gd}$) of the fabricated HEMT exhibited ~1.7 THz·V, which is reasonably good for the devices on 200-mm-diameter Si(111).

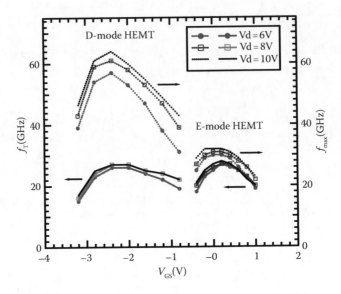

**FIGURE 3.17**  Device $f_T$ and $f_{max}$ as a function of $V_g$ and $V_D$ for both unrecessed- and recessed-gate AlGaN/GaN HEMTs on 200-mm-diameter Si(111).

Recently, researchers at the Nanyang Technological University have also achieved JFoM = 5.41 THz·V for 0.3-μm T-gate [57] and 8.32 THz·V for 0.15-μm T-gate [58] AlGaN/GaN HEMTs without any additional field plate on 100-mm-diameter high-resistivity Si substrate by $(NH_4)_2S_x$ sulfur passivation.

### 3.2.5  UNIFORMITY STUDIES

The uniformity of the electrical properties of AlGaN/GaN HEMT structure and device parameters is essential to get higher device yield. Christy et al. [14] reported a radial uniformity by measuring the Hall samples taken from ½ of a 200-mm-diameter Si substrate. Figure 3.18a shows the schematic cross-section of the AlGaN/GaN HEMT structure with AlN spacer layer on 200-mm-diameter Si substrate. About 18 Hall samples were prepared and measured the Hall parameters (sheet resistance, 2DEG mobility, and sheet carrier density) at room temperature. The radial distribution of Hall parameters is symmetrical when a line is drawn in the center of the wafer (See Figure 3.18b). Due to the symmetrical nature of Hall parameters, a ¼ of a 200-mm wafer was taken for device fabrication and in-wafer uniformity study (See Figure 3.19). Such approach has also been studied and reported using AlGaN/GaN HEMTs 100-mm-diameter sapphire substrate [7]. The uniformity of the measured parameters reported in this work was calculated using the formula, Uniformity (%) = (1 − σ/x) 100, where, σ is the standard deviation of the measured data and $x$ is the average of the measured data. Figure 3.20 shows the contour mapping of the 2DEG profile for full 200-mm-diameter AlGaN/GaN HEMTs on Si which was measured by mercury probe method. The 2DEG carrier concentration profile peaks at a depth of ~21 nm, which is in agreement with the grown barrier thickness measured by cross-sectional

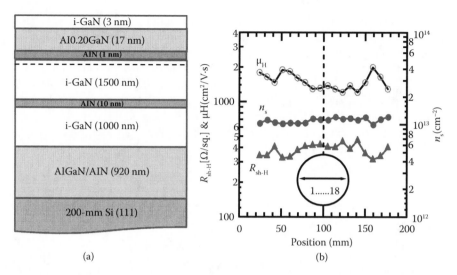

(a)                    (b)

**FIGURE 3.18** Schematic cross-section of the AlGaN/GaN HEMT structure with 0.8-μm-thick AlN spacer layer on 200-mm-diameter Si substrate. (b) The radial distribution of Hall parameters measured at room temperature from the Hall samples of 200-mm-diameter Si(111).

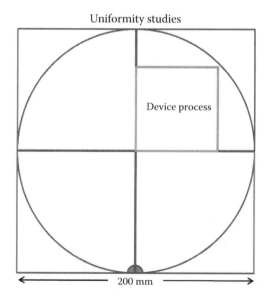

**FIGURE 3.19** Schematic diagram to show 200-mm-diameter AlGaN/GaN HEMT wafer. The selected region (first quadrant) of the wafer for the fabrication of submicron devices for uniformity studies.

STEM image (see Figure 3.8). The average 2DEG depth is 21.44 ± 0.28 nm. Three wafers with similar device structures were also statistically analyzed. The wafer-to-wafer uniformity of 2DEG depth is also greater than 98%. Apart from the uniformity of Hall parameters, barrier thickness, and background doping concentration of buffer, Arulkumaran et al. [15] has also investigated the uniformity of the device DC parameters. The average $I_{ON}/I_{OFF}$ ratio of the fabricated HEMTs are >9 × 10⁶ obtained from 20 devices of ¼ of a 200-mm-diameter Si (See Figure 3.21).

Figure 3.22 shows the contour mapping of (a) $I_{Dmax}$, (b) $g_{mmax}$, (c) threshold voltage ($V_{th}$), and (d) 2DEG depth of AlGaN/GaN HEMTs on a ¼ of 200-mm-diameter Si substrate. The average $I_{Dmax}$ and $g_{mmax}$ values of 685 mA/mm and 202 mS/mm

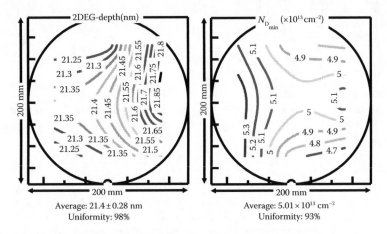

Average: 21.4 ± 0.28 nm
Uniformity: 98%

Average: 5.01 × 10¹³ cm⁻²
Uniformity: 93%

**FIGURE 3.20** Contour mapping of the 2DEG profile (Left: 2DEG peak location from the surface; Right: lowest value of the carrier concentration [$N_D$] in the buffer GaN) for full 200-mm-diameter AlGaN/GaN HEMTs on Si, which was measured by mercury probe method.

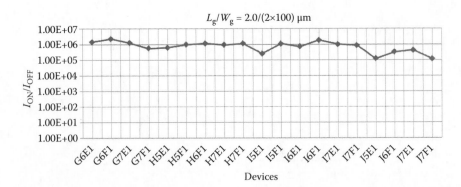

**FIGURE 3.21** The measure $I_{ON}/I_{OFF}$ ratio of AlGaN/GaN HEMTs on 20-mm-diameter Si substrate. About 20 devices were selected for uniformity measurements in the ¼ of 200-mm-diameter Si(111) substrate.

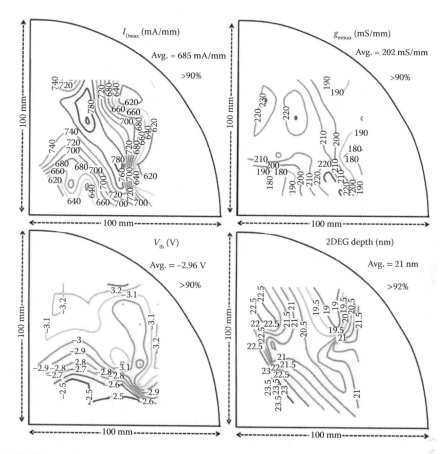

**FIGURE 3.22** The contour mapping of (a) $I_{Dmax}$, (b) $g_{mmax}$, (c) threshold voltage ($V_{th}$), and 2DEG depth of AlGaN/GaN HEMTs on a ¼ of 200-mm-diameter Si(111) substrate.

with uniformities of >90% were observed on a ¼ of the 200-mm wafer. The uniformity of the device parameters is >90%, which is consistent with the uniformity of the material properties such as $\mu_H$ (average = 1600 cm²/V·s), $n_s$ (average = 1.11 × 10¹³ cm⁻²), and $R_{sh-H}$ (average = 387 Ω/sq). From the measured Hall parameters, it was shown that the grown AlGaN/AlN/GaN heterostructures have good uniformity across the 200-mm-diameter Si(111) substrate. The nonuniformity of $\mu_H$, $n_s$, and $R_{sh-H}$ are <10% over a full 200-mm-Si substrate (see Figure 3.18). The average $V_{th}$ value of the devices is –2.96 V with a uniformity of >90%, which is consistent with the uniformity of 2DEG depth measured by C–V measurements. Christy et al. [14] reported the overall epilayer thickness uniformity (3.71 μm with standard deviation of 0.053 μm) and Al content (23.1% with standard deviation of 0.71%) uniformity in the AlGaN barrier.

Figure 3.23 shows the measured sheet resistance ($R_{sh-P}$) of AlGaN/AlN/GaN heterostructures using van der Pauw patterns on the device processed sample and the $R_{sh-H}$ values measured in the van der Pauw Hall samples. The trend of $R_{sh-P}$ values follows closely the trend of $R_{sh-H}$ measured by Hall measurements at room

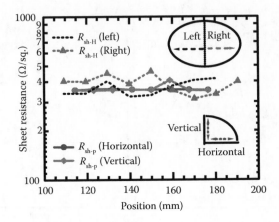

**FIGURE 3.23**  Measured sheet resistance ($R_{sh-P}$) values from the *van der Pauw* patterns in the device processed sample and the $R_{sh-H}$ values measured in the *van der Pauw* Hall samples of AlGaN/GaN heterostructures on 200-mm-diameter Si(111).

temperature. The processed quarter of the 200-mm GaN HEMT sample's average $R_{sh-P}$ values are 380 ± 14 Ω/sq with a uniformity >90%, which is consistent with the $R_{sh-H}$ values (387 ± 44 Ω/sq) obtained from Hall samples prepared from a full 200-mm GaN HEMT wafer (See Figure 3.18b). From the experimental data, inference can be made that the remaining three quarters of the 200-mm wafer can have similar uniformity as the selected quarter of the 200-mm GaN HEMT wafer. Overall, the radial distribution of $R_{sh-H}$ over a full 200-mm GaN HEMT wafer and $R_{sh-P}$ over a quarter of the 200-mm GaN HEMT wafer is >90%, which is very much comparable to the AlGaN/GaN HEMT structures on a 100-mm Si substrate [4]. The $BV_{Buff}$ of AlGaN/AlN/GaN HEMTs with an average value of 170 V has also been measured over a quarter of the 200-mm diameter Si substrate with >88% uniformity. The highest $BV_{Buff}$ on this HEMT structure is 200 V. Further increase of $BV_{Buff}$ is also possible by increasing the buffer GaN thickness [5,9].

### 3.2.6  SUMMARY

There are only a few research groups that have successfully demonstrated crack-free AlGaN/GaN HEMT structures on 200-mm-diameter Si(111) substrate with uniform electrical and structural properties. The bowing values of AlGaN/GaN HEMTs on 200-mm-diameter Si substrate are in the range between 20 and 90 μm. These thicknesses depend on the starting substrate thicknesses as well as the buffer GaN. AlGaN/GaN HEMTs with 2.0-μm gate and 0.3-μm gate were fabricated and their device characteristics were also measured and compared with the measured electrical and structural properties. The realization of OFF-state breakdown voltage beyond 1100 V for 3.8-μm-thick device with low ON-resistance and high FoM are the key challenges for AlGaN/GaN HEMTs on 200-mm-diameter Si technology to be competitive with other technologies such as SiC field effect transistors (FETs). The radial distribution of sheet resistance by van der Pauw patterns, Hall parameters, and 2DEG profile with uniformity >90% over a full 200-mm-diameter Si(111) sub-strates has been demonstrated. The recessed-gate E-mode AlGaN/GaN HEMTs were also realized on 200-mm-diameter Si(111) substrate. Good uniformity of >90% has been achieved with the GaN HEMT parameters (buffer leakage current, drain current density, extrinsic transconductance, threshold voltage), which are in agree-ment with the electrical parameters of the grown AlGaN/GaN heterostructures on 200-mm-diameter Si(111) substrate. The recessed-gate and unrecessed-gate AlGaN/GaN HEMTs exhibited an $f_T/f_{max}$ ratio of 28/32 GHz and $f_T/f_{max}$ ratio of 28/64 GHz. The demonstrated results shows the feasibility of achieving good uniformity AlGaN/GaN HEMTs on 200-mm-diameter Si(111) substrate for low-cost high-power and high-speed switching device applications.

## 3.3   NON-GOLD PROCESS TECHNOLOGY

GaN device processing in dedicated III-V fabs typically uses Ni/Au or Mo/Au lift-off metallization schemes to form ohmic and gate contacts. GaN device processing in a high-volume production Si CMOS fab requires the implementation of non-gold metallization schemes and the use of dry etch patterning instead of lift-off. To date, the fabrication of non-gold ohmic contacts with $R_c$ value below 1.0 $\Omega$-mm remains a challenge [24,59]. Recently, AlGaN/GaN HEMTs on a large Si substrates (up to 200-mm diameter) have been investigated to develop a high-volume low-cost fabri-cation process using the existing 200-mm silicon CMOS line [24]. The established conventional gold-based ohmic scheme (Ti/Al/Ni/Au) cannot be used in a typi-cal Si CMOS line because of contamination issues caused by gold. Moreover, the usage of gold inevitably results in higher cost, which is commercially undesirable. Therefore, the development of CMOS-compatible non-gold ohmic contacts for the fabrication of GaN HEMTs is of utmost importance. In this section, we discuss the non-gold ohmic contacts, gate, and transmission metal, which are suitable to use in the CMOS processing line. Then the influence of selected non-gold metal stacks in both AlGaN/GaN and InAlN/GaN HEMT structures and its device characteristics will also be discussed. Apart from DC characteristics, small-signal characteristics of the submicron gate devices will also be discussed in detail.

### 3.3.1   GaN HEMTs with Non-Gold Metal Stack

To establish a low-cost process technology, researchers have fabricated sub-micron gate to 1.5-μm-gate AlGaN/GaN HEMTs using "non-gold" or "Au-free" ohmic contacts. Table 3.4 summarizes the best reported $R_c$ values for "gold-" and "non-gold-" based ohmic contacts for AlGaN/GaN HEMTs on Si. Malmros et al. [67] have achieved an extremely low contact resistance ($R_c$) of ~0.06 $\Omega$-mm using a non-gold Ta/Al/Ta (10/280/20 nm) ohmic contact to a low Al content ($x_{Al}$) barrier of $Al_xGa_{1-x}N$ in an AlGaN/GaN HEMT structure. However, the low $x_{Al}$ of 0.14 makes the AlGaN/GaN HEMT structure suffer from poor 2DEG properties. At the same time, they reported a higher $R_c$ of ~0.28 $\Omega$-mm for a higher Al content $Al_{0.25}Ga_{0.75}N$/GaN HEMT structure. Lee et al. [63] and De Jaeger et al. [24] have tried to use non-gold ohmic contacts of Ti/Al/W (60/100/30 nm) for the conventional

**TABLE 3.4**

**Contact Resistance Values for Different GaN HEMT Structures on Si Substrate Using Gold-Based and CMOS-Compatible Non-Gold Ohmic Contacts**

| Research Group | HEMT on Si | Metal Stack | Annealing Temperature [°C] | $R_c$ (Ω-mm) | |
|---|---|---|---|---|---|
| | | | | Gold | Non-Gold |
| Nitronex [4] | AlGaN/GaN | Ti/Al/Ni/Au | 825 | 0.45 | |
| NTU [21] | AlGaN/GaN | Ti/Al/Ni/Au | 825 | 0.12 | — |
| NTU [50] | AlGaN/AlN/ GaN | Ti/Al/Ni/Au | 825 | 0.33 (Recess) | — |
| ETH-Z [60] | AlGaN/GaN | Ti/Al/Ni/Au | 850 | 0.45 | — |
| HKUST [61] | AlGaN/GaN | Ti/Al/Ni/Au | 850 | 0.70 | — |
| IEMN [62] | AlGaN/GaN | Ti/Al/Ni/Au | 900 | 0.50 | — |
| IMEC [24] | AlGaN/GaN/ AlGaN | Ti/Al/Ti/TiN | 550 | — | 1.25 (Recess) |
| IMEC [59] | AlGaN/GaN/ AlGaN | Ti/Al/W | 800 | — | 0.65 |
| MIT [63] | AlGaN/GaN | Ti/Al/W | 870 | — | 0.49 (Recess) |
| NTU[64] | AlGaN/GaN | Ta/Si/Ti/Al/ Ni/Ta | 800 | — | 0.24 |
| NUS [65] | AlGaN/GaN | Ti/Al | 650 | — | 5.7 |
| SMART [66] | AlGaN/AlN/ GaN | Ti/Al/Ni/Pt | 975 | — | 0.57 |

$Al_{0.26}Ga_{0.74}N/GaN$ HEMT structure and Ti/Al/Ti/TiN contacts (20/100/20/60 nm) for an $Al_{0.25}Ga_{0.75}N/GaN/Al_{0.18}Ga_{0.82}N$ double heterostructure HEMT, respectively, with ohmic-recess etching, which increases the complexity of the fabrication process and its reproducibility. Even with the ohmic-recess process, they achieved $R_c$ values of 0.49 Ω-mm [63] and 1.25 Ω-mm [24].

Hove et al. [59] reported non-gold Ti/Al/W (20/100/20 nm) ohmic contacts without a recess-etching process and achieved an $R_c$ of only 0.65 Ω-mm. Hence, further reduction of $Rc$ is still essential for the demonstration of high-speed switching devices with low $R_{ON}$ similar to the conventional gold-based ohmic contacts on conventional un-doped $Al_{0.26}Ga_{0.74}N/GaN$ HEMTs [21]. Arulkumaran et al. [64] proposed and demonstrated a Ta/Si/Ti/Al/Ni/Ta (5/2/15/140/30/25 nm) ohmic contact on a conventional undoped $Al_{0.26}Ga_{0.74}N/GaN$ HEMTs with a record low $R_c$ of 0.24 Ω-mm. The motivation for selecting Ta and Si is that Ta has a metal work function (~4.0–4.8 eV, thickness dependent) close to the electron affinity of GaN (~4.1 eV), whereas Si has been widely used as an $n$-type dopant for III-V compound materials (GaAs, InP) including III-nitrides. Thus, the incorporation of Ta and Si in the ohmic metallization is believed to be a potential candidate to further reduce the $R_c$. In addition, both Ta and Si are commonly used in conventional Si CMOS fabrication line. The conventional AlGaN/GaN HEMT structure consists of a 1.4-μm-thick transition layer, a 0.8-μm-thick UID GaN buffer, an 18-nm-thick UID-$Al_{0.26}Ga_{0.74}N$ barrier, and a 2-nm-thick UID-GaN cap from bottom to top. There was no doping in the barrier

(AlGaN) and cap (GaN) layers. At room temperature, the 2DEG sheet carrier density and 2DEG mobility were measured as $1.1 \times 10^{13}$ cm$^{-2}$ and 1450 cm$^2$/V·s, respectively. The repeatability of non-gold ohmic contacts were also studied and reported by Li et al. [68]. The optimized annealing temperature was in the range between 800 and 850°C for 30 seconds in a N$_2$ atmosphere for Ta/Si/Ti/Al/Ni/Ta ohmic contact.

Figure 3.24a shows the resistance versus transfer length model (TLM) gaps of gold-based (Ti/Al/Ni/Au) and non-gold (Ta/Si/Ti/Al/Ni/Ta) ohmic contacts on a conventional AlGaN/GaN HEMTs. The non-gold ohmic contact exhibited the lowest $R_c$ of 0.24 Ω-mm and specific contact resistivity of $4.9 \times 10^{-6}$ Ω-cm$^2$ [64]. For the case of gold-based ohmic contact (Ti/Al/Ni/Au), researchers have already achieved $R_c$ <0.18 Ω-mm for the conventional AlGaN/GaN HEMTs on Si substrate [21]. Although the gold-based ohmic contacts exhibiting low $R_c$ values, the surface morphology is rough. Figure 3.24 shows optical micrograph of (b) conventional gold-based and (c) non-gold ohmic contacts. The surface morphology of non-gold ohmic contact is smooth with good edge definition when compared with gold-based ohmic contact's surface morphology [63,64,68]. The RMS surface roughness of non-gold ohmic contact is around 5.5 nm. This simple ohmic scheme also avoids the need to use other complicated techniques such as an ohmic-recess [49] or a regrown ohmic contact [69], which will complicate the manufacturing process. Recently, Liu et al. [66] reported low $R_c$ of 0.57 Ω-mm for undoped AlGaN/GaN HEMTs with 1-nm-thick AlN spacer layer using non-gold metal stack (Ti/Al/Ni/Pt) by annealing at very high temperature of 975°C. The RMS surface roughness was also improved from 131.9 nm for gold (Ti/Al/Ni/Au) ohmic contact to 4.6 nm for non-gold (Ti/Al/Ni/Pt) ohmic contact, as measured by AFM. More recently, Arulkumaran et al. [70] have also demonstrated record-low $R_c$ of 0.34 Ω-mm with smooth surface morphology for lattice matched In$_{0.17}$Al$_{0.82}$N/GaN HEMT structure on Si substrate using non-gold metal stack Ta/Si/Ti/Al/Ni/Ta (5/5/20/120/40/30 nm). The low $R_c$ with the non-gold

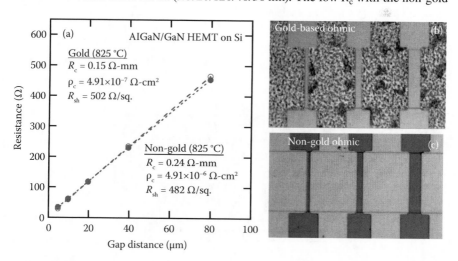

**FIGURE 3.24** Total resistance versus transfer length model (TLM) gaps of (a) gold-based (Ti/Al/Ni/Au) and (b) non-gold (Ta/Si/Ti/Al/Ni/Ta) ohmic contacts on a conventional AlGaN/GaN HEMTs.

ohmic contact is due to the formation of a low-metal-work-function Ti$_x$Si$_y$ alloy at the metal–semiconductor interface as well as the intermixing of Ti$_x$Si$_y$ with the bottom Ta metal layer. The formation of the alloy was confirmed by powder X-ray diffraction, HR-TEM, and energy-dispersive X-ray spectroscopy analysis [68]. The developed low $R_c$ non-gold ohmic contact to the conventional undoped AlGaN/GaN HEMT structure is shown to be promising for the mass production of low-cost GaN HEMTs on Si substrates in well-established silicon CMOS fabrication lines.

After the ohmic contact formation on AlGaN/GaN HEMT structure with non-gold metal scheme, the 2-μm-gate was formed by conventional optical lithography using the optimized CMOS-compatible non-gold gate Ni/Al/Ta (50/400/50 nm) metal stack. Finally, the non-gold transmission metal stack Ti/Al/Ta (50/800/30 nm)was formed followed by final passivation with 120-nm-thick PECVD grown Si$_3$N$_4$. Figure 3.25 shows the schematic cross-section of the fabricated AlGaN/GaN HEMTs on Si substrate.

Figure 3.26 shows (a) $I_{DS}$–$V_{DS}$ and (b) transfer characteristics of 2-μm-gate AlGaN/GaN HEMTs with non-gold metal stacks. The device with non-gold metal stacks exhibited $I_{Dmax}$ of 515 mA/mm, $g_{mmax}$ of 167 mS/mm, and $V_{th}$ of −2.5 V. For OFF-state breakdown voltage ($BV_{gd}$) measurements, the gate bias was maintained at −5.0 V with Si substrate grounded [21,71]. The samples were immersed in Fluorinert™(FC-40) to avoid any atmospheric surface flashover at the gate-drain region while $BV_{gd}$ measurements [21]. Figure 3.27 shows the I$_{Dmax}$, g$_{mmax}$, and three-terminal OFF-state BV$_{gd}$ of AlGaN/GaN HEMTs for different L$_{gd}$. Table 3.5 shows the measured device parameters to compare with the reported devices fabricated using CMOS-compatible non-gold metal stack. The $BV_{gd}$ linearly increased up to $L_{gd}$ = 8 μm. For the $L_{gd}$ = 10 μm, the $BV_{gd}$ started saturating due to the increase of buffer/substrate leakage current. The GaN HEMTs with $L_{gd}$ = 8 μm and 10 μm exhibited a $BV_{gd}$ of 640 and 670 V with specific ON-resistance ($R_{DS[ON]}$) = 0.92 and 1.17 mΩ-cm$^2$, respectively [71]. The obtained breakdown strength from the slope is ~0.8 MV/cm. The measured $BV_{gd}$ is almost equivalent to the $BV_{gd}$ values measured in the GaN HEMTs with gold-based ohmic contacts [21]. Due to the occurrence of low $Rc$ (~0.24 Ω-mm) HEMTs exhibited low static-$R_{DS[ON]}$ values. The power device FoM = $BV_{gd}^2/R_{DS[ON]}$ is as high as 4.45 × 10$^8$v$^2$.Ω$^{-1}$-cm$^{-1}$. This is

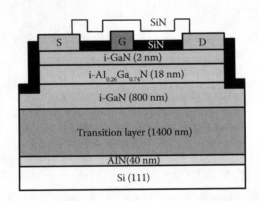

**FIGURE 3.25** Schematic cross-section of the fabricated AlGaN/GaN HEMTs on Si substrate.

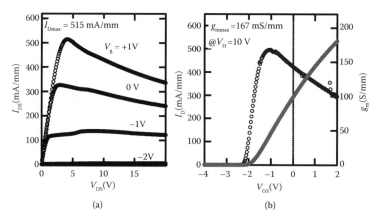

**FIGURE 3.26**   This figure shows (a) $I_{DS}$–$V_{DS}$ and (b) transfer characteristics of the fabricated AlGaN/GaN HEMTs on Si substrate with non-gold metal stack.

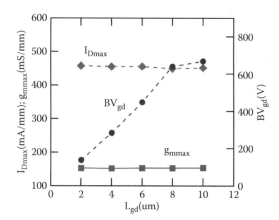

**FIGURE 3.27**   $I_{Dmax}$, $g_{mmax}$ and $BV_{gd}$ of AlGaN/GaN HEMTs for different $L_{gd}$

the state-of-the-art value reported for the undoped AlGaN/GaN HEMTs on Si substrate using CMOS-compatible non-gold metal stack [71]. Due to a thin-GaN buffer layer (2.2 µm), the $BV_{gd}$ is limited. Without compromising the $R_{DS[ON]}$, enhanced $BV_{gd}$ can be achievable beyond 1.4 kV by increasing the buffer GaN thickness from 2.2 to 6 µm [3,7,10]. Liu et al. [66] reported the $BV_{gd}$ of 427 V with $R_{DS[ON]}$ of 0.77 mΩ-cm$^2$ for AlGaN/AlN/GaN HEMTs on Si substrate with non-gold ohmic contacts. The calculated power device FoM is ~2.37 × 10$^8$ V$^2$·Ω$^{-1}$·cm$^{-2}$. More recently, Imec team achieved the $R_c$ of 0.62 Ω-mm with Ti/Al/TiN (Ti/Al ratio is 0.2) ohmic contact for AlGaN/GaN/AlGaN HEMT annealed at low temperature of 550 °C [72]. Figure 3.28 shows the comparison chart for the specific ON-resistance $R_{DS[ON]}$ versus $BV_{gd}$ of AlGaN/GaN HEMTs on a Si substrate with conventional III-V gold-based ohmic contacts and non-gold ohmic contacts. For the high-power switching devices, the dynamic ON-resistance should also be low. Hence, the devices can operate in a higher speed with low switching losses. The demonstration of $BV_{gd}$

**TABLE 3.5  Measured Device Parameters for the GaN HEMTs on Si Fabricated with CMOS-Compatible Non-Gold Metal Stack**

| Group | Epi Structure | Ohmic Metal Stack | Anneal Temperature (°C) | $R_c$ (Ω-mm) | $d_{Buff}$ (µm) | $L_{gd}$ (µm) | $BV_{gd}$ (V) | $R_{DS(ON)}$ (mΩ-cm²) | FoM = $BV_{gd}^2/R_{DS(ON)}$ (×10⁸·V²·Ω⁻¹·cm⁻²) | Breakdown Strength (MV/cm) |
|---|---|---|---|---|---|---|---|---|---|---|
| IMEC [59] | AlGaN/ GaN/ AlGaN | Ti/Al/ Ti/TiN | 550 | 1.25 | 2.65 | 10 | 750 | 2.90 | 1.94 | 0.75 |
| MIT [63] | AlGaN/ GaN | Ti/ Al/W | 870 | 0.49 | 1.8 | 3 | 87 | 0.05 | 1.51 | 0.29 |
| NUS [65] | AlGaN/ GaN | Ti/Al | 650 | 5.7 | 4.8 | 5 | 800 | 3.00 | 2.13 | 1.60 |
| SMART [66] | AlGaN/ AlN/GaN | Ti/Al/ Ni/Pt | 975 | 0.6 | 2.5 | 5 | 427 | 0.77 | 2.37 | 0.71 |
| NTU [71] | AlGaN/ GaN | Ta/Si/ Ti/Al/ Ni/Ta | 800 | 0.24 | 2.2 | 8 | 640 | 0.92 | 4.45 | 0.80 |
| | | | | | | 10 | 670 | 1.17 | 3.84 | 0.67 |

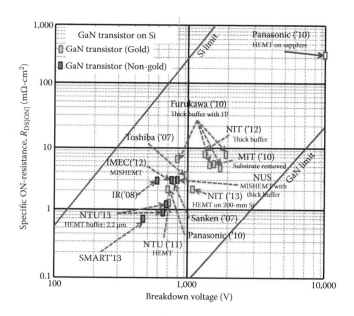

**FIGURE 3.28** Benchmarking of specific ON-resistance $R_{DS[ON]}$ versus $BV_{gd}$ for AlGaN/GaN HEMTs on a Si substrate with conventional III-IV gold-based ohmic contacts and non-gold ohmic contacts.

of 640 V with specific ON-resistance as low as 0.92 mΩ-cm$^2$ and the high FoM (~4.45 × 10$^8$ V$^2$·Ω$^{-1}$·cm$^{-2}$) for 2.4-μm-thick AlGaN/GaN HEMTs on Si substrate with CMOS-compatible non-gold ohmic, Schottky, and transmission line is one of the major steps toward the realization of low-cost efficient high-power devices [64,71]. In addition, the non-gold metal stack fabricated AlGaN/GaN HEMTs with $L_{gd} = 5$ μm exhibited the dynamic-$R_{DS[ON]}$ of 0.58 mΩ-cm$^2$ for $V_{gs0} = -5$ V and $V_{ds0} = 40$ V, which is equivalent to the static-$R_{DS[ON]}$ of similar-dimension GaN HEMTs on Si substrate [71]. Table 3.5 shows the measured DC parameters for the AlGaN/GaN HEMTs on Si substrate fabricated with CMOS-compatible non-gold metal stacks. These experimental results are indicating that the AlGaN/GaN HEMTs on Si substrate with CMOS-compatible non-gold metal stack are also feasible to achieve high OFF-state $BV_{gd}$ with low-dynamic-$R_{DS[ON]}$, which make them very attractive for high-speed high-voltage power-switching device applications.

### 3.3.2 Submicron Gate AlGaN/GaN HEMTs with Non-Gold Ohmic Contact

Until now, researchers have only reported >1.5-μm-gate-length AlGaN/GaN HEMTs using "non-gold" or "Au-free" metal stacks for low-cost high-power electronics [24,59,63,66,67,71]. However, for high-frequency applications, submicron-gate HEMTs are necessary. To form a submicron T-gate, a first layer of 120-nm-thick SiN was deposited by PECVD after the ammonium sulfide ((NH$_4$)$_2$S$_x$) treatment [57,58,73]. The 0.15-μm footprint was opened and the SiN was etched by CF$_4$/O$_2$

plasma. Then a 0.5-μm gate head was subsequently formed with a CMOS-compatible non-gold metal stack of Ni/Al/Ta (100/400/30 nm). Finally, the non-gold metal-stack Ti/Al/Ta (50/800/30 nm) as an interconnect metal was deposited followed by final passivation with 120-nm-thick PECVD-grown SiN [64]. Figure 3.29 shows the schematic cross-section of the submicron gate (0.15 μm) AlGaN/GaN HEMTs with non-gold metal stacks [64]. Two different AlGaN barrier thickness (18-nm and 5-nm) devices were fabricated and its characteristics were compared. The metal schemes used for this study (i.e., ohmic contact, Schottky contact, and interconnect metal lines) are commonly used in the silicon fabrication process line.

Figure 3.30 shows the $I_{DS}$–$V_{DS}$ characteristics of (a) 18-nm-thick AlGaN barrier and (b) 5-nm-thick AlGaN barrier 0.15-μm-gate-length AlGaN/GaN HEMTs $[L_{sg}/L_g/L_{gd}/W_g = 0.8/0.15/1.7/(2 \times 75)$ μm] on Si with non-gold metal stacks. Figure 3.31 shows the transfer characteristics of 18-nm-thick AlGaN barrier [64] and 5-nm-thick AlGaN barrier GaN HEMTs on Si substrate with non-gold metal stacks. The HEMTs

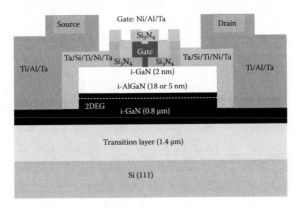

**FIGURE 3.29** Schematic cross-section of the submicron gate (0.15 μm) AlGaN/GaN HEMTs with non-gold metal stacks.

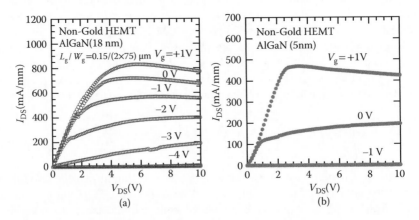

**FIGURE 3.30** $I_{DS}$–$V_{DS}$ characteristics of (a) 18-nm-thick AlGaN barrier and (b) 5-nm-thick AlGaN barrier 0.15-μm-gate-length AlGaN/GaN HEMTs $[L_{sg}/L_g/L_{gd}/W_g = 0.8/0.15/1.7/(2 \times 75)$ μm] on Si with non-gold metal stacks.

with 18-nm-thick AlGaN barrier exhibited a $I_{Dmax}$ of 830 mA/mm, a $g_{mmax}$ of 263 mS/mm, and a $V_{th}$ of −3.75 V. Subsequently, the HEMTs with 5-nm-thick AlGaN barrier exhibited a $I_{Dmax}$ of 470 mA/mm, a $g_{mmax}$ of 310 mS/mm, and a $V_{th}$ of −0.82 V. The Ni/Al/Ta Schottky gate exhibited a barrier height of 0.88 eV and a reverse gate-leakage current (−20 V) of $3.8 \times 10^{-3}$ mA/mm.

Figure 3.32 shows the small-signal gain versus frequency for the CMOS-compatible non-gold metal stack AlGaN/GaN HEMTs with (a) 18-nm-thick AlGaN

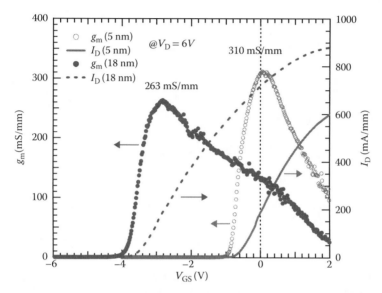

**FIGURE 3.31** Transfer characteristics of 18-nm-thick AlGaN barrier and 5-nm-thick AlGaN barrier AlGaN/GaN HEMTs on Si substrate with non-gold metal stacks.

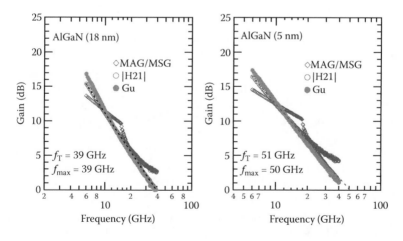

**FIGURE 3.32** Small-signal microwave gain versus frequency for the complementary metal–oxide–semiconductor-compatible non-gold metal stack AlGaN/GaN HEMTs with (a) 18-nm-thick AlGaN barrier and (b) 5-nm-thick AlGaN barrier.

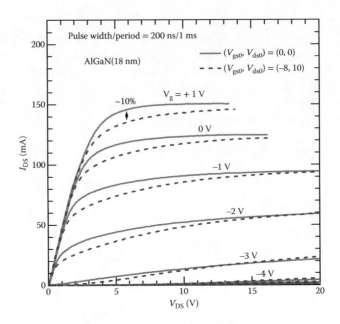

**FIGURE 3.33**   Pulsed $I_{DS}$–$V_{DS}$ characteristics (pulse width = 200 ns; pulse period = 1 ms) of AlGaN/GaN HEMTs with a CMOS-compatible non-gold metal stacks.

barrier [64] and (b) 5-nm-thick AlGaN barrier. The HEMT with 18-nm-thick- and 5-nm- thick-AlGaN barrier exhibited an $f_T$ of 39 and 51 GHz and an $f_{max}$ of 39 and 50 GHz, respectively. This is believed to be the first reported microwave performance of sub-micron AlGaN/GaN HEMTs using CMOS-compatible non-gold metal stack ohmic contacts. Figure 3.33 shows the pulsed $I$–$V$ characteristics (pulse width = 200 ns; pulse period = 1 ms) of AlGaN/GaN HEMTs with a CMOS-compatible non-gold metal stack. Very little current collapse (<10%) was observed for the gate- and drain-quiescent biases ($V_{gs0} = -8$ V, $V_{ds0} = 10$ V). Similarly, very low current collapse (≤10%) AlGaN/GaN HEMTs with gold-based ohmic metal stacks were also realized and reported elsewhere [56,57,69]. The three-terminal $BV_{gd}$ for the HEMTs with 18-nm-thick- and 5-nm-thick-AlGaN barrier was 90 V and 83 V, respectively. The J-FoM ($= f_T \times BV_{gd}$) of HEMTs are in the range between 3.51 THz·V and 3.83 THz·V for different gate-drain spacing devices ($L_{gd} = 1.7$– 3.5 μm). The Johnson's FoM for 5-nm-thick AlGaN barrier HEMTs is as high as 4.23 THz·V. The obtained values are comparable or better than that of the 0.15-μm gate-length GaN HEMTs (3.3 THz·V) on Si(111) fabricated using a non-CMOS-compatible gold metal stack [74].

In summary, submicron-gate-length AlGaN/GaN HEMTs have been successfully demonstrated on high-resistivity Si substrate with DC and microwave performances using a CMOS-compatible non-gold metal stack. These results demonstrate the feasibility of non-gold metallization process to achieve submicron gate AlGaN/GaN HEMTs on Si substrate for high-frequency applications.

### 3.3.3 Submicron Gate InAlN/GaN HEMTs with Non-Gold Ohmic Contact

InAlN/GaN HEMTs have great advantage over conventional AlGaN/GaN HEMTs due to the larger bandgap discontinuity ($\Delta Ec \sim 0.68$ eV) [16,70,75,76], which results in two to three times higher 2DEG density in the order of $\sim 2.73 \times 10^{13}$ cm$^{-2}$. Moreover, AlGaN/GaN HEMTs also face the strain-induced reliability due to large ($\sim 18\%$) lattice mismatch between AlGaN barrier and GaN buffer layer [75]. Lattice-matched In$_x$Al$_{1-x}$N/GaN HEMT ($x \sim 0.17$) mitigates strain-induced reliability due to the absence of piezoelectric polarization, which helps to improve the overall device characteristics. By inserting a thin layer of AlN between InAlN barrier layer and GaN active buffer layer, the alloy disorder scattering is significantly reduced, which in turn enhances the transport properties in the channel. The optimized heterostructure with gate length scaling and efficient process technique can provide improved DC and RF performances. Recently, Yue et al. [77] achieved very high $f_T/f_{max} = 400/33$ GHz with 30-nm-gate-length InAlN/GaN HEMT on SiC substrate with regrown n$^+$-GaN and Au-based ohmic contacts. However, very few reports are available on InAlN/GaN HEMTs fabricated on Si substrate. Furthermore, most reported InAlN/GaN HEMTs were fabricated using conventional III–V gold-based ohmic metal schemes, which are not compatible to existing Si process line [77–79]. Hence, non-gold ohmic contacts with low contact resistance ($R_c$) are essential for InAlN/GaN HEMTs to be manufacturable in the matured Si process line.

Table 3.6 summarizes some of the best reported $R_c$ values for "non-gold-" based ohmic contacts for InAlN/GaN HEMTs on Si and SiC. Various non-gold ohmic metal stacks (Ti/Al/Ni/W [16], Hf/Al/Ta [80], Ta/Al/Ta [81], Ti/Al/Ni [82]) have been used by researchers to achieve low ohmic contact on InAlN/GaN HEMT. Researchers at Chalmers University also achieved $R_c = 0.64$ $\Omega$-mm using Ta/Al/Ta metal stack on InAlN HEMT on SiC [81]. To further improve the $R_c$, a non-gold ohmic metal scheme using Ta with a thin layer of Si layer for AlGaN/GaN HEMTs on Si substrate was proposed and discussed in the Section 3.3.1. The record-low $R_c$ of 0.36 $\Omega$-mm so

**TABLE 3.6  Benchmarking Different Non-Gold Ohmic Contacts to InAlN/GaN HEMTs**

| Affiliation | InAlN/AlN/GaNHEMT on | Ohmic Metal | Anneal Temperature (°C)/Time (seconds) | $R_c$ ($\Omega$-mm) |
|---|---|---|---|---|
| IMRE [16] | Si | Ti/Al/Ni/W | 900/60 | 0.56 |
| NTU [70] | Si | Ta/Si/Ti/Al/Ni/Ta | 825/30 | 0.36 |
| NUS [80] | Si | Hf/Al/Ta | 600/60 | 0.59 |
| Chalmers [81] | SiC | Ta/Al/Ta | 550/— | 0.64 |
| ULM [82] | Sapphire | Ti/Al/Ni and Ta/Cu/Ta | 900/— | 1.60 |

far achieved with annealing temperature at 825°C for 30 seconds using Ta/Si/Ti/Al/Ni/Ta metal stack [70].

The HEMT structure was grown by MOCVD with a 9-nm-thick $In_{0.17}Ga_{0.83}N$ barrier, 1-nm-thick AlN spacer layer, 1000-nm-thick $i$-GaN buffer and 100-nm-thick NL on high-resistivity Si (111) substrate (see Figure 3.34). Recently, Tripathy et al. [16] successfully demonstrated the growth and its characteristics of InAlN/AlN/GaN on 200-mm-diameter conducting Si(111) substrate using MOCVD system. The grown structure exhibited room temperature 2-DEG mobility of 786 $cm^2$/V·s and sheet carrier density of $2.74 \times 10^{13}$ $cm^{-2}$. After the formation of mesa isolation using dry etching by $BCl_3/Cl_2$ plasma, Ta/Si/Ti/Al/Ni/Ta (5/5/20/120/40/30 nm) ohmic metal was deposited and annealed at 825°C for 30 seconds in an $N_2$ environment with a RTA system. A foot print of ~0.17-μm Schottky T-gate with 0.5-μm gate head was formed with a metal stack of Ni/Au (150/400 nm) using EBL. Subsequently, the non-gold metal-stack Ti/Al/Ta (50/800/30 nm) was also formed as an interconnect metal [64,70]. Finally, the devices were passivated with 120-nm-thick PECVD-grown SiN. Figure 3.34 shows the (a) schematic cross-section of the fabricated InAlN/GaN HEMTs on silicon and (b) HR-TEM image of the formed T-gate.

Figure 3.35 shows the current–voltage characteristics of non-gold and gold-based ohmic contacts for InAlN/AlN/GaN HEMTs on Si substrate. Inset of Figure 3.34 shows the total resistance versus TLM gaps. The conventional gold-based ohmic contact (Ti/Al/Ni/Au) on InAlN/AlN/GaN HEMT structure exhibited an average $R_c$ as low as 0.33 Ω-mm and an average specific contact resistivity ($\rho_c$) of $3.27 \times 10^{-6}$ Ω-$cm^2$. The non-gold ohmic contacts exhibited an average $R_c$ as low as 0.36 Ω-mm with $\rho_c$ of $4.47 \times 10^{-6}$ Ω-$cm^2$, which is comparable with the gold-based ohmic contacts. The achieved $R_c$ is believed to be the lowest ever reported for non-gold ohmic contacts on InAlN/AlN/GaN HEMTs on Si substrate and it is also lower than that of InAlN/AlN/GaN HEMTs on SiC substrate [81]. Smooth surface morphology and good edge definition has also been observed, which is similar to non-gold ohmic contacts on

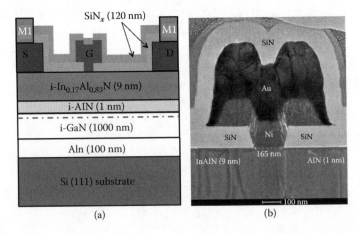

(a)                                         (b)

**FIGURE 3.34** (a) Schematic cross-section of InAlN/GaN HEMTs on Si with non-gold Ohmic contacts (b) cross-sectional HR-TEM image of T-gate formed InAlN/GaN HEMTs with non-gold ohmic contact.

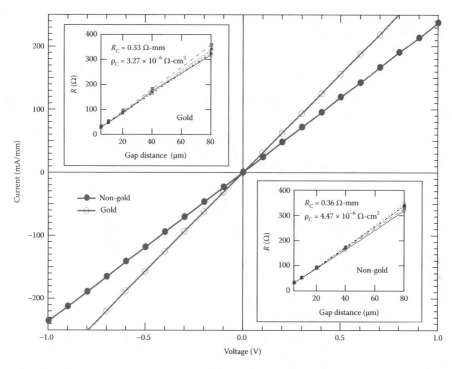

**FIGURE 3.35**  Current–voltage characteristics of non-gold and gold-based ohmic contacts for InAlN/AlN/GaN HEMTs on Si substrate. Inset: the total resistance versus TLM gaps.

AlGaN/GaN HEMT system [64,68]. For a comparison, different gold-based ohmic contacts to InAlN/AlN/GaN HEMTs on different substrates are listed in Table 3.7.

Figure 3.36 shows the typical (a) $I_{DS}$–$V_{DS}$ and (b) transfer characteristics of 0.17-μm-gate-length InAlN/GaN HEMTs [$L_{sg}/L_g/L_{gd}/W_g$ = 0.8/0.17/1.7/(2 × 75) μm] on Si substrate with non-gold ohmic contacts. The HEMTs exhibited an $I_{Dmax}$ of 1110 mA/mm and a $g_{mmax}$ of 353 mS/mm with a good channel pinch-off [78]. The $I_{Dmax}$ value is slightly smaller than the reported 1170 mA/mm for 50-nm-gate length with source-drain spacing of 1.0 μm [81]. However, the reported device suffered from short channel effect and exhibited high output conductance beyond $V_D$ = 4 V. The reverse gate leakage current of the device measured at $V_G$ = −15 V was 1.6 × 10$^{-2}$ mA/mm, which is typical for Ni/Au Schottky gate InAlN/GaN HEMTs. The device exhibited a Schottky barrier height of 0.81 eV with an ideality factor of 1.38.

Figure 3.37a shows the pulsed $I_{DS}$–$V_{DS}$ characteristics (width = 200 ns, period = 1 ms)of InAlN/GaN HEMTs on Si substrate with gate-lag ($V_{gs0}$ = −5 V, $V_{ds0}$ = 0 V) and drain-lag ($V_{gs0}$ = −5 V, $V_{ds0}$ = 10 V) conditions. Very small (~8%) drain current ($I_D$) collapse was observed from the gate-lag (−5,0) and drain-lag (−5,10) at $V_D$ = 6 V. The observation of small $I_D$ collapse in gate-lag measurement is due to the lattice-matched InAlN/GaN HEMT structure, thus the absence of piezoelectric polarization. The strain-free InAlN barrier layer has also been verified by Leach et al. [88]. In addition, the current collapse by surface related traps is also suppressed by the optimized SiN

**TABLE 3.7  Benchmarking Different Gold-Based Ohmic Contacts to InAlN/ GaN HEMTs**

| Affiliation | InAlN/ AlN/GaN HEMT on | Ohmic Metal | Treatments Prior to Metallization | Anneal Temperature (°C)/Time (seconds) | $R_c$ (Ω-mm) |
|---|---|---|---|---|---|
| NTU [78] | Si | Ti/Al/Ni/ Au | — | 825/30 | 0.33 |
| NIT [79] | Si | Ti/Al/Ni/ Au | — | 800/30 | 0.60 |
| ETH [83] | Si | Ti/Al/Au | — | 800/30 and 850/30 (two-time annealed) | 0.36 |
| UND [77] | SiC | Ti/Au | Regrown n+-GaN by MBE | Nonalloyed | 0.16 |
| ISSE [84] | SiC | Ti/Al/Ni/ Au | Ohmic recess (SiCl$_4$ plasma) | 600/— | 0.70 |
| IEMN [85] | SiC | Ti/Al/Ni/ Au | — | 900/30 | 0.15 |
| UIUC [86] | SiC | Mo/Al/Mo/ Au | Ohmic recess (SiCl$_4$ plasma) | 650/30 | 0.15 |
| MIT [87] | SiC | Si/Ge/Ti/ Al/Ni/Au | — | 820/30 | 0.35 |

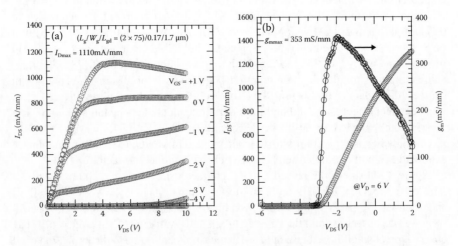

**FIGURE 3.36**  (a) $I_{DS}$–$V_{DS}$ and (b) transfer characteristics of 0.17-μm gate length InAlN/ GaN HEMTs [$L_{sg}/L_g/L_{gd}/W_g$ = 0.8/0.17/1.7/(2 × 75) μm] on Si with non-gold ohmic contacts.

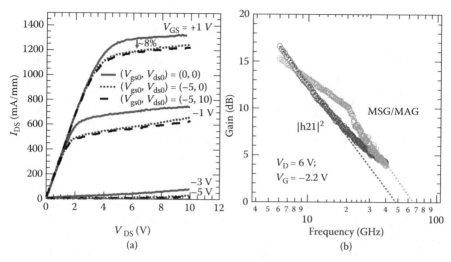

**FIGURE 3.37**  (a) Pulsed $I_{DS}$–$V_{DS}$ characteristics (width = 200 ns; period = 1 ms) of 0.17 μm InAlN/GaN HEMTs on Si substrate with gate-lag ($V_{gs0}$ = –5 V, $V_{ds0}$ = 0 V) and drain-lag ($V_{gs0}$ = –5 V, $V_{ds0}$ = 10 V) conditions. (b) Small-signal microwave performance measured at $V_g$ = –2.2 V and $V_D$ = 6 V for 0.17 μm InAlN/GaN HEMTs on Si with non-gold ohmic contact.

passivation with ammonium sulfide pretreatment [57,58,73,78]. Figure 3.37b shows the small-signal microwave performance of 0.17-μm InAlN/GaN HEMTs measured at $V_g$ = –2.2 V and $V_D$ = 6 V. The HEMT exhibited an $f_T$ of 48 GHz and an $f$max of 66 GHz without de-embedding. So far, this is the first report with microwave characteristics in the literature using non-gold metal stack on InAlN/GaN HEMTs.

In summary, submicron (0.17-μm) gate $In_{0.17}Al_{0.83}$N/AlN/GaN HEMTs on Si substrate have been successfully demonstrated with promising device DC and microwave performances. The preliminary results show the feasibility of using non-gold metal stack as a low $R_c$ ohmic contact to achieve improved performances of submicron gate InAlN/AlN/GaN HEMTs on Si substrate for high-frequency applications. This will enable a wider choice of high-performance III-Nitride devices to be available for low-cost and high-volume applications.

## 3.4  CONCLUSIONS

Nitride-based semiconductors have undoubtedly become the choice of technology for many applications spanning from optoelectronic to electronics applications. In order for this technology to be more widely adopted in commercial markets, it is imperative that the production cost will have to be lowered substantially while maintaining the high performance. To achieve that, there are two key technology hurdles, which will need to be overcome. These are (1) growth of GaN and its related materials on large-diameter silicon substrate and (2) development of process technology using the existing completely depreciated silicon CMOS line.

In 2012, Imec, Belgium, demonstrated a fully CMOS-compatible and noble metal-free LED integration flow. They have demonstrated working LED devices with sizes up to $2 \times 2$ mm$^2$ on 200-mm GaN-on-Si wafer [89]. The optical performance of the GaN LEDs fabricated on 200-mm-diameter Si is similar to that of the available LED on sapphire substrates. They are now in progress to improve the efficiency for mass production of LEDs using GaN-on-Si technology. In 2012, M/s. Toshiba, Japan, and M/s. Bridgelux have also jointly started mass production of low-power LED chips using large diameter GaN-on-Si technology [90]. The planned capacity is 10 million units per month. The produced chips can provide 112 lm at 350 mA. Recently, a France company M/s. Aleda has also demonstrated cost-disruptive MicroLEDs on GaN on 200-mm diameter Si substrates [91]. At the same time, the industries are also putting significant effort to commercialize the devices for automobile, solar energy convertors, compact AC–DC power supplies for computer, high-efficiency power conversion for uninterrupted power supplies (UPS) to reduce the conversion losses. The developed non-gold metal stack will also be useful for the supply of cost-effective chips using CMOS compatible process line. One of the main concerns is the bowing of the wafer, which could be a hindrance to wafer process. However, with the advancement in the growth technologies, the bowing values of AlGaN/GaN HEMTs on 200-mm-diameter Si substrate are in the acceptable range down to 50 µm. As a result, more commercial products (e.g., low voltage small size power adaptors for laptop, UPS system, high-efficiency power convertors, etc.) using GaN HEMTs fabricated on 200-mm-diameter silicon substrates have been unveiled.

As growth technologies for GaN on large silicon substrates are making excellent progress, GaN-based HEMTs fabricated on 200-mm-diameter Si substrates with encouraging performances have also been reported [11–16,48]. AlGaN/GaN HEMTs with 2.0-µm-gate and 0.3-µm-gate were fabricated and their device characteristics were also measured and compared with the measured electrical (2DEG mobility >1500 cm$^2$/V·s and sheet carrier density >1 $\times$ 10$^{13}$ cm$^{-2}$) and structural properties. For example, the average edge- and screw-dislocation densities of the AlGaN/GaN HEMT structure on 200-mm diameter Si were in the order of $1.5 \times 10^9$ and $5.3 \times 10^8$ cm$^{-2}$, respectively, which is equivalent to the dislocation density of AlGaN/GaN HEMT structure on small diameter (75- to 150-mm) Si substrate [12,13,15]. Though the structural and electrical properties are in equal to the AlGaN/GaN HEMT structure on small diameter Si substrate, the realization of OFF-state breakdown voltage >1100 V using thick GaN buffer layer for low dynamic ON-resistance with high FoM still remains the key challenges. In order for AlGaN/GaN HEMTs on 200-mm diameter Si technology to be competitive with other technologies such as SiC FETs, better growth and device processes are still needed.

For lower chip cost, it is necessary to have very high device yield per wafer. This will require high uniformity in the starting raw epitaxial wafer and the subsequent fabricated devices. In this aspect, it is encouraging that reported results have demonstrated that very high yield can be achieved both in growth and device fabrication. The radial distribution of sheet resistance by van der Pauw patterns, Hall parameters, and 2DEG profile has been reported to achieve >90% over a full 200-mm-diameter Si(111) substrates. The recessed-gate E-mode AlGaN/GaN HEMTs were also realized on 200-mm-diameter Si(111) substrate. The uniformity of the GaN HEMT parameters are also >90%, which are in good agreement with the electrical parameters of the grown AlGaN/GaN heterostructures on 200-mm-diameter Si(111) substrate. The demonstrated results shows the feasibility of achieving good uniformity AlGaN/GaN HEMTs on 200-mm-diameter Si(111) substrate. To utilize the existing CMOS-compatible silicon wafer processing lines, non-gold ohmic contact with low contact resistance and low leakage current achievable dielectric gate material is of paramount importance. For example, $R_c$ <0.2 $\Omega$-mm, low specific ON-resistance, high breakdown voltage (>1100 V), and low drain current collapse or dispersion, which are normally coming from the buffer (GaN or AlGaN), barrier (AlGaN or InAlN), and surface-related traps.

In conclusion, this chapter has reviewed the status of the two key challenging technologies namely the growth of GaN HEMTs on large diameter silicon substrates and the non-gold ohmic contacts. These two key technology challenges have to be overcome to facilitate the realization of high-volume low-cost GaN-based commercial products. The achieved results reported thus far have indicated that these goals are feasible in the near future and one will see that nitride semiconductors will continue to transform the electronic landscapes just as GaN has transformed the lighting industry and improve the way we live.

## REFERENCES

1. Philips, The LED lighting revolution: A summary of the global savings potential, May 2012, accessed July 4, 2015, http://www.lighting.philips.com/pwc_li/main/connect/Assets/tools-literature/LED-lighting-revolution-booklet.pdf.
2. Semiconductor TODAY, GaN-on-Si opportunity for extending the life of CMOS silicon fabs?, Compounds & Advanced Silicon, vol. 10, December 2013–January 2014, accessed July 4, 2015, http://www.semiconductor-today.com/features/PDF/SemiconductorToday_December-Jan2014-GaN-on-Si.pdf.
3. Semiconductor TODAY, GaN-on-Si LEDs to grow at 69% CAGR from 1% market share in 2013 to 40% in 2020, 4 December 2013, http://www.semiconductor-today.com/news_items/2013/DEC/IHS_041213.shtml.
4. J. D. Brown, R. Borges, E. Piner, A. Vescan, S. Singhal and R. Therrien, "AlGaN/GaN HFETs fabricated on 100-mm GaN on Si(111) substrates", *Solid-State Electronics*, vol. 46, p.1535, Feb. 2002.
5. S. Arulkumaran, T. Egawa, S. Matsui and H. Ishikawa, "Enhancement of breakdown voltage by AlN buffer layer thickness in AlGaN/GaN HEMTs on 4 in. diameter silicon", *Applied Physics Letters*, vol. 86, p.123503, Mar. 2005.
6. A. Dadgar, C. Hums, A. Diez, J. Bläsing and A. Krost, "Growth of blue GaN LED structures on 150-mm Si(111)", *Journal of Crystal Growth*, vol. 297, pp.279–282, Dec. 2006.
7. S. Arulkumaran, T. Egawa and H. Ishikawa, "Studies of AlGaN/GaN high-electron-mobility transistors on 4-in. diameter silicon and sapphire substrates", *Solid-State Electronics*, vol. 49, p.1632, Oct. 2005.
8. A. R. Boyd, S. Degroote, M. Leys, F. Schulte, O. Rockenfeller, M. Luenenbuerger, M. Germain, J. Kaeppeler and M. Heuken: "Growth of GaN/AlGaN on 200-mm diameter Si(111) wafers by MOCVD", *Physica Status Solidi (C)*, vol. 6, p.s1045, June 2009.
9. S. L. Selvaraj, T. Suzue and T. Egawa, "Breakdown enhancement of AlGaN/GaN HEMTs on 4-in silicon by improving the GaN quality on thick buffer layers", *IEEE Electron Device Letters*, vol. 30, p.587, June 2009.

10. Y. Umeda, A. Suzuki, Y. Anda, M. Ishida, T. Ueda and T. Tanaka: "Blocking voltage boosting technology for GaN transistors by widening depletion layer in Si substrate", IEEE International Electron Device Meeting, pp.20.5.1–20.5.4, Dec. 2010.

11. K. Chen, H. Liang, M. V. Hove, K. Geens, B. D. Jaeger, P. Srivastava, X. Kang, P. Favia, H. Bender, S. Decoutere, J. Dekoster, J. A. Borniquel, S. W. Jun and H. Chung, "AlGaN/ GaN/AlGaN double heterostructures grown on 200-mm silicon (111) substrates with high electron mobility", *Applied Physics Express*, vol. 5, p.011002, Jan. 2012.

12. S. Arulkumaran, G. I. Ng, S. Vicknesh, H. Wang, K. S. Ang, J. P. Y. Tan, V. K. Lin, S. Todd, G.-Q. Lo and S. Tripathy, "Direct current and microwave characteristics of sub-micron AlGaN/GaN HEMTs on 8-in Si(111) substrate", *Japan Journal of Applied Physics*, vol. 51, p.111001, Oct. 2012.

13. S. Tripathy, V. K. X. Lin, S. B. Dolmanan, J. P. Y. Tan, R. S. Kajen, L. K. Bera, S. L. Teo, M. Krishna Kumar, S. Arulkumaran, G. I. Ng, S. Vicknesh, S. Todd, W. Z. Wang, G. Q. Lo, H. Li, D. Lee and S. Han, "AlGaN/GaN two-dimensional-electron gas hetero-structures on 200-mm diameter Si(111)", *Applied Physics Letters*, vol. 101, p.082110, Aug. 2012.

14. D. Christy, T. Egawa, Y. Yano, H. Tokunaga, H. Shimamura, Y. Yamaoka, A. Ubukata, T. Tabuchi and K. Matsumoto, "Uniform growth of AlGaN/GaN high electron mobility transistors on 200-mm silicon (111) substrate", *Applied Physics Express*, vol. 6, p.026501, Feb. 2013.

15. S. Arulkumaran, G. I. Ng, S. Vicknesh, C. M. Manojkumar, M. J. Anand, H. Wang, K. S. Ang, S. L. Selvaraj, W. Z. Wang, G. –Q. Lo and S. Tripathy, "Uniformity studies of AlGaN/GaN HEMTs on 200-mm diameter Si(111) substrate", CS MANTECH Conference, May 13–16, 2013, New Orleans, LA, pp.289–292.

16. S. Tripathy, L. M. Kyaw, S. B. Dolmanan, Y. J. Ngoo, Y. Liu, M. K. Bera, S. P. Singh, H. R. Tan, T. N. Bhat and E. F. Chor, "InxAl1-xN/AlN/GaN high electron mobility transistor structures on 200 mm diameter Si(111) substrates with Au-free device processing", *ECS Journal of Solid State Science and Technology*, vol. 3, no. 5, pp.Q84–Q88, Mar. 2014.

17. J. W. Chung, B. Lu and T. Palacios, "On-wafer seamless integration of GaN and Si (100) electronics", Compound Semiconductor Integrated Circuit Symposium (CISC), 2009, pp.1–4.

18. J. W. Chung, E. L. Piner and T. Palacios, "N-face GaN/AlGaN HEMTs fabricated through layer transfer technology", *IEEE Electron Device Letters*, vol. 30, no. 2, pp.113–116, Feb. 2009.

19. Y. Wu, J.-M. Matt, M. L. Moore and S. Heikman, "A 97.8% efficient GaN HEMT boost converter with 300-W output power at 1 MHz", *IEEE Electron Device Letters*, vol. 29, p.824, Aug. 2008.

20. S. Iwakami, O. Machida, M. Yanagihara, T. Ehara, N. Kaneko, H. Goto and A. Iwabuchi, "20 mΩ, 750 V high-power AlGaN/GaN HFETs on Si substrate", *Japan Journal of Applied Physics*, vol. 46, p.L587, June 2007.

21. S. Arulkumaran, S. Vicknesh, G. I. Ng, S. L. Selvaraj and T. Egawa, "Improved power device figure-of-merit (4×108 V2Ω-1cm-2) in AlGaN/GaN HEMTs on high-resistivity 4-in. Si", *Applied Physics Express*, vol. 4, p.08410, Aug. 2011.

22. M. A. Briere, "GaN based power devices: Cost-effective revolutionary performance", www.irf.com: Power Semiconductor Materials, p.29, accessed July 4, 2015, http:// www.irf.com/pressroom/articles/560PEE0811.pdf

23. Transphorm Inc., "Transphorm Releases First JEDEC-Qualified 600 Volt GaN on Silicon Power Devices", Press Release March 14, 2013, accessed July 4, 2015, http://www.transphormusa.com/news/transphorm-releases-first-jedec-qualified-600-volt-gan-silicon-power-devices.

24. B. D. Jaeger, M. V. Hove, D. Wellekens, X. Kang, H. Liang, G. Mannaert, K. Geens and S. Decoutere, "Au-free CMOS-compatible AlGaN/GaN HEMT processing on 200-mm Si substrates", 24th International Symposium on Power Semiconductor Devices (ISPSD), June 2012, p.49.
25. H. M. Manasevit, F. M. Erdmann, and W. I. Simpson, "The use of metalorganics in the preparation of semiconductor materials", *Journal of the Electrochemical Society: Solid State Science*, vol. 118, no. 11, pp.1864–1868, 1971.
26. T. Takeuchi. H. Arnano, K. Hirarnatsu, N. Sawaki and I. Akasaki, "Growth of single crystalline GaN film on Si substrate using 3C-SiC as an intermediate layer", *Journal of Crystal Growth*, vol. 115, no. 1–4, pp.634–635, Dec. 1991.
27. A. Watanabe, T. Takeuchi, K. Hirosawa, H. Amano, K. Hiramatsu and I. Akasaki, "The growth of single crystalline GaN on a Si substrate using AlN as an intermediate layer", *Journal of Crystal Growth*, vol. 128, no. 1–4, pp.391–396. Mar. 1993.
28. H. Ishikawa, K. Yamamoto, T. Egawa, T. Soga, T. Jimbo and M. Umeno, "Thermal stability of GaN on (111) Si substrate", *Journal of Crystal Growth*, vol. 189–190, pp. 178–182, June 1998.
29. Y.-D Lin, S. Yamamoto, C.-Y. Huang, C.-L. Hsiung, F. Wu, K. Fujito, H. Ohta, J. S. Speck, S. P. DenBaars and S. Nakamura "High quality InGaN/AlGaN multiple quantum wells for semi-polar InGaN green laser diodes", *Applied Physics Letters*, vol. 3, no. 8, pp.082001, July 2010.
30. M. Adachi, Y. Yoshizumi, Y. Enya, T. Kyono, T. Sumitomo, S. Tokuyama, S. Takagi, K. Sumiyoshi, N. Saga, T. Ikegami, M. Ueno, K. Katayama and T. Nakamura, "Low threshold current density InGaN based 520–530 nm green laser diodes on semi-polar (20-21) free-standing GaN substrates", *Applied Physics Express*, vol. 3, no. 12, p.121001, Dec. 2010.
31. J. Sun, H. Fatima, A. Koudymov, A. Chitnis, X. Hu, H.-M. Wang, J. Zhang, G. Simin, J. Yang and M. Khan, "Thermal management of AlGaN–GaN HFETs on Sapphire using flip-chip bonding with epoxy underfill", *IEEE Electron Device Letters*, vol. 24, no. 6, pp.374–376, June 2003.
32. A. Kim, "Trends in LED manufacturing", Philips Lumileds, accessed July 4, 2015, http://theconfab.com/wp-content/uploads/2014/confab_2014_andrew_kim.pdf.
33. R. S. Pengelly, S. M. Wood, J. W. Milligan, S. T. Sheppard and W. L. Pribble, "A review of GaN on SiC high electron-mobility power transistors and MMICs", *IEEE Transactions on Electron Devices*, vol. 60, no. 6, pp.1764–1783, Feb. 2012.
34. P. R. Hageman, J. J. Schermer and P. K. Larsen, "GaN growth on single-crystal diamond substrates by metalorganic chemical vapour deposition and hydride vapour deposition", *Thin Solid Films*, vol. 443, no. 1–2, pp.9–13, Oct. 2003.
35. G. W. G. van Dreumel, P. T. Tinnemans, A. A. J. van den Heuvel, T. Bohnen, J. G. Buijnsters, J. J. ter Meulen, W. J. P. van Enckevort, P. R. Hageman and E. Vlieg, "Realising epitaxial growth of GaN on (001) diamond", *Journal of Applied Physics*, vol. 110, no. 1,p.013503, July 2011.
36. K. Hirama, Y. Taniyasu and M. Kasu, "Epitaxial growth of AlGaN/GaN high-electron mobility transistor structure on diamond (111) surface", *Japan Journal of Applied Physics*, vol. 51, p.090114, 2012.
37. P. C. Chao, K. Chu and C. Creamer, "A new high power GaN-on-diamond HEMT with low-temperature bonded substrate technology", CS MANTECH Conference, May 13–16, 2013, New Orleans, LA, pp.179–182.
38. J. Pomeroy, M. Bernardoni, A. Sarua, A. Manoi, D. C. Dumka, D. M. Fanning and M. Kuball, "Achieving the best thermal performance for GaN-on-diamond", IEEE Compound Semiconductor Integrated Circuit Symposium (CSICS), Monterey, CA, 2013, pp.1–4.

39. F. Semond, P. Lorenzini, N. Grandjean and J. Massies, "High-electron-mobility AlGaN/GaN heterostructures grown on Si(111) by molecular-beam epitaxy", *Applied Physics Letters*, vol. 78, no. 3, pp.335–337, 2001.

40. K. Radhakrishnan, N. Dharmarasu, Z. Sun, S. Arulkumaran and G. I. Ng, "Demonstration of AlGaN/GaN HEMTs on 100-mm diameter silicon by plasma assisted MBE", *Applied Physics Letters*, vol. 97, no. 23, pp.1–3, Dec 2010.

41. N. Dharmarasu, K. Radhakrishnan, M. Agrawal, L. Ravikiran, S. Arulkumaran, K. E. Lee and G. I. Ng, "Demonstration of AlGaN/GaN high-electron-mobility-transistors on 100-mm diameter Si (111) by ammonia molecular beam epitaxy", *Applied Physics Express*, vol. 5, no. 9, p. 091003, Aug. 2012.

42. J. Laroche, K. Ip, M. Breen, W. Hoke, Y. Cao, J. Bettencourt, D. Guenther, G. Gebara, T. Kennedy, Schultz, O. Laboutin, C. Fong, T. Trimble, W. Johnson, T. Kazior and J. Comeau, "GaN HEMT fabrication in a 200-mm Si foundry environment: The time has come", 225th ECS meeting, 1505 & ECS Transactions, vol. 61 no. 4, May 2014, pp.29–32.

43. L. M. Kyaw, S. B. Dolmanan, M. K. Bera, Y. Liu, H. R. Tan, T. N. Bhat, Y. Dikme, E. F. Chor and S. Tripathy, "Influence of RuOx gate thermal annealing on electrical characteristics of $Al_xGa_{1-x}N$/GaN HEMTs on 200-mm silicon", *ECS Solid State Letters*, vol. 3, no. 2, pp.Q5–Q8, 2014.

44. S. Arulkumaran, T. Egawa, H. Ishikawa and T. Jimbo, "Characterization of different-Al-content $Al_xGa_{1-x}N$/GaN heterostructures and HEMTs on sapphire", *Journal of Vacuum Science and Technology B*, vol. 21, p.888, March 2003.

45. A. Ubukata, K. Ikenaga, N. Akutsu, A. Yamaguchi, K. Matsumoto, T. Yamazaki and T. Egawa, "GaN growth on 150-mm-diameter (111) Si substrates", *Journal of Crystal Growth*, vol. 298, pp.198–201, Jan. 2007.

46. A. Ubukata, Y. Yano, H. Shimamura, A. Yamaguchi, T. Tabuchi and K. Matsumoto, "High-growth-rate AlGaN buffer layers and atmospheric-pressure growth of low-carbon GaN for AlGaN/GaN HEMT on the 6-in.-diameter Si substrate metal-organic vapor phase epitaxy system", *Journal of Crystal Growth*, vol. 370, pp.269–272, May 2013.

47. D. S. Lee, J. Su, B. Krishnan, G. D. Papasouliotis and A. Paranjpe, "Production readiness of AlGaN/GaN HEMT on 6"/8" Si", *ECS Transactions*, vol. 58, no. 4, pp.311–314, Feb. 2013.

48. T. Egawa, "Heteroepitaxial growth and power devices using AlGaN/GaN HEMT on 200 mm Si (111) substrate", IEEE Compound Semiconductor Integrated Circuit Symposium (CSICS), Monterey, CA, 2013, pp.1–4.

49. T. N. Bhat, S. B. Dolmanan and S. Tripathy, "Structural and optical properties of $Al_xGa_{1-x}N$/GaN high electron mobility transistor structures grown on 200-mm diameter Si (111) substrates," *Journal of Vaccum Science and Technology* B, vol. 32, no. 2, p.021206, Feb. 2014.

50. S. Arulkumaran, G. I. Ng, S. Vicknesh, Z. H. Liu and M. Bryan, "Improved recess-ohmics in AlGaN/GaN HEMTs with AlN spacer layer on silicon substrate", *Physica Status Solidi C*, vol. 7, p.2412, Oct. 2010.

51. S. Arulkumaran, Z. H. Liu and G. I. Ng, "Effect of gate-source and gate-drain Si3N4 passivation on current collapse in AlGaN/GaN HEMTs on silicon", *Applied Physics Letters*, vol. 90, p.173504, Apr, 2007.

52. G. I. Ng and S. Arulkumaran, Nano-semiconductors: Devices and technology, in *Chapter 14: GaN HEMTs Technology and Applications*, October 24, 2011, CRC Press, Boca Raton, FL, p.389.

53. L. M. Kyaw, L. K. Bera, Y. Liu, M. K. Bera, S. P. Singh, S. B. Dolmanan, H. R. Tan, T. N. Bhat, E. F. Chor and S. Tripathy, "Probing channel temperature profiles in $Al_xGa_{1-x}N$/GaN high electron mobility transistors on 200mm diameter Si(111) by optical spectroscopy," *Applied Physics Letters*, vol. 105, no. 7, p.073504, Aug. 2014.

54. S. Arulkumaran, Z. H. Liu, G. I. Ng, T. Aggerstam, J. Bu, R. Zeng, M. Sjodin, K. Radhakrishnan C. L. Tan and S. Lourdudoss, "Enhancement of both DC and microwave characteristics of AlGaN/GaN HEMTs by furnace annealing", *Applied Physics Letters*, vol. 88, no. 2, pp.023502–023502-3, Jan. 2006.
55. S. Arulkumaran, Z. H. Liu, G. I. Ng, W. C. Cheong, R. Zeng, J. Bu, H. Wang, K. Radhakrishnan and C. L. Tan, "Temperature dependent microwave performance of AlGaN/GaN HEMTs on high-resistivity silicon substrate, *Thin Solid Films*, vol. 515, no. 10, pp.4517–4521, Mar. 2007.
56. S. Haffouz, H. Tang, J. A. Bardwell, E. M. Hsu, J. B. Webb and S. Rolfe, "AlGaN/GaN field effect transistors with C-doped GaN buffer layer as an electrical isolation template grown by molecular beam epitaxy", *Solid-State Electronics*, vol. 49, no. 5, pp.802–807, May 2005.
57. S. Arulkumaran, G. I. Ng and S. Vicknesh, "Enhanced breakdown voltage with high Johnson's figure- of-merit in 0.3-μm T-gate AlGaN/GaN HEMTs on silicon by (NH4)2Sx treatment", IEEE Electron Device Letters, vol. 34, no. 11, pp.1364–1366, Nov. 2013.
58. K. Ranjan, S. Arulkumaran, G. I. Ng and S. Vicknesh, "High Johnson's figure-of-merit (8.32 THz·V) in 0.15-μm conventional T-gate AlGaN/GaN HEMTs on silicon", *Applied Physics Express*, vol. 7, no. 4, p. 044102, Mar. 2014.
59. M. V. Hove, S. Boulay, S. R. Bahl, S. Stoffels, X. Kang, D. Wellekens, K. Geens, A.Delabie and S. Decoutere "CMOS process-compatible high-power low-leakage AlGaN/GaN MISHEMT on silicon", *IEEE Electron Device Letters*, vol. 33, no. 5, pp.667–669, May 2012.
60. H. F. Sun, A. R. Alt, H. Benedickter and C. R. Bolognesi, "100 nm Gate AlGaN/GaN HEMTs on silicon with fT=90 GHz", *Electronics Letters*, vol. 45, no. 7, pp. 376–377, Mar. 2009.
61. S. Jia, Y. Cai, D. Wang, B. Zhang, K. M. Lau, and K. J. Chen, "E-mode AlGaN/GaN HEMTs on Si substrate", *IEEE Transactions on Electron Devices*, vol. 53, no. 6, pp.1474–1477, June 2006.
62. A. Minko, V. Hoel, S. Lepilliet, G. Dambrine, J. C. De Jaeger, Y. Cordier, F. Semond, F. Natali and J. Massies. "High microwave and noise performance of 0.17μm AlGaN/GaN HEMTs on high-resistivity silicon substrates," *IEEE Electron Device Letters*, vol. 25, no. 4, pp.167–169, Apr. 2004.
63. H. S. Lee, D. S. Lee and T. Palacios, "AlGaN/GaN high-electron-mobility transistors fabricated through a Au-free technology," *IEEE Electron Device Letters*, vol. 32, no.5, pp.623–625, May 2011.
64. S. Arulkumaran, G. I. Ng, S. Vicknesh, W. Hong, K. S. Ang, C. M. Manojkumar, K. L. Teo and K. Ranjan, "Demonstration of submicron-gate AlGaN/GaN high-electron-mobility transistors on silicon with complementary metal–oxide–semiconductor-compatible non-gold metal stack", *Applied Physics Express*, vol. 6, no. 1, p.016501, Jan. 2013.
65. X. Liu, C. Zhan, K.W. Chan, W. Liu, L. S. Tan, K. J. Chen, and Y. C. Yeo, "AlGaN/GaN-on-silicon metal–oxide–semiconductor HEMT with breakdown voltage of 800V and on-state resistance of 3 mΩ-cm² using a CMOS compatible gold-free process", *Applied Physics Express*, vol. 5, p.066501, May 2012.
66. Z. H. Liu, M. Sun, H.-S. Lee, M. Heuken and T. Palacios, "AlGaN/AlN/GaN high-electron-mobility transistors fabricated with Au-free technology", *Applied Physics Express*, vol. 6, no. 9, p.096502, Sep. 2013.
67. A. Malmros, H. Blanck and N. Rorsman, "Electrical properties, microstructure, and thermal stability of Ta-based ohmic contacts annealed at low temperature for GaN HEMTs," *Semiconductor Science and Technology*, vol. 26, no. 7, p.075006, July 2011.

68. Y. Li, G.I. Ng, S. Arulkumaran, C. M. Manojkumar, K. S. Ang, H. Wang, G. Ye, R. Hofstetter and M. J. Anand, "Investigations of CMOS-compatible non-gold Ta/Si/Ti/ Al/Ni/Ta ohmic contact for AlGaN/GaN HEMT on Si with low contact resistance", 10th International Conference on Nitride Semiconductors, August 25–30, 2013, Washington, DC

69. K. Shinohara, A. Corrion, D. Regan, I. Milosavljevic, D. Brown, S. Burnham, P. J. Willadsen, C. Butler C, A. Schmitz, D. Wheeler, A. Fung, M. Micovic, "220 GHz $f_T$ and 400 GHz $f_{max}$ in 40-nm GaN DH-HEMTs with re-grown ohmic," IEEE Electron Devices Meeting Technical Digest, December 6–8, 2010, San Fransico, CA, pp.30.1.1–30.1.4.

70. S. Arulkumaran, G. I. Ng, K. Ranjan, C. M. Manoj Kumar, S. C. Foo, K. S. Ang, S. Vicknesh, S. B. Dolmanan and S. Tripathy, "Record low contact resistance for InAlN/GaN HEMTs on Si with non-gold metal", Solid State Materials and Devices (SSDM) 2014, Tsukuba, Japan, Sept. 2014.

71. M.J. Anand, G. I. Ng, S. Arulkumaran, H. Wang, Y. Li, S. Vicknesh and T. Egawa, "Low specific ON-resistance and high figure-of-merit AlGaN/GaN HEMTs on Si substrate with non-gold metal stacks", 71st Annual Device Research Conference (DRC), Notre Dame, IN, June 23–26, 2013, pp.1–2.

72. A. Firrincieli, B. De Jaeger, S. You, D. Wellekens, M. V. Hove and S. Decoutere, "Au-free low temperature Ohmic contacts for AlGaN/GaN power devices on 200-mm Si substrates", *Japanese Journal of Applied Physics*, vol. 53, p.04EF01, 2014.

73. S. Vicknesh, S. Arulkumaran and G. I. Ng, "'Effective suppression of current collapse in both E- and D-mode AlGaN/GaN HEMTs on Si by [(NH4)2Sx] passivation', Proceedings of IEEE MTT-S International Microwave Symposium Digest, Canada, June 2012, pp.1–2.

74. S. Yoshida, M. Tanomura, Y. Murase, K. Yamanoguchi, K. Ota, K. Matsunaga and H. Shimawaki, "A 76 GHz GaN-on-silicon power amplifier for automotive radar systems", Proceedingsof IEEE MTT-S International Microwave Symposium Digest, 2009, p.665.

75. J. Kuzmik, "Power electronics on InAlN/(In)GaN: Prospect for a record performance", *IEEE Electron Device Letters*, vol. 22, no. 11, pp.510–512, Nov. 2001.

76. J. Xie, X. Ni, M. Wu, J. H. Leach, U. Ozgur and H. Morkoc, "High electron mobility in nearly lattice-matched AlInN/AlN/GaN heterostructure field effect transistors', *Applied Physics Letters*, vol. 91, no. 13, pp.132116, Sep. 2007.

77. Y. Yue, Z. Hu, J. Guo, B. S. Rodriguez, G. Li, R. Wang, F. Faria, B. Song, X. Gao, S. Guo, T. Kosel, G. Snider, P. Fay, D. Jena and H. Xing, "Ultrascaled InAlN/GaN high electron mobility transistors with cutoff frequency of 400 GHz", *Japan Journal of Applied Physics*, vol. 52, no. 8S, pp.08JN14, May 2013.

78. S. Arulkumaran, K. Ranjan, G. I. Ng, C. M. Manoj Kumar, S. Vicknesh, S. B. Dolmanan and S. Tripathy, "High-frequency microwave noise characteristics of InAlN/GaN HEMTs on Si(111) substrate", *IEEE Electron Device Letters*, vol. 35, no. 10, pp.992–994, Oct. 2014.

79. A. Watanabe, J. J. Freedsman, R. Oda, T. Ito and T. Egawa, "Characterization of InAlN/ GaN high- electron-mobility transistors grown on Si substrate using graded layer and strain-layer superlattice", *Applied Physics Express*, vol. 7, no. 4, pp.041002, Apr. 2014.

80. Y. Liu, L. M. Kyaw, M. K. Bera, S. P. Singh, Y. J. Ngoo, G. Q. Lo, E. F. Chor, "Low thermal budget Au-free Hf-based ohmic contacts on InAlN/GaN heterostructures", *ECS Transactions*, vol. 61, no. 4, pp.319–327, May 2014.

81. A. Malmros, P. Gamarra, M. Thorsell, M.-A. Di F-Poisson, C. Lacam, M. Tordjman, R. Aubry, H. Zirath and N. Rorsman, "Evaluation of an InAlN/AlN/GaN HEMT with Ta-based ohmic contacts and PECVD SiN passivation", *Physica Status Solidi C*, vol. 11, no. 3–4, pp.924–927, Apr. 2014.

82. M. Alomari, D. Maier, J.-F. Carlin, N. Grandjean, M. A. D.-Poisson, S. Delage and E. Kohn, "Au-free ohmic contacts for high temperature InAlN/GaN HEMTs", *ECS Transactions*,vol. 25, no. 12, pp.33–36, Oct. 2009.

83. H. Sun, A. R. Alt, H. Benedickter, C. R. Bolognesi, E. Feltin, J.-F. Carlin, M. Gonschorek, and N. Grandjean, Ultrahigh-speed AlInN/GaN high electron mobility transistors grown on (111) high-resistivity silicon with FT = 143 GHz, *Applied Physics Express*, vol. 3, no. 6, pp.094101, Sep. 2010.

84. G. Pozzovivo, J. Kuzmik, C. Giesen, M. Heuken, J. Liday, G. Strasser, D. Pogany, "Low resistance ohmic contacts annealed at 600 °C on a InAlN/GaN heterostructure with SiCl4-reactive ion etching surface treatment", *Physica Status Solidi C*, vol. 6, no. 52, pp.S999–S1001, June 2009.

85. O. Jardel, G. Callet, J. Dufraisse, M. Piazza, N. Sarazin, E. Chartier, M. Oualli, R. Aubry, T. Reveyrand, J.-C. Jacquet, M. A. Di Forte Poisson, E. Morvan, S. Piotrowicz and S. L. Delage, "First demonstration of AlInN/GaN HEMTs amplifiers at K band", 2012 IEEE/MTT-S International Microwave Symposium Digest, Montreal, QC, Canada, 17–22 June 2012, pp.1–3.

86. J. Lee, M. Yan, B. Ofuonye, J. Jang, X. Gao, S. Guo and I. Adesida, "Low resistance Mo/Al/Mo/Au ohmic contact scheme to InAlN/AlN/GaN heterostructure", *Physica Status Solidi A*, vol. 208, no. 7, pp.1538–1540, July 2011.

87. D. S. Lee, X. Gao, S. Guo, D. Kopp, P. Fay and T. Palacios, "300-GHz InAlN/GaN HEMTs with InGaN back barrier," *IEEE Electron Device Letters*, vol. 32, no. 11, pp.1525–1527, 2011.

88. J. H. Leach, M. Wu, X. Ni, X. Li, U. Ozgur and H. Morkoc, "Effect of lattice mismatch on gate lag in high quality InAlN/AlN/GaN HFET structures", *Physica Status Solidi A*, vol. 207, no. 1, pp.211–216, 2010.

89. Imec, More Efficient Switching and Lighting with GaN Technology, Annual report 2012, http://annualreport.imec.be/Domains/Energy/GaN-technology/page.aspx/1199.

90. Toshiba, Toshiba to Start Sales of White LED Packages, Press releases, 2012, accessed July 4, 2015, http://www.toshiba.co.jp/about/press/2012_12/pr1401.htm.

91. Aledia, Aledia makes its first LEDs on 8-inch silicon wafers using cost-disruptive microwire technology, accessed July 4, 2015, http://www.aledia.com/en/news/aledia-makes-its-first-leds-on-8-inch-silicon-wafers-using-cost-disruptive-microwire-tech-nology-1/.

# 4 GaN-HEMT Scaling Technologies for High Frequency Radio Frequency and Mixed Signal Applications

*Keisuke Shinohara*

## CONTENTS

## ABSTRACT

This chapter focuses on recent advancements in device scaling technology of GaN-based high electron mobility transistors (HEMTs) for high-frequency RF and mixed signal applications. First, unique material properties of GaN-on-SiC technology suitable for high-power and high-efficiency performance in microwave and millimeter-wave power amplifier monolithic microwave integrated circuits will be described. Then, discusses frequency limitations of conventional GaN-HEMTs, state-of-the-art device scaling technologies that boosted GaN-HEMT frequency performance to 500 GHz range, monolithic E/D-mode integration process, and potential applications enabled by the deeply-scaled GaN transistors will be discussed.

## 4.1  INTRODUCTION

Millimeter-wave is a radio wave with a frequency between 30 GHz and 300 GHz. Millimeter-wave radio offers several advantages over lower frequency microwave radio. Its wide bandwidth enables high data transmission rates of up to 10 Gbit/s. The highly directional and narrow radiation pattern from millimeter-wave radios allows many radios to be deployed without causing interference. Its short wavelength (1–10 mm) reduces the size of antennas and allows for the construction of a smaller and lighter apparatus. When used for imaging applications, images with high resolution can be obtained due to the short wavelength. Atmospheric absorption by oxygen and water exists in various parts of the millimeter-wave spectrum, allowing a repeated use of radios in a short range. By taking advantage of their unique characteristics, a wide variety of applications using millimeter-wave radios have emerged. These applications include satellite communications (35, 60, 94 GHz), wireless LAN (60 GHz), point-to-point backhaul system (70–80 GHz), 10 Gbit/s wireless link for uncompressed HDTV signal transmission (120 GHz), body scanners for airport security (24–30 GHz), automotive radars (77, 79 GHz), passive imaging system for aircraft safe landing (94 GHz), radio astronomy, environmental remote sensing, and so on. Consequently, the demand for compact, low-cost, and high-performance millimeter-wave components is significantly increasing.

Key performances required for high-frequency monolithic microwave integrated circuits (MMICs) are high-power, high-efficiency, high-linearity, low-noise, and high-integration characteristics. GaAs-based metal-semiconductor field effect transistors (MESFETs), HEMTs and heterojunction bipolar transistors (HBTs) are transistor technologies commonly used in MMICs due to their superior electrical properties over Si such as high electron mobility and high saturation velocity. Recent advancement in Si CMOS and SiGe HBT technologies, however, improved their high-frequency performance while offering a high-volume, low-cost solution with a high level of integration, enabling complex millimeter-wave phased array radio frequency integrated circuit (RFIC) [1]. Low-voltage and low-noise performance of InP-based HEMTs are suitable for low-power and low-noise amplifier (LNA) applications. More recently, MMICs operating in sub-THz regime have been successfully demonstrated based on ultra-high frequency performance realized in highly-scaled InP HEMTs and HBTs [2–5].

Since the first demonstration of GaN-based MESFETs in 1993 [6] and HEMTs in 1994 [7], tremendous progress on GaN transistor technology has been made in a broad range of technical areas from substrate materials, epitaxial crystal growth, heterostructure design, device process, MMIC design, to packaging. The unique properties of GaN-based material systems offer unprecedented flexibility in designing advanced electron devices. The availability of AlGaN/GaN heterostructures enabled

HEMTs that utilize high-density two-dimensional electron gas (2DEG) accumulated in the GaN channel layer through spontaneous and piezoelectric polarization effect [8]. The high electron mobility has been obtained through progress in the material growth technique such as molecular beam epitaxy (MBE) and metalorganic chemical vapor deposition (MOCVD). Combined with a high breakdown voltage, GaN HEMTs enable power amplifiers (PAs) with high power added efficiency (PAE) that have significantly higher output power and power density than is presently available from amplifier circuits based on other material systems such as Si, GaAs, or InP. In addition, thermal characteristics are enhanced using high thermal conductivity SiC substrates. Based on these advantages for PA applications, the technology has been commercialized in a wide range of applications from pulsed radars to cable television (CATV) amplifier modules and mobile base stations. Recently, remarkable progress has been made in high-frequency E/D-mode GaN HEMT technologies that expand the GaN design space to address higher-frequency higher-efficiency PAs, ultra-LNAs with high input power survivability, ultra-linear mixers, and increased output-power digital-to-analog converts. This chapter reviews the advantages of GaN-based material systems suitable for high-frequency high-power applications, and recent advancement in GaN HEMT scaling technologies and device performances, and their potential applications.

## 4.2 MATERIAL PROPERTIES OF GaN-BASED MATERIAL SYSTEM

Table 4.1 summarizes material properties of commonly used microwave semiconductors, showing advantages of GaN-based material system for high-frequency and high-power applications. III-N semiconductors, with a wide range of band-gap energy varying from 0.7 eV (InN) to 3.4 eV (GaN) to 6.1 eV (AlN), and availability of their ternary (AlGaN, InAlN) and quaternary (InAlGaN) alloys enable flexible design of heterostructures. HEMT epitaxial structures based on III-N heterostructures typically consist of a wide band-gap top barrier material such as AlGaN or InAlN and a GaN channel layer. A high-density two-dimensional electron gas (2DEG) is induced at an interface through spontaneous and piezoelectric polarization effects without a need for intentional doping [8]. A large conduction band offset between the AlGaN (or InAlN) barrier layer and the GaN channel layer effectively increases electron confinement in the channel. The high 2DEG density of $1-2 \times 10^{13}$ cm$^{-2}$ combined with their high electron mobility in the range of 1500–2000 cm$^2$/V·s leads to a low channel sheet resistance, resulting in exceptionally low device on-resistance. The high saturation velocity enables not only high current densities leading to a high output power density but also high-frequency operation. Owing to the high critical electric field of GaN, which is >10× higher than Si, AlGaN/GaN HEMTs possesses high breakdown voltage, which allows a large drain bias voltage, leading to a high-power density with a high output impedance per unit RF power. This results in easier matching and lower loss matching circuits. In addition to the unique electrical properties of GaN-based material system, availability of semi-insulating 4H- and 6H-SiC substrates with high thermal conductivity greatly helps to spread heat generated by self-heating during high-power operation, allowing transistors to operate at high-power densities and high efficiency while reducing the requirement for cooling.

**TABLE 4.1**
**Material Properties of Microwave Semiconductors**

|  | Si | InP | GaAs | SiC | InN | GaN | AlN | Diamond |
|---|---|---|---|---|---|---|---|---|
| Egap (eV) | 1.1 | 1.34 | 1.43 | 3.3 (4H) | 0.63 | 3.4 | 6.1 | 5.5 |
| Electron mobility (cm²/V·s) | 1,350 | 12,000* | 8,500* | 900 | 3,300 | 2,000* | 1,100 | 1,900 |
| 2DEG density (×10¹³ cm⁻²) | N/A | 0.3* | <0.2 | N/A | N/A | >2 | N/A | N/A |
| Electron effective mass | 0.26 | 0.08 | 0.067 | 0.29 | 0.11 | 0.2 | 0.4 | 1.4 |
| Saturation velocity (×10⁷ cm/s) | 1 | 3.3 | 1 | 2 | 3.5 | 1.5–2.5 | 1.5 | 1.9 |
| Critical electric field (MV/cm) | 0.3 | 0.5 | 0.4 | 3 | 1 | 3.3 | 6–15 | 10 |
| Thermal conductivity (W/cm·K) | 1.3 | 0.7 | 0.5 | 4.9 | 1.2 | 2 | 2 | 6–20 |
| Relative dielectric constant | 12 | 12.5 | 13 | 9.8 | 15.3 | 9.5 | 9 | 5.7 |

* Measured on InAlAs/InGaAs, AlGaAs/InGaAs, AlGaN/GaN HEMT structures.

## 4.3 CURRENT STATUS OF GaN POWER AMPLIFIER PERFORMANCE

PAs based on GaN-on-SiC HEMT technology operating in microwave and millimeter-wave frequency ranges have been successfully demonstrated. State-of-the-art power levels have been reported with total output powers of 912 W at 2.9 GHz [9], 550 W at 3.6 GHz [10] under pulsed conditions, and continuous wave (CW) output powers of 179 W at 2 GHz [11], 38 W at 10 GHz [12], and 20.7 W at 27 GHz [13]. PA MMICs operating in W-band frequency range have also been reported with a pulsed output power of 3.2 W at 86 GHz [14], and a CW output power of 2.1 W at 93.5 GHz with an associated PAE of 19% [15]. Figure 4.1 compares CW output power of PA ICs reported for various semiconductor transistor technologies as a function of operating frequency. Owing to high-current and high-voltage handling capability, GaN HEMT amplifiers show more than 5–10 times higher output power than any other technologies in the frequency range of up to 100 GHz. The output power of GaN HEMT amplifiers decreases with increasing frequency by ~$1/f$ up to 30 GHz, and by ~$1/f^2$ between 30 and 100 GHz. The ~$1/f^2$ dependence of the output power above

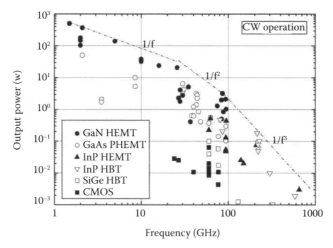

**FIGURE 4.1**   Output power of power amplifier ICs under CW operation based on various semiconductor transistor technologies.

30 GHz is attributed to a combination of the total gate periphery that approximately scales with $1/f$ and the maximum operational drain voltage that is proportional to the lateral device dimension (source-drain distance ~$1/f$) when the devices are scaled for millimeter-wave frequency operations (>30 GHz). The slower slope ~$1/f$ below 30 GHz is a result of a limited heat dissipation density in large gate periphery devices used in this frequency range to maintain a sufficiently low junction temperature ($T_j$). The maximum operating frequency of GaN PA MMICs reported so far is 96 GHz, which was limited by the frequency performance of GaN HEMTs; $f_T/f_{max}$ = 97/230 GHz was reported for 140-nm gate GaN HEMTs used in the W-band MMICs [16]. On the other hand, InP HEMTs and InP HBTs, the technologies that have the highest cutoff frequencies $f_T/f_{max}$ of >600/>1200 GHz (InP HEMT) and 400/700 GHz (InP HBT), demonstrate the highest power performance beyond 100 GHz; for instance, $P_{out}$ = 90 mW at 220 GHz [17], 180 mW at 214 GHz [18], 10 mW at 300 GHz [19], and 3 mW at 653.5 GHz [20]. At such high frequencies (>200 GHz), insufficient gain (maximum stable gain [MSG]/maximum available gain [MAG]) across a span of a load-line imposes a severe limitation [21], leading to a more rapid reduction in output power with frequency at a slope of roughly $1/f^3$ (Figure 4.1).

High PAs with a high PAE is critically important in most applications because they greatly reduce power consumption, cooling requirement, and therefore, size and weight of the system. RF waveform engineering enabled by harmonic load-pull measurement systems greatly enhanced design flexibility of high efficiency PAs [22]. Very high efficiency GaN HEMT PAs with 84.9% PAE and a 3.3 W/mm power density at 2.14 GHz have been reported by designing inverse class-F second- and third-harmonic terminations [23]. The first class-E switched-mode PAs operating in X-band demonstrated 61% PAE at 3.5 W of output power [24]. High-efficiency X-band class-E amplifiers with a PAE of 72%, which deliver an output power density of 3.2 W/mm were also reported [25]. Figure 4.2 shows PAE of GaN HEMT amplifiers as a function of frequency. The high-efficiency performance of GaN HEMT PAs

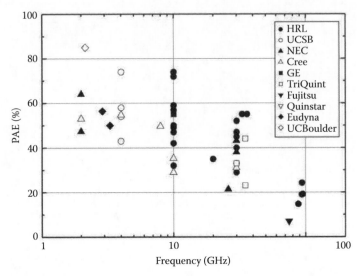

**FIGURE 4.2** Power-added efficiency (PAE) of GaN-HEMT power amplifiers operating at various frequency ranges.

have been limited in the frequency range of <10 GHz mainly due to an insufficient gain and a limited cutoff frequency necessary for high-efficiency switched-mode operation at higher frequencies. At W-band, the highest PAE reported to date is 24% [16]. Figure 4.3 shows a comparison of the output power and efficiency performance between various GaN, GaAs, and InP amplifiers operating in W-band [15,16]. GaN HEMT amplifiers have comparable PAE and about 3–4 times more output power than InP HEMT amplifiers in this frequency range. To take full advantages of GaN-based material systems for high-power and high-efficiency characteristics at higher frequencies, it is critical to further boost the frequency performance of GaN HEMTs.

Owing to their excellent electrical properties (high electron mobility, high saturation velocity, and high 2DEG density) of GaN-based material system, cutoff frequencies ($f_T$/$f_{max}$) of GaN HEMTs can potentially exceed 500 GHz through aggressive device scaling. Realization of such high-speed GaN transistors makes it possible to build GaN PA MMICs that have sufficient gain and output power above 100 GHz. Increasing PAE of PA MMICs is of critical importance because dissipated power during high-power operation generates heat that limits output power and eventually affects device reliability. In that context, switched-mode amplifier configurations such as class-D, E, F, and $F^{-1}$ offer very high efficiencies compared with class-A/B amplifiers. In class-E amplifiers, where GaN HEMTs are used as switches, a high $f_T$ of 5–10× higher than the operating frequency, a low device on-resistance, and a high breakdown voltage are required. Therefore, 500-GHz-class GaN transistors scaled for a low on-resistance while maintaining a high breakdown voltage enable high-power high-efficiency class-E amplifiers operating in millimeter-wave frequencies. Furthermore, thermal management of GaN-based electronics is a key to enable the technology to reach its full potential. There have been a lot of researches and developments addressing the thermal management issues at both device and packaging

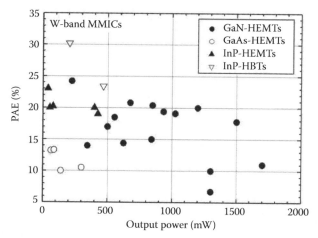

**FIGURE 4.3** Comparison of PAE versus output power among GaN-, GaAs-, and InP-based power amplifiers in the W-band.

levels. They include use of a high thermal conductivity substrate such as diamond, employing a novel packaging technology that efficiently spread heat out from device areas [26] or introducing liquid cooling [27].

Section 4.4 discusses frequency limitations in conventional GaN devices, key challenges for device scaling, and novel scaling technologies that enabled state-of-the-art 500 GHz-class GaN HEMTs, and introduces anticipated applications enabled by the emerging highly scaled GaN technologies.

## 4.4 RECENT PROGRESS IN HIGH-FREQUENCY GaN HEMT TECHNOLOGY

For high-frequency RF applications, a cutoff frequency $f_T$, and a maximum oscillation frequency $f_{max}$, a breakdown voltage $BV$, an on-resistance $R_{on}$, and a maximum drain current $I_{dmax}$ are key device performance parameters. Device scaling has successfully increased $f_T$ and $f_{max}$ of GaN transistors but simultaneously deteriorated $BV$ due to associated dimension scaling. Low $BV$ greatly restricts the dynamic range of the circuit and represents a severe limitation. The high breakdown field of GaN (3.4 MV/cm) has been the main motivation for GaN transistors designed for PAs. Figure 4.4 compares Johnson figure of merit (JFoM) which is defined as the product of $f_T$ and $BV$ among various high-speed device technologies. GaN HEMTs demonstrate the highest JFoM—about five times higher than that of InP-based HEMTs and HBTs. Recent advancements in GaN device scaling technologies led to the highest $f_T$ of 454 GHz in GaN HEMTs with a simultaneous $BV$ of 10 V, maintaining the superior JFoM power performance [28]. In addition to JFoM, $f_{max}$ is an important performance parameter for high-frequency PA applications where transistors must have sufficient power gain at the operating frequency. Proportional scaling of intrinsic and parasitic delay components in GaN HEMTs has significantly enhanced both $f_T$ and $f_{max}$ up to 450 and 600 GHz range, respectively (Figure 4.5). A low device $R_{on}$ is particularly important when GaN HEMTs are used in high-efficiency PAs, LNAs, or low-loss

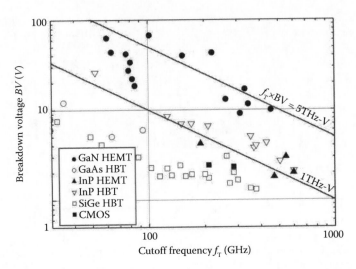

**FIGURE 4.4** Comparison of Johnson's figure-of-merit among various high-speed device technologies.

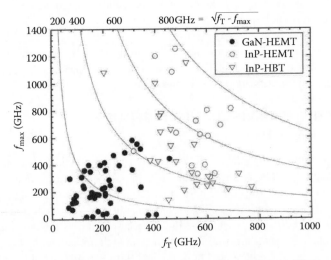

**FIGURE 4.5** Comparison of cutoff frequencies ($f_T$) and maximum oscillation frequencies ($f_{max}$) among III-N HEMTs, InP HEMTs, and InP HBTs.

RF switches. When a device is operated under a large signal condition, an actual $R_{on}$ is sometimes higher than the static $R_{on}$ due to electron trapping in deep levels in the buffer and/or the surface states of the AlGaN/GaN materials [29,30]. This device on-resistance under a large signal operation is called "dynamic" $R_{on}$, which affects high-power and efficiency performance of GaN HEMT PAs. Reduction of deep level traps in the buffer layer by optimizing epitaxial growth conditions [31], and surface passivation [32] or electric field engineering using gate field plates [33,34] to reduce the trapping effects are key technologies to minimize dynamic $R_{on}$.

## 4.5 GaN DEVICE SCALING TECHNOLOGIES

Figure 4.6 shows a cross-sectional GaN HEMT structure that illustrates important device scaling features for high-frequency operation and its equivalent circuit model. The cutoff frequency ($f_T$) of the HEMT can be expressed by a sum of intrinsic and parasitic delay times using the following equation [35]:

$$\tau = \frac{1}{2f_T} = \frac{C_{gs} + C_{gd}}{g_m} + C_{gd} \cdot (R_s + R_d) \cdot \left[ 1 + \left( 1 + \frac{C_{gs}}{C_{gd}} \right) \frac{g_d}{g_m} \right]$$

where $C_{gs}$, $C_{gd}$, $g_m$, $g_d$, $R_s$, and $R_d$ represent the gate-to-source and the gate-to-drain capacitances, transconductance, output conductance, source and drain resistances, respectively. The first term ($C_{gs} + C_{gd}$)/$g_m$ is equivalent to an intrinsic electron transit time = ($L_g + \Delta L_g$)/$v_{ave}$, where $L_g$ and $\Delta L_g$ represent a physical gate length and an excess gate length associated with drain depletion and gate fringing field, and $v_{ave}$ is an average electron velocity. The second term $C_{gd} \cdot (R_s + R_d)$ represents a charging delay due to the parasitic source and drain resistance. The last term is a delay originated with $g_d$, which becomes significant when the device shows a severe

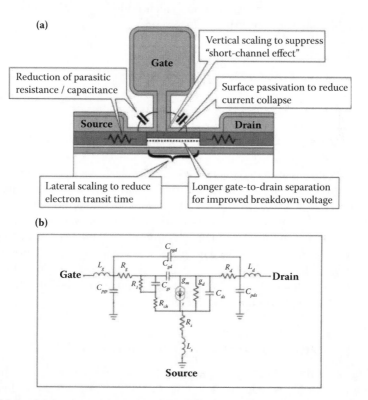

**FIGURE 4.6** Schematic cross section of a GaN HEMT illustrating important device scaling features for (a) high-frequency operation and (b) an equivalent circuit model.

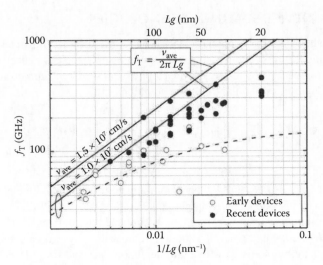

**FIGURE 4.7** Scaling behavior of cutoff frequency ($f_T$) of GaN HEMTs with the gate length ($L_g$).

short-channel effect. The maximum oscillation frequency ($f_{max}$) can be expressed with the following equation [35]:

$$f_{max} \cong \frac{f_T}{2\sqrt{(R_i + R_s + R_g)g_d + 2\pi f_T R_g C_{gd}}}$$

where $R_i$ and $R_g$ represent the input and gate resistances, respectively. Optimization of $f_{max}$ requires a device design by which device parameters such as $R_g$, $g_d$, and $C_{gd}$ are minimized while improving $f_T$.

A guiding principle in improving HEMT frequency response is scaling the gate length ($L_g$) to reduce an electron transit time. Figure 4.7 shows $f_T$ as a function of $1/L_g$ measured for short-gate GaN HEMTs. Distinct deviation from the linear scaling behavior observed for $L_g$ <100 nm in early devices indicates that parasitic delay components start to dominate the total delay time at the short gate length. Therefore, proportional scaling of intrinsic and parasitic delay components enables true scaling of GaN technology to sub-50-nm gate length. In the following Sections 4.5.1 through 4.5.6, frequency limitations and novel scaling technologies to address these limitations are discussed in detail.

### 4.5.1 Vertically-Scaled HEMT Epitaxial Structure

To obtain improved HEMT frequency response through device scaling, the electron transit time is reduced by scaling the gate length ($L_g$). Accordingly, the thicknesses of HEMT epitaxial layers are reduced to prevent the "short-channel effect"—degraded drain current modulation by the gate voltage seen in very short gate HEMTs—which causes a negative threshold voltage ($V_{th}$) shift, a decreased transconductance ($g_m$), an increased drain-induced barrier lowering (DIBL), and an increased output conductance ($g_d$). Increased $g_d$ degrades both $f_T$ and $f_{max}$ as can be seen in the above equations. The short-channel effect is characterized by the channel aspect ratio ($L_g/d$, where $d$ is the gate-to-channel distance), and $L_g/d \geq 5$ is typically required to prevent a severe $V_{th}$ shift and $g_m$ degradation (Figure 4.8) [36,37]. Spontaneous and piezoelectric polarization in GaN/AlGaN heterostructures induces a two-dimensional electron gas (2DEG) in the GaN channel with a high electron sheet density ($n_s$) of $>1\times10^{13}$ cm$^{-2}$ without the need of modulation doping when the epitaxial layers are grown on a polar c-plane substrate [8]. The 2DEG electron sheet density is, therefore, determined by the thickness and aluminum content in the AlGaN top barrier layer. Because of the largest polarization effect and barrier height of AlN among $Al_xGa_{1-x}N$ ($0 < x \leq 1$) barrier materials, it is the most effective to use as a top barrier material to minimize the barrier thickness while achieving a high $n_s$ and simultaneously suppressing the gate leakage [38,39]. The $In_{0.17}Al_{0.83}N$ alloy is lattice matched to GaN, and a higher 2DEG sheet density about twice of typical AlGaN/GaN HEMTs can be obtained due to its larger spontaneous polarization effect. Several research groups developed high-frequency HEMTs using the InAlN top barrier [40–42], and reported the highest $f_T$ of 400 GHz in this material system [43]. Such a high 2DEG sheet density realized in AlGaN/GaN and InAlN/GaN systems can lead to insufficient electron confinement. A back barrier structure beneath the 2DEG channel layer improves the confinement and effectively suppresses the short channel effect. An AlGaN back barrier structure was first introduced by Chen et al. (Figure 4.9) [44], and its effective suppression of the short-channel effect in $Al_{0.3}Ga_{0.7}N/GaN/Al_{0.04}Ga_{0.96}N$ double heterojunction HEMT structure was demonstrated by Micovic et al. [45]. Alternatively, InGaN layer inserted in the GaN buffer layer just below the 2DEG layer produces a conduction band barrier and effectively acts as a back barrier [46]. The conventional HEMT structures are built on Ga-polar AlGaN/GaN heterostructures where the 2DEG is formed below the hetero-interface. On the other

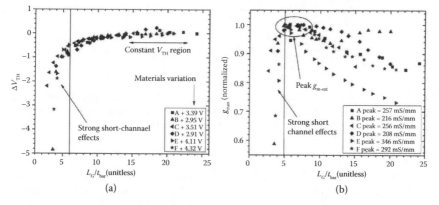

**FIGURE 4.8** (a) Threshold voltage ($V_{th}$) and (b) normalized transconductance ($g_m$) as a function of channel aspect ratio. (G.H. Jessen et al., *IEEE Trans. Eelecron. Devices*, 2589, 2007. © 2007 IEEE.)

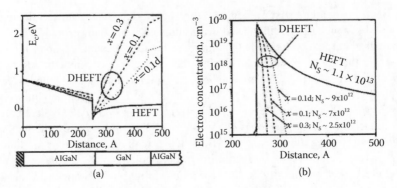

**FIGURE 4.9** (a) Conduction band diagrams and (b) electron concentration profiles for AlGaN/GaN/AlGaN double heterostructure for improved carrier confinement. (Reprinted with permission from C.Q. Chen, J.P. Zhang, V. Adivarahan, A. Koudymov, H. Fatima, AlGaN/GaN/AlGaN double heterostructure for high-power III-N field-effect transistors, *Appl. Phys. Lett.*, 82(25), 4593–4595, 2003. Copyright © 2003, American Institute of Physics.)

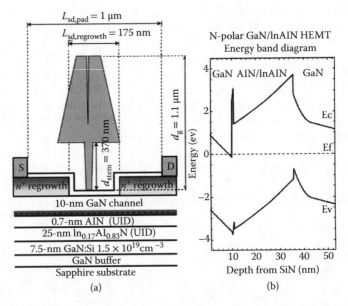

**FIGURE 4.10** (a) Epitaxial structure and (b) energy band diagram for N-polar GaN/InAlN MIS-HEMT grown by MOCVD on an off-axis sapphire substrate. (D. Denninghoff et al., N-polar GaN/InAlN MIS-HEMT with 400-GHz fmax, *DRC Tech. Dig.*, 151–152, 2012. Copyright © 2012 IEEE.)

hand, N-polar AlGaN/GaN heterostuctures induce the 2DEG above the hetero-interface due to a flipped polarity, and this built-in back barrier greatly improves electron confinement. Researchers at USCB have developed high-frequency HEMTs with 400 GHz $f_{max}$ based on this unique N-polar heterostructures as shown in Figure 4.10 [47,48].

## 4.5.2  E/D-Mode Device Integration

Enhancement-mode HEMTs are attractive because they eliminate the need of a negative voltage supply when used in MMICs and also can build logic circuits by monolithically integrating with D-mode HEMTs such as direct-coupled field-effect transistor logic (DCFL) ring oscillator circuits. Figure 4.11a and b illustrate vertically-scaled AlN/GaN/AlGaN double-heterojunction (DH) HEMT epitaxial structures grown on a semi-insulating SiC substrate by MBE and their corresponding band diagrams. Two kinds of top barrier structures consisting of (a) GaN (2.5 nm)/AlN (3.5 nm) and (b) $Al_{0.5}Ga_{0.5}N$ (2.5 nm)/AlN (2.0 nm) are designed for D- and E-mode device operation. Both structures have a 20-nm-thick GaN channel and an $Al_{0.08}Ga_{0.92}N$ back barrier. The AlN top barrier layer enables minimization of the gate-to-channel distance ($d$) (6 nm for D-mode and 4.5 nm for E-mode) while maintaining a high 2DEG density and a low gate leakage current. The $V_{th}$ of the transistors was controlled by $d$ and the Al content in the (Al)GaN cap layer that determines the surface barrier height ($\Omega_B$) of the Pt Schottky gate. A high 2DEG density ($n_s$) of 1.2(D)/1.1(E)×10$^{13}$ cm$^{-2}$ and high electron mobility ($\mu$) of 1200(D)/1250(E) cm2/V·s are measured after surface passivation with SiN. A monolithic integration of the E/D-mode epitaxial layers was successfully demonstrated using selective area epitaxial regrowth by MBE (Figure 4.11c) [49]. TriQuint and University of Notre Dame team reported monolithic integration of E/D-mode InAlN/AlN/GaN HEMTs using a gate recess approach [50–52]. The vertically-scaled HEMT structure depicted in the inset of Figure 4.12 is grown on a 6H-SiC substrate by MOCVD. Following a nucleation layer, a GaN buffer is grown. The vertically scaled barrier layer consisted of an AlN transition layer followed by a thin $In_{0.18}Al_{0.82}N$ layer lattice matched to GaN. A very low sheet resistance of 220 Ω/ sq. is obtained. The gate Schottky contacts are formed on InAlN and AlN layers for D-mode and E-mode HEMTs, respectively. Based on these technologies 501-stage

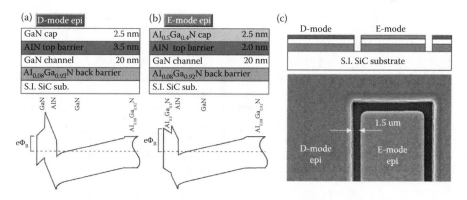

**FIGURE 4.11**  Vertically scaled double heterojunction (DH) HEMT epitaxial structures with (a) GaN (2.5 nm)/AlN(3.5 nm) top barrier for D-mode operation and (b) $Al_{0.5}Ga_{0.5}N$(2.5 nm)/AlN(2.0 nm) top barrier for E-mode operation. (c) E/D-mode epitaxial integration by selective area regrowth by MBE. (K. Shinohara et al., Scaling of GaN HEMTs and Schottky diodes for submillimeter-wave MMIC applications, *IEEE Trans. Electron Devices*, 60(10), 1982, 2013. Copyright © 2013 IEEE.)

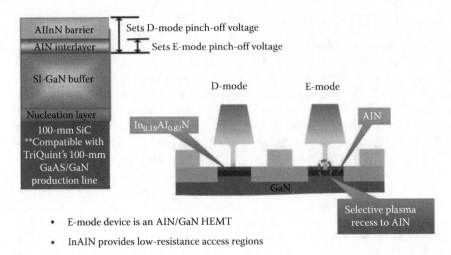

FIGURE 4.12  Monolithic integration of $In_{0.18}Al_{0.82}N$/AlN/GaN D-mode HEMT and gate recessed AlN/GaN E-mode HEMT. (P. Saunier, GaN for Next Generation Electronics, Compound Semiconductor Integrated Circuit Symposium (CSICS), 2014. Copyright © 2014 IEEE.)

GaN-based E/D-mode DCFL ring oscillator circuits consisting of >1000 transistors are demonstrated for mixed signal and logic applications [51,53].

### 4.5.3  REGROWN OHMIC TECHNOLOGY

Parasitic $RC$ charging time is reduced by minimizing source and drain resistances. Conventional alloyed ohmic contacts typically exhibit a high contact resistance of >0.4 $\Omega$·mm due to the high potential barrier of AlGaN, limiting the reduction of parasitic resistance in proportion to the device size scaling. $n^+$-GaN ohmic regrowth is one of the key aspects of GaN HEMT scaling [54–56]. An ohmic contact regrowth technique that allows direct contact of $n^+$-GaN to 2DEG enables an extremely low interface resistance owing to high $n_s$ in GaN HEMT structures [57]. A combination of a high 2DEG sheet density ($n_s$) and high electron mobility ($\mu$) results in a low 2DEG sheet resistance (typically, 300–400 $\Omega$ /sq.) that is comparable to InGaAs HEMTs, reducing resistances in the source-gate and gate-drain access regions. Figure 4.13 compares two n$^+$-GaN ohmic structures: Figure 4.13a shows a regrown $n^+$-GaN ohmic layer directly contacting the 2DEG, where electrons are supplied from the 3D $n^+$-GaN source to the 2DEG channel near the gate (3D–2D), Figure 4.13b shows an $n^+$-GaN ohmic layer regrown on top of the (Al)GaN/AlN barrier, where electrons are supplied from the 2DEG source to the 2DEG channel (2D–2D). The access resistance ($R_{ac}$), defined as a total resistance from the ohmic metal to the edge of the gate, is 0.101 and 0.127 $\Omega$·mm for 3D–2D and 2D–2D contacts, respectively. These are the lowest values ever reported for GaN HEMTs and are comparable to the state-of-the-art InP-based transistors [58,59]. An extremely low interfacial resistance ($R_{int}$) between the $n^+$-GaN and the 2DEG of only 0.026 $\Omega$·mm [57,56] almost reaches its theoretical limit [$\sim h/(2q^2 \cdot n_s^{1/2}) = 0.036$ $\Omega$·mm] [60].

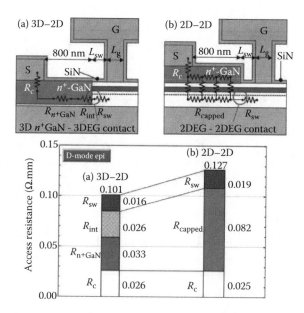

**FIGURE 4.13** Comparison of two regrown n⁺-GaN ohmic structures. (a) 3 dimensional n⁺-GaN source to 2DEG channel contact (3D–2D). (b) 2DEG source to 2DEG channel contact (2D–2D). Access resistance ($R_{ac}$) components for two regrown n⁺-GaN ohmic contacts are shown below. (K. Shinohara et al., Scaling of GaN HEMTs and Schottky diodes for submillimeter-wave MMIC applications, *IEEE Trans. Electron Devices*, 60(10), 1982, 2013. Copyright © 2013 IEEE.)

## 4.5.4 LATERAL SCALING USING SELF-ALIGNED-GATE TECHNOLOGY

When the gate length ($L_g$) is scaled down to a sub-50-nm range and becomes much smaller than the source-to-drain distance ($L_{ds}$), HEMT frequency performance becomes limited by an extrinsic delay associated with depletion of the 2DEG channel at the drain-side of the gate. The delay is called "drain-delay" (a transit time for electrons through the depleted region), which can be minimized by reducing the gate-to-drain distance ($L_{gd}$). As the critical device dimensions shrinks to nanometer-scale, it is essential to develop self-aligned process, similar to manufacturable Si CMOS technology, which allows for precise dimension control, high device yield and uniformity. Combined with n+GaN ohmic contact regrowth technique, self-aligned-gate (SAG) technology by which an ultrashort gate is self-aligned to the source-drain contacts using SiN sidewall spacers was successfully demonstrated (Figure 4.14b and c). Aggressive S-D scaling increases $f_T$ and maximum drain current but simultaneously deteriorates *BV* due to an increased electric field confined in the short gate-to-drain (G-D) spacing. Too short $L_{gd}$ also results in a high $C_{gd}$ and a high $g_d$ which decrease $f_{max}$. To realize a simultaneously high $f_T$, $f_{max}$ and BV, an asymmetric gate structure with a short $L_{gs}$ and a moderate $L_{gd}$ is preferred. A novel self-aligned-gate process that allows for an independent control of $L_{gs}$ and $L_{gd}$ was demonstrated for optimum high-speed and high breakdown voltage performances (Figure 4.14d).

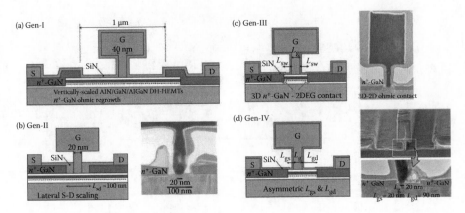

**FIGURE 4.14**    Technology cross-sections of GaN HEMT scaling technology for various generation. (K. Shinohara et al., Scaling of GaN HEMTs and Schottky diodes for submillimeter-wave MMIC applications, *IEEE Trans. Electron Devices*, 60(10), 1982, 2013. Copyright © 2013 IEEE.)

### 4.5.5   OPTIMUM T-SHAPED GATE DESIGN FOR MINIMIZED PARASITIC CAPACITANCE

The gate resistance ($R_g$) is a key device parameter that affects $f_{max}$ and a noise performance of HEMTs. A T-shaped gate structure has been used to reduce $R_g$ while minimizing the gate length ($L_g$). With aggressive device scaling, a parasitic capacitance ($C_p$) associated with T-shaped gate geometry becomes no longer negligible and a careful design of T-gate structure is required. A $C_p$ was calculated using 3D full-wave electromagnetic field simulation. Figure 4.15 illustrates the simulated T-gate structure that has a lateral gate head length of 400nm (only the right half of the T-gate is shown), where the height of the gate foot ($H_{gf}$) is varied from 25 to 200 nm. The $n^+$-GaN ohmic contact was buried in the HEMT epi layers, which greatly reduces $C_p$ between the gate-foot and the $n^+$-GaN. The electric field due to the parasitic coupling of the T-shaped gate geometry is represented by a contour plot where the color indicates the magnitude of the electric field. A high field region was formed between the bottom of the gate foot and the $n^+$-GaN contact. Figure 4.15b shows simulated $C_p$ for three structures shown by the illustrations; T-gate, gate-head only, and gate-foot only, when dielectric constant of the medium equals to 1 (air). Since the charge distribution in the T-gate differs from that in gate-head only and gate-foot only cases, the simulated $C_p$ for the T-gate is not a simple sum of these two components. However, by comparing the $C_p$ values of the gate-head and the gate-foot structures, we can see each contribution to the total $C_p$. For $H_{gf}$ below 100 nm, $C_p$ is mostly determined by the high capacitance associated with the gate head. At $H_{gf} = 200$ nm, $C_p$'s arising from the gate-head and the gate-foot become comparable, and a further increase in $H_{gf}$ does not significantly reduce the total $C_p$.

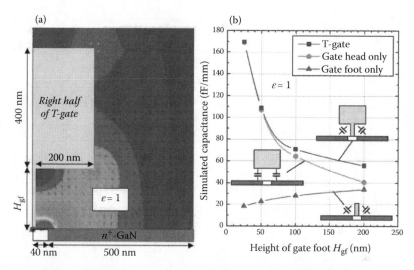

**FIGURE 4.15** Parasitic capacitance simulation of T-shaped gate. (a) Simulated T-gate structure (shown only right half of T-gate) and electric field distribution around the gate. (b) Simulated capacitance as a function of gate-foot height for three structures; T-gate, gate head only, gate foot only. (K. Shinohara et al., Scaling of GaN HEMTs and Schottky diodes for submillimeter-wave MMIC applications, *IEEE Trans. Electron Devices*, 60(10), 1982, 2013. Copyright © 2013 IEEE.)

### 4.5.6 Surface Passivation for Low Dynamic R<sub>ON</sub>

The polarization effects in GaN HEMTs lead to surface states that have an unfavorable impact on device performance. When the nitride semiconductor surface is not passivated, positively charged surface donor states can trap electrons and form a "virtual gate" that depletes the 2DEG, significantly reducing drain current. This phenomenon is called "current collapse" which degrades power performance of GaN PAs under a high voltage operation. Surface passivation technology has been developed to reduce the trapping effects and prevent the formation of the virtual gate. SiN has been the most widely used passivation layer. The physical mechanism of surface passivation by the SiN film is considered to be either reduction of trap-state density [61,62] or the change of surface donor states [63]. An alternative $Al_2O_3$ dielectric layer deposited by atomic layer deposition technique is shown to improve pulsed I-V and RF power performance over PECVD SiN [64,65].

## 4.6 DEVICE PERFORMANCE AND SCALING BEHAVIOR

### 4.6.1 E/D-Mode Device Operation

Figure 4.16 shows output and transfer characteristics of deeply scaled self-aligned-gate GaN-HEMTs with 3D–2D (Figure 4.14b) and 2D–2D (Figure 4.14c) ohmic contacts. Both devices have a gate length ($L_g$) of 60 nm and a source-drain distance

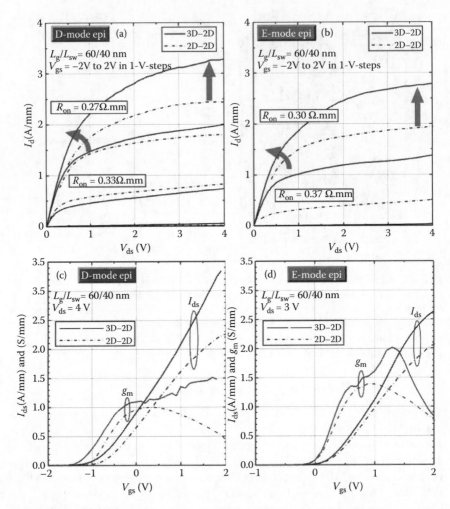

**FIGURE 4.16** (a,b) DC output and (c,d) transfer characteristics of deeply scaled E/D-mode GaN HEMTs. (K. Shinohara et al., Scaling of GaN HEMTs and Schottky diodes for submillimeter-wave MMIC applications, *IEEE Trans. Electron Devices*, 60(10), 1982, 2013. Copyright © 2013 IEEE.)

($L_{sd}$) of 140 nm and were fabricated on depletion (D) and enhancement (E)-mode epitaxial structures shown in Figure 4.11. The gate is placed symmetrically in between source and drain, where the gate-source distance ($L_{gs}$) and the gate-drain distance ($L_{gd}$) are 40 nm, respectively. The $R_{on}$ for the 3D–2D contact (0.27 Ω·mm (0.30 Ω·mm)) is smaller than that for the 2D–2D contact (0.33 Ω·mm (0.37 Ω·mm)) by 18% (19%) for depletion (enhancement)-mode device. The reduction of $R_{on}$ is consistent with the improved device access resistance ($R_{ac}$) as shown in Figure 4.13. Maximum drain current ($I_{dmax}$) is dramatically increased by 34% (45%) for D (E)-mode device. The increased $I_{dmax}$ in the 3D–2D contact is a result of an enhanced $g_m$ at high $I_{ds}$. It is reported that in deeply-scaled FETs highly-doped source/drain (S/D) can significantly improve device performance by enhancing electron supply in the source [66,67]. A heavily-doped 3D $n^+$-GaN source in direct contact with 2DEG near the gate mitigates the limited electron supply from the source, i.e., "source-starvation," resulting in suppression of $g_m$ roll-off at high $I_{ds}$ which was typically observed in previous devices with the 2D–2D contact. Exceptionally high $g_m$ values of 1.5 S/mm for the D-mode device and 2.0 S/mm for the E-mode device are achieved in GaN-based HEMTs by aggressive lateral device scaling.

### 4.6.2 Short Channel Effect

Figure 4.17 shows the dependence of the threshold voltage ($V_{th}$) on the gate length ($L_g$) for self-aligned-gate D- and E-mode devices with $L_{gs} = L_{gd} = 40$ nm. The $V_{th}$ shift ($\Delta V_{th}$) is –1.5 V for the 20-nm D-mode device and –0.4 V for the 20-nm E-mode device. The smaller $\Delta V_{th}$ obtained in the E-mode devices illustrates effectiveness of the thinner top barrier for suppression of the short-channel effect. As the lateral S-D dimension shrinks, electrostatic isolation between gate and drain is reduced, causing an increased DIBL. Figure 4.18a shows sub-threshold characteristics of a 60-nm E-mode device with a gate-drain distance ($L_{gd}$) of 70 nm (= $L_{gs}$). A subthreshold slope at $V_{ds} = 1$V is 144 mV/dec. DIBL defined at 10 mA/mm is 63 mV/V. Figure 4.18b shows dependence of DIBL on $L_g$ for three different $L_{gd}$'s of 30, 50, and 70 nm. DIBL rapidly increases with decreasing $L_g$ as a result of the increased short-channel effect. At $L_{gd} = 70$ nm, however, the increase of DIBL is suppressed as compared to $L_{gd} = 30$ and 50 nm cases, and DIBL is still as small as 100 mV/V at $L_g = 20$ nm. This result suggests that a moderate $L_{gd}$ (>70 nm) is required to maintain a good isolation between gate and drain, leading to good pinch-off characteristics, a suppressed $\Delta V_{th}$, and a lower $g_d$ for the 20-nm devices.

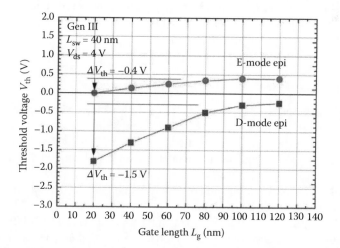

**FIGURE 4.17**  Threshold voltage ($V_{th}$) versus $L_g$ for E/D-mode GaN HEMTs. Vertical epitaxial scaling using an AlN top barrier suppressed the short channel effect. (K. Shinohara et al., Scaling of GaN HEMTs and Schottky diodes for submillimeter-wave MMIC applications, *IEEE Trans. Electron Devices*, 60(10), 1982, 2013. Copyright © 2013 IEEE.)

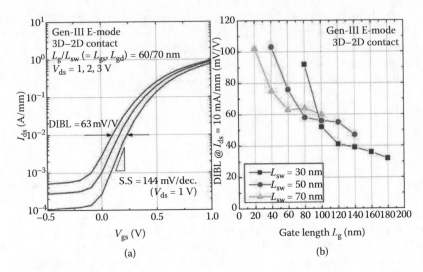

**FIGURE 4.18** (a) Subthreshold characteristics of 60-nm GaN HEMTs. (b) Dependence of drain-induced barrier lowering (DIBL) on $L_g$ for different gate-to-drain distance ($L_{gd} = L_{sw}$) (K. Shinohara et al., Scaling of GaN HEMTs and Schottky diodes for submillimeter-wave MMIC applications, *IEEE Trans. Electron Devices*, 60(10), 1982, 2013. Copyright © 2013 IEEE.)

### 4.6.3 BREAKDOWN VOLTAGE

Figure 4.19 shows three terminal off-state breakdown voltage ($BV_{off}$) plotted as a function of $L_{gd}$ for devices with $L_g = 20$–80 nm. The $BV_{off}$ is defined as a $V_{ds}$ at which catastrophic device breakdown occurs at an off-state $V_{gs}$ condition. At $L_{gd}$ below 100 nm, $BV_{off}$ increased almost linearly with increasing $L_{gd}$ with a slope of 3.25 MV/cm, close to the critical breakdown field of GaN (3.4 MV/cm). For $L_{gd}$ above 100 nm, $BV_{off}$ no longer scaled proportionally with $L_{gd}$—field shaping with a field plate structure may be necessary to maintain scaling for larger devices.

### 4.6.4 RF PERFORMANCE AND FREQUENCY LIMITATIONS

As a result of proportional scaling of device dimensions and parasitic resistances and capacitances, the highest $f_T$ of 454 GHz with a simultaneous $f_{max}$ of 444 GHz is obtained for 20-nm self-aligned-gate HEMTs with 3D–2D ohmic contacts [28] (Figure 4.20). Small signal device model parameters extracted from measured S-parameters show good agreement with measured DC parameters. Figure 4.21 shows peak $f_T$ and $f_{max}$ as a function of $L_g$ for D- and E-mode devices, showing high gate length scalability down to 20 nm.

Next, the impact of lateral S-D scaling on RF performance of GaN HEMTs is discussed. Figure 4.22a shows output characteristics and $f_T/f_{max}$ contour plots for 40-nm D-mode HEMT with an unscaled S-D ($L_{gs} = L_{gd} = 500$ nm) (Figure 4.14a). $f_T$ showed a peak value of 220 GHz at $V_{ds} = 2$ V, which is close to the saturation voltage ($V_{sat}$), and $I_{ds} = 0.2$ A/mm, and decreases monotonically with an increase in $V_{ds}$ and $I_{ds}$. This trend has been typically observed for the S-D unscaled devices [10,34]. $f_{max}$ showed a

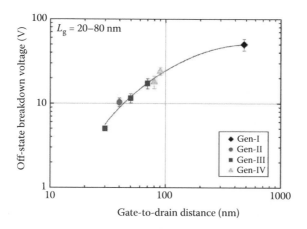

**FIGURE 4.19** Off-state breakdown voltage of scaled GaN HEMTs as a function of the gate-to-drain distance ($L_{gd}$) (K. Shinohara et al., Scaling of GaN HEMTs and Schottky diodes for submillimeter-wave MMIC applications, *IEEE Trans. Electron Devices*, 60(10), 1982, 2013. Copyright © 2013 IEEE.)

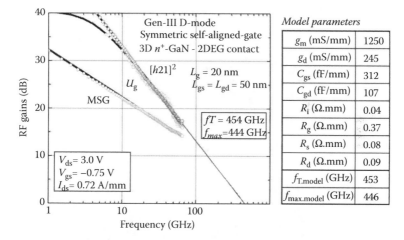

**FIGURE 4.20** Measured and modeled RF gains versus frequency (0.5–65 GHz) for 20-nm D-mode self-aligned-gate HEMT with a scaled source-to-drain distance of 120 nm. (K. Shinohara et al., Scaling of GaN HEMTs and Schottky diodes for submillimeter-wave MMIC applications, *IEEE Trans. Electron Devices*, 60(10), 1982, 2013. Copyright © 2013 IEEE.)

peak value of 400 GHz at a higher $V_{ds}$ of 6 V and a similar $I_{ds}$ of 0.24 A/mm. Figure 4.22b shows output characteristics and $f_T/f_{max}$ contour plots for 20-nm D-mode HEMT with an aggressive S-D scaling ($L_{gs} = L_{gd} = 40$ nm) (Figure 4.14c). Unlike the unscaled device, the device showed a continuous increase of $f_T$ with $V_{ds}$ above the $V_{sat}$ (~0.5 V) until reaching a maximum value of 310 GHz at $V_{ds}$=5 V and $I_{ds}$ = 0.7 A/mm. A peak $f_{max}$ of 377 GHz was obtained at a similar bias condition. The data demonstrate that

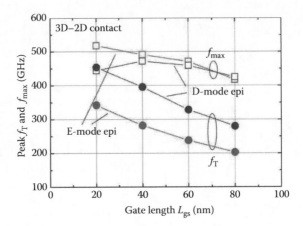

**FIGURE 4.21** $f_T$ and $f_{max}$ versus $L_g$ for deeply scaled E/D-mode GaN HEMTs (K. Shinohara et al., Scaling of GaN HEMTs and Schottky diodes for submillimeter-wave MMIC applications, *IEEE Trans. Electron Devices*, 60(10), 1982, 2013. Copyright © 2013 IEEE.)

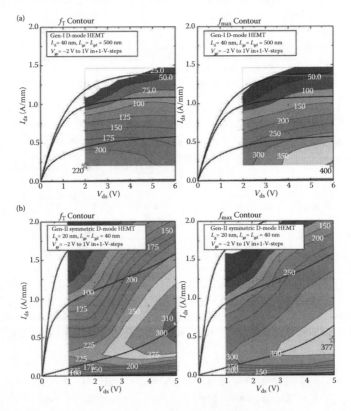

**FIGURE 4.22** $f_T$ and $f_{max}$ contour plots for source-drain (a) unscaled ($L_{gs} = L_{gd} = 500$ nm) and (b) scaled GaN HEMTs ($L_{gs} = L_{gd} = 40$ nm). (K. Shinohara et al., Scaling of GaN HEMTs and Schottky diodes for submillimeter-wave MMIC applications, *IEEE Trans. Electron Devices*, 60(10), 1982, 2013. Copyright © 2013 IEEE.)

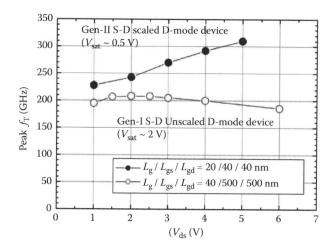

**FIGURE 4.23** $f_T$ and $f_{max}$ contour plots for source-drain (a) unscaled ($L_{gs} = L_{gd} = 500$ nm) and (b) scaled GaN HEMTs ($L_{gs} = L_{gd} = 40$ nm). (K. Shinohara et al., Scaling of GaN HEMTs and Schottky diodes for submillimeter-wave MMIC applications, *IEEE Trans. Electron Devices*, 60(10), 1982, 2013. Copyright © 2013 IEEE.)

aggressive S-D scaling not only increases the peak $f_T$ value but also widens the $I_{ds}$-$V_{ds}$ range where both high $f_T$ and $f_{max}$ are obtained. Figure 4.23 plots peak $f_T$ versus $V_{ds}$ for the two devices.

To understand the unique dependence of $f_T$ on $V_{ds}$ in the scaled device and clarify frequency limitations of devices, delay time analysis was performed based on Moll's method [68]. In Figure 4.24, total delay time ($=1/2\bar{\pi}\cdot f_T$) consisting of parasitic charging time, channel charging time, and electron transit time was plotted as a function of voltage across the channel [$= V_{ds} - I_{ds}\cdot(R_s + R_d)$]. The parasitic charging time [$= C_{gd}\cdot(R_s + R_d)$] in the scaled device was shorter than the unscaled device due to a reduced $R_s$ and $R_d$ as observed in the reduced $R_{on}$. The channel charging time was determined from the dependence of total intrinsic delay time on $1/I_{ds}$ as detailed in Ref. [69]. The electron transit time ($\tau_{transit}$) is the sum of the gate transit time ($\tau_{gate}$) and drain delay ($\tau_{drain}$) as expressed by $\tau_{transit} = \tau_{gate} + \tau_{drain} = L_g/v_e + \Delta L_{gd}/(\alpha\cdot v_e)$ ($\alpha = 2\sim3$), where $v_e\Delta L_{gd}$, and $\alpha$ represent the electron velocity, width of drain depletion, and a factor given by the effect of image charges in the carrier transport in the depletion region, respectively [70]. The $\tau_{transit}$ of the unscaled device monotonically increased with a slope of 0.054 ps/V. This increase was mainly attributed to the increased $\tau_{drain}$ due to wider $\Delta L_{gd}$ at higher voltages. The scaled device, on the other hand, showed a monotonic decrease in $\tau_{transit}$ with increasing voltage (-0.057 ps/V). A similar trend was first reported in Ref. [71]. The data suggest that $\Delta L_{gd}$ and therefore $\tau_{drain}$, which typically increase with voltage in the unscaled devices, were suppressed in the scaled device. More importantly, the observed decrease of $\tau_{transit}$ with increasing voltage indicates an increasing $v_e$. An enhanced average electron velocity ($v_{ave}$) of $1.5 \times 10^7$ cm/s was indeed measured at $V_{ds} = 5$ V, which was 36% higher than $1.1 \times 10^7$ cm/s at $V_{ds} = 1$ V, from the dependence of total delay time on $L_g$ as shown in Figure 4.25.

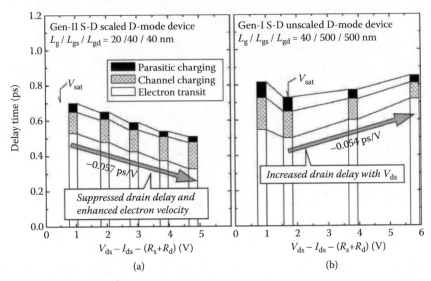

**FIGURE 4.24** Delay time analysis for (a) source-drain scaled ($L_{gs}$ = $L_{gd}$ = 40 nm) and (b) unscaled GaN HEMTs ($L_{gs}$ = $L_{gd}$ = 500 nm). (K. Shinohara et al., Scaling of GaN HEMTs and Schottky diodes for submillimeter-wave MMIC applications, *IEEE Trans. Electron Devices*, 60(10), 1982, 2013. Copyright © 2013 IEEE.)

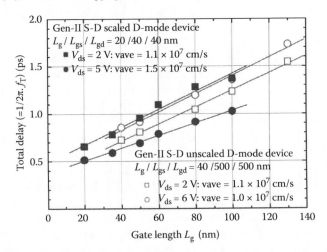

**FIGURE 4.25** Electron velocity enhancement at higher drain-source voltage in source-drain scaled HEMTs. (K. Shinohara et al., Scaling of GaN HEMTs and Schottky diodes for submillimeter-wave MMIC applications, *IEEE Trans. Electron Devices*, 60(10), 1982, 2013. Copyright © 2013 IEEE.)

This velocity enhancement implies velocity overshoot due to a higher lateral electric field achieved in the scaled device with a very short $L_{gd}$. Kodama et al. simulated the velocity overshoot effect in laterally scaled GaN transistors using Monte Carlo simulation, demonstrating an effectiveness of $L_{gd}$ scaling on the overshoot velocity [72]. Although aggressive lateral S-D scaling significantly reduces $R_{on}$ and increases

$f_T$, the devices suffer from a low $BV$ (Figure 4.19) and a high $g_d$ (Figure 4.22b) caused by DIBL due to the short $L_{gd}$ (Figure 4.18), limiting $f_{max}$.

An asymmetric self-aligned-gate technology addresses this trade-off by enabling independent control of $L_{gs}$ and $L_{gd}$ by accurately forming sidewall spacers with different thicknesses at the source and drain-side of the gate (Figure 4.14d). Figure 4.26a and b shows output characteristic and $f_T/f_{max}$ contour plots for 20-nm asymmetric self-aligned-gate HEMT with a short $L_{gs}$ of 30 nm and a longer $L_{gd}$ of 80 nm, which was designed for a simultaneous high-speed and high-$BV$ performance. A higher $BV_{off}$ of 17 V than that (10 V) of the symmetric device is a result of the longer $L_{gd}$ (Figure 4.19). It is noteworthy that there are two peaks in the $f_T$ contour plot—317 GHz at $V_{ds} = 1$ V and $I_{ds} = 0.33$ A/mm; and 329 GHz at $V_{ds} = 4$ V and $I_{ds} = 0.53$ A/mm. This feature can be explained as follows. The first peak at $V_{ds} = 1$ V corresponds to the peak observed at around $V_{sat}$ in unscaled device (Figure 4.22a). Since the device has a

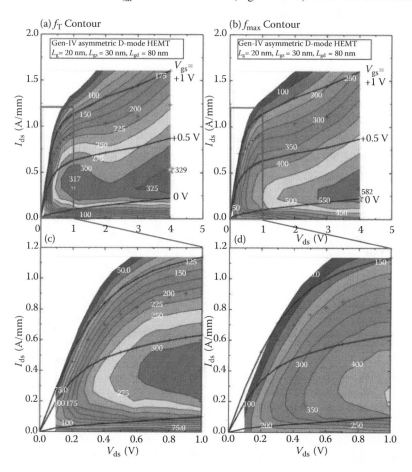

**FIGURE 4.26** (a), (c) $f_T$ and (b), (d) $f_{max}$ contour plots for deeply scaled asymmetric self-aligned-gate GaN HEMTs ($L_{gs} = 20$ nm, $L_{gs} = 30$ nm, $L_{gd} = 80$ nm). (K. Shinohara et al., Scaling of GaN HEMTs and Schottky diodes for submillimeter-wave MMIC applications, *IEEE Trans. Electron Devices*, 60(10), 1982, 2013. Copyright © 2013 IEEE.)

longer $L_{gd}$ than that of the symmetric-scaled device (Figure 4.22b), $f_T$ decreases with $V_{ds}$ due to the increased drain delay until the drain depletion is terminated by the 3D $n^+$-GaN drain ohmic contact. Then, $f_T$ increases again with $V_{ds}$ due to an increased electron velocity. The increased $L_{gd}$ improved electrostatic isolation between G-D, resulting in a reduced $g_d$ and $C_{gd}$. The record-high $f_{max}$ of 582 GHz with a simultaneous $f_T$ of 310 GHz was obtained at $V_{ds} = 4$ V and $I_{ds} = 0.25$ A/mm. RF characteristics measured at the peak $f_T$ condition ($V_{ds} = 4$ V, $V_{gs} = 0.25$ V, $I_{ds} = 0.53$ A/mm) are shown in Figure 4.27. The small signal model showed a significant increase in the voltage gain (=$g_m/g_d$), 10.9 versus 5.1, and a reduction in $C_{gd}$, 89 fF/mm versus 107 fF/mm, as compared to the symmetric device with a shorter $L_{gd} = 50$ nm (see Figure 4.20). High $f_{max}$ obtained in a wide range of $I_{ds}$ and $V_{ds}$ makes it very promising to realize high-power submillimeter-wave GaN amplifier MMICs.

### 4.6.5  LOW NOISE PERFORMANCE

The minimum noise figure ($NF_{min}$) can be well approximated by the semiempirical equation given by Fukui.

$$NF_{min} = 1 + k \cdot \frac{f}{f_T} \cdot \left[ g_m \cdot (R_g + R_s) \right]^{1/2}$$

where $k$ is a fitting parameter [73]. Deeply scaled GaN-HEMTs have a great potential for low-power LNA applications because of their high $f_T/f_{max}$ and high $g_m$ realized at low $V_{ds}$ as well as a low $R_s$. Figure 4.28 plots $NF_{min}$ and associated gain ($G_a$) at 20 GHz

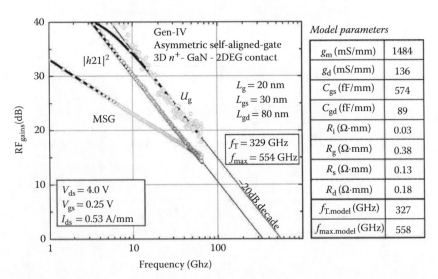

**FIGURE 4.27**  Measured and modeled RF gains versus frequency (0.5–65 GHz) for 20-nm asymmetric self-aligned-gate HEMT with $L_{gs} = 30$ nm and $L_{gd} = 80$ nm. (K. Shinohara et al., Scaling of GaN HEMTs and Schottky diodes for submillimeter-wave MMIC applications, *IEEE Trans. Electron Devices*, 60(10), 1982, 2013. Copyright © 2013 IEEE.)

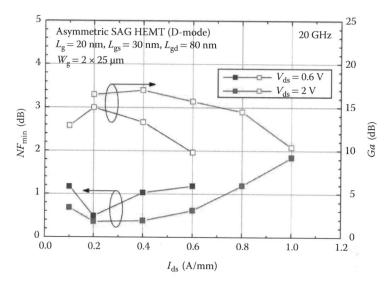

**FIGURE 4.28** $NF_{min}$ and $G_a$ at 20 GHz as a function of drain current density for 20-nm asymmetric self-aligned-gate HEMT with $L_{gs} = 30$ nm and $L_{gd} = 80$ nm. (K. Shinohara et al., Scaling of GaN HEMTs and Schottky diodes for submillimeter-wave MMIC applications, *IEEE Trans. Electron Devices*, 60(10), 1982, 2013. Copyright © 2013 IEEE.)

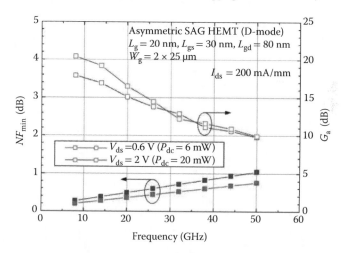

**FIGURE 4.29** $NF_{min}$ and $G_a$ at 20 GHz as a function of frequency for 20-nm asymmetric self-aligned-gate HEMT with $L_{gs} = 30$ nm and $L_{gd} = 80$ nm. (K. Shinohara et al., Scaling of GaN HEMTs and Schottky diodes for submillimeter-wave MMIC applications, *IEEE Trans. Electron Devices*, 60(10), 1982, 2013. Copyright © 2013 IEEE.)

as a function of drain current density measured for 20-nm asymmetric self-aligned-gate HEMT with a gate width ($W_g$) of 2 × 25 µm. The lowest $NF_{min}$ of 0.36 dB (0.49 dB) with a $G_a$ of 16.5 dB (15 dB) was obtained at $I_{ds} = 0.2$ A/mm and $V_{ds} = 2$ V (0.6 V). Figure 4.29 shows dependence of $NF_{min}$ and $G_a$ as a function of frequency at the optimum $I_{ds} = 0.2$ A/mm. The device exhibited an $NF_{min}$ of 0.76 dB (1.05 dB) with a $G_a$ of

9.9 dB (9.8 dB) at 50 GHz at an extremely low DC power dissipation of 20 mW (6 mW). The excellent noise performance with a high gain obtained at a low DC power enables low DC power GaN LNA with high input power survivability and a wide dynamic range, which had been a big challenge for narrow band-gap In(Ga)As-HEMTs.

## 4.7   FUTURE PROSPECTS OF THE TECHNOLOGY

Recent advancement of aggressive device scaling technologies in GaN-HEMTs enabled an unprecedented combination of high-frequency, low-resistance, high-breakdown, and low-noise characteristics. This new capability offers many practical advantages in RF MMICs such as extremely efficient switch-mode PAs, PA MMICs operating in G-band and higher frequencies, low DC power consumption LNAs with high input power survivability, and robust low-loss RF switches. In addition, large-scale monolithic integration of high-speed E/D-mode GaN-HEMTs realizes complex GaN-based mixed-signal and digital logic circuits.

## REFERENCES

1. G.M. Rebeiz, "Millimeter-Wave SiGe RFICs for Large-Scale Phased Arrays," Bipolar-BiCMOS Circuits and Technology Meeting (BCTM), 56, 2014.
2. R. Lai, W.R. Deal, X.B. Mei et al., "Fabrication of InP HEMT Devices with Extremely High Fmax," International Conference on InP and Related Materials, Versailles, France, 2008.
3. W.R. Deal, "InP HEMT for Sub-Millimeter Wave Space Applications: Status and Challenges," International Conference on Infrared, Millimeter, and Terahertz waves (IRMMW-THz), Tucson, AZ, 2014.
4. M. Seo, M. Urteaga, J. Hacker, and et al., "A 600GHz InP HBT Amplifier Using Cross-Coupled Feedback Stabilization and Dual Differential Power Combining," 2013 International Microwave Symposium, Seattle, WA, June 2013.
5. J. Hacker, M. Urteaga, M. Seo, and et al., "InP HBT Amplifier MMICs Operating to 0.67 THz," 2013 International Microwave Symposium, Seattle, WA, June 2013.
6. M.A. Khan, T.N. Kuznia, A.R. Bhattaraia, and D.T. Olson, "Metal semiconductor field effect transistor based on single crystal GaN," *Appl. Phys. Lett.*, 62, 1786, 1993.
7. M.A. Khan, J.N. Kuznia, D.T. Olson, and et al., "Microwave performance of a 0.25 um AlGaN/GaN heterostructure filed effect transistor," *Appl. Phys. Lett.*, 65, 1121, 1994.
8. O. Ambacher, J.Smart, J.R. Shealy et al., "Two-dimensional electron gases induced by spontaneous and piezoelectric polarization charges in N- and Ga-face AlGaN/GaN heterostructures," *J. Appl. Phys.*, 85(6), 3222, 1999.
9. E. Mitani, M. Aojima, A. Maekawa, and S. Sano, "An 800-W AlGaN/GaN HEMT for S-band High-Power Application," 2007 CS MANTECH Technical Digest, 213, 2007.
10. Y.-F. Wu, S.M. Wood, R.P. Smith et al., "An Internally-matched GaN HEMT Amplifier with 550-watt Peak Power at 3.5 GHz," International Electron Device Meeting (IEDM), 2006.
11. Y. Okamoto, Y. Ando, K. Hataya et al., "A 149 W Recessed-Gate AlGaN/GaN FP-FET," International Microwave Symposium (MTT-S), 3, 1351, 2004.
12. S.T. Shemard, R.P. Smith, W.L. Pribble et al., "High Power Hybrid and MMIC Amplifiers Using WideBandgap Semiconductor Devices on Semi-insulating SiC Substrates," Device Research Conference (DRC), 175, 2002.
13. Y. Murase, A. Wakejima, T. Inoue et al., "CW 20-W AlGaN/GaN FET Power Amplifier for Quasi-Millimeter Wave Applications," Compound Semiconductor Integrated Circuit Symposium (CSICS), 2007.

14. J. Schellenberg, B. Kim, and T. Phan, "W-Band, Broadband 2W GaN MMIC," International Microwave Symposium (MTT-S), 2013.
15. M. Micovic, A. Kurdoghlian, A. Margomenos et al., "92-96 GHz GaN Power Amplifiers," International Microwave Symposium (MTT-S), 2012.
16. D.F. Brown, A. Williams, K. Shinohara et al., "W-Band Power Performance of AlGaN/GaN DHFETs with Regrown n+ GaN Ohmic Contacts by MBE," IEEE Electron Device Meeting (IEDM), 2011.
17. V. Radisic, D. Scott, A. Cavus, and C. Monier, "220-GHz High-Efficiency InP HBT Power Amplifiers," *IEEE Trans. Microwave Theory and Techniques* 62(12), 3001, 2014.
18. T.B. Reed, Z. Griffith, P. Rowell, M. Field, and M. Rodwell, "A 180 mW InP HBT Power Amplifier MMIC at 214 GHz," Compound Semiconductor Integrated Circuit Symposium (CSICS), 2013.
19. H.G. Yu, S.H. Choi, S. Jeon, and M. Kim, "300 GHz InP HBT amplifier with 10 mW output power," *Electronics Lett.*, 50(5), 377, 2014.
20. V. Radisic, K.M.K.H. Leong, X. Mei, S. Sarkozy, W. Yoshida, and W.R. Deal, "Power amplification at 0.65 THz using InP HEMTs," *IEEE Trans. Microwave Theory and Techniques* (MTT-S), 60(3), 724, 2012.
21. Z. Griffith, M. Urteaga, P. Rowell, R. Pierson, M. Field, "Multi-finger 250nm InP HBTs for 220GHz mm-Wave Power," International Conference on InP and Related materials, Santa Barbara, CA, 2012.
22. P.J. Tasker, "RF Waveform Measurement and Engineering," Compound Semiconductor Integrated Circuit Symposium (CSICS), 2009.
23. M. Roberg, J. Hoversten, and Z. Popovic, "GaN HEMT PA with over 84% power added efficiency," *Electronics Lett.*, 46(23), 1553, 2010.
24. J.S. Moon, H. Moyer, P. Macdonald et al., "High efficiency X-band class-E GaN MMIC high-power amplifiers," IEEE Topical Conference on Power Amplifiers for Wireless and Radio Applications (PAWR), 9, 2012.
25. A. Margomenos, M. Micovic, A.Kurdoghlian et al., "X Band Highly Efficient GaN Power Amplifier Utilizing Built-In Electroformed Heat Sinks for Advanced Thermal Management," International Microwave Symposium (MTT-S), 2013.
26. D.D. Dumka and T.M. Chou, "Evaluation of Thermal Resistance of AlGaN/GaN Heterostructure on Diamond Substrate," IEEE Intersociety Conference on Thermal and Thermomechanical Phenomena in Electronic Systems (ITherm), 1210, Orland, FL, 2014.
27. C.T. Creamer, K.K. Chu, P.C. Chao et al., "Microchannel Cooled, High Power GaN-on-Diamond MMIC," Lester Eastman Conference on Higher Performance Devices (LEC), 2014.
28. K. Shinohara, D.C. Regan, Y. Tang et al., "Scaling of GaN HEMTs and Schottky diodes for submillimeter-wave MMIC applications," *IEEE Trans. Electron Devices*, 60(10), 1982, 2013.
29. S.C. Binari, P.B. Klein, and T.E. Kazior, "Trapping effects in GaN and SiC Microwave FETs," *Proc. IEEE*, 90, 1048–1058, 2002.
30. G. Meneghesso, G. Verzellesi, R. Pierobon et al., "Surface-related drain current dispersion effects in AlGaN/GaN HEMTs," *IEEE Trans. Electron Devices*, 51, 1554–1561, 2004.
31. H. Fujimoto, W. Saito, A. Yoshioka, T. Nitta, Y. Kakiuchi, and Y. Saito, "Wafer Quality Target for Current-Collapse-Free GaN-HEMTs in High Voltage Applications," CSMANTECH Conference, Chicago, IL, 2008.
32. S. Arulkumaran, T. Egawa, H. Ishikawa, T. Jimbo, and Y. Sano, "Surface passivation effects on AlGaN/GaN high-electron-mobility transistors with SiO2, Si3N4, and silicon oxynitride," *Appl. Phys. Lett.*, 84(4), 613, 2004.
33. Y.-F. Wu, A. Saxler, M. Moore et al., "30-W/mm GaN HEMTs by Field Plate Optimization," *IEEE Electron Device Lett.*, 25(3), 117, 2004.

34. Y.-F. Wu, M. Moore, A. Saxler, T. Wisleder, and P. Parikh, "40-W/mm Double Field-plated GaN HEMTs," *DRC Tech. Dig.*, 151, 2006.

35. P.J. Tasker and B. Hughes, "Importance of source and drain resistance to the maximum fT of millimeter-wave MODFET's," *IEEE Elec. Dev. Lett.*, 10(7), 291-293, 1989.

36. Y. Awano, M. Kosugi, K. Kosemura, T. Mimura, and M. Abe, "Short-channel effects in subquarter-micrometer-gate HEMT's: Simulation and experiment," *IEEE Trans. Electron Devices*, 36(10), 2260, 1989.

37. G.H. Jessen, R.C. Fitch, J.K. Gillespie et al., "Short-Channel Effect Limitations on High-Frequency Operation of AlGaN/GaN HEMTs for T-Gate Devices," *IEEE Trans. Eelecron. Devices*, 2589, 2007.

38. I.P. Smorchkova, S. Keller, S. Heikman et al., "Two-dimensional electron-gas AlN/GaN heterostructures with extremely thin AlN barriers" *Appl. Phys. Lett.*, 77(24), 3998, 2000.

39. Y. Cao and D. Jena, "High-mobility window for two-dimensional electron gases at ultrathin AlN/GaN heterojunctions" *Appl. Phys. Lett.*, 90(18), 182112, 2007.

40. H. Sun, A.R. Alt, H. Benedickter et al., "205-GHz (Al,In)N/GaN HEMTs," *IEEE Elec. Dev. Lett.*, 31(9), 957–959, 2010.

41. D.S. Lee, X. Gao, S. Guo, D. Kopp, P. Fay, and T. Palacios, "300-GHz InAlN/GaN HEMTs with InGaN Back Barrier," *IEEE Elec. Dev. Lett.*, 32(11), 1525, 2011.

42. Y. Yue, Z. Hu, J. Guo a et al., "InAlN/AlN/GaN HEMTs With Regrown Ohmic Contacts and fT of 370 GHz," *IEEE Elec. Dev. Lett.*, 33(7), 988–990, 2012.

43. Y. Yue, Z. Hu, J. Guo et al., "Ultrascaled InAlN/GaN HEMTs with fT of 400 GHz," *Jpn. J. Appl. Phy.*, 52(8), 08JN14, 2013.

44. C.Q. Chen, J.P. Zhang, V. Adivarahan, A. Koudymov, H. Fatima, "AlGaN/GaN/AlGaN double heterostructure for high-power III-N field-effect transistors," *Appl. Phys. Lett.*, 82(25), 4593–4595, 2003.

45. M. Micovic, P. Hashimoto, M. Hu et al., "GaN double heterojunction field effect transistor for microwave and millimeterwave power applications," *IEDM Tech. Dig.*, 807–810, 2004.

46. T. Palacios, A. Chakraborty, S. Heikman, S. Keller, S.P. DenBaars, and U.K. Mishra, "AlGaN/GaN high electron mobility transistors with InGaN back-barriers," *IEEE Elec. Dev. Lett.*, 27(1), 13–15, 2006.

47. Nidhi, S. Dasgupta, D.F. Brown, S. Keller, J.S. Speck, and U.K. Mishra, "N-polar GaN-based highly scaled self-aligned MIS-HEMTs with state-of-the-art fT·Lg product of 16.8 GHz-μm," *IEDM Tech. Dig.*, 955–957, 2009.

48. D. Denninghoff, J. Lu, M. Laurent, E. Ahmadi, S. Keller, U.K. Mishra, "N-polar GaN/InAlN MIS-HEMT with 400-GHz fmax," *DRC Tech. Dig.*, 151–152, 2012.

49. D.F. Brown, K. Shinohara, A. Williams et al., "Monolithic Integration of Enhancement- and Depletion-mode AlN/GaN/AlGaN DHFETs by Selective MBE Regrowth," *IEEE Trans. Electron Devices*, 58(4), 1063–1067, 2011.

50. B. Song, B. Sensale-Rodrigueza, R. Wang et al., "Monolithically integrated E/D-mode InAlN HEMTs with ft/fmax >200/220 GHz," *DRC Tech. Dig.*, 2012.

51. M.L. Schuette, A. Ketterson, B. Song et al., "Gate-recessed integrated E/D GaN HEMT technology with fT/fmax > 300 GHz," *IEEE Electron Device Lett.*, 34(6), 741, 2013.

52. P. Saunier, "GaN for Next Generation Electronics," Compound Semiconductor Integrated Circuit Symposium (CSICS), 2014.

53. A.L. Corrion, K. Shinohara, D. Regan et al., "High-speed 501-stage DCFL GaN ring oscillator circuits," *IEEE Electron Device Lett.*, 34(7), 846, 2013.

54. Y.-F. Wu, D. Kapolnek, P. Kozodoy et al., "AlGaN/GaN MODFETs with Low Ohmic Contact Resistances by Source/Drain n+ Regrowth," IEEE International Symposium on Compound Semiconductors (ISCS), 431, 1998.

55. I. Milosavljevic, K. Shinohara, D. Regan et al., "Vertically scaled GaN/AlN DH-HEMTs with regrown n+GaN ohmic contacts by MBE," *DRC Tech. Dig.*, 159–160, 2010.

56. J. Guo, G. Li, F. Faria et al., "MBE-regrown Ohmics in InAlN HEMTs with a regrowth interface resistance of 0.05 Ω-mm," *IEEE Elec. Dev. Lett.*, 33(4), 525–527, 2012.

57. K. Shinohara, D. Regan, A. Corrion et al., "Self-aligned-Gate GaN-HEMTs with heavily-doped n+-GaN ohmic contacts to 2DEG," *IEDM Tech. Dig.*, 617–620, 2012.

58. D.-H. Kim, B. Brar, and J.A. del Alamo, "fT = 688 GHz and fmax = 800 GHz in Lg = 40 nm In0.7Ga0.3As MHEMTs with gm_max > 2.7 mS/μm," IEDM Tech. *Dig.*, 319–322, 2011.

59. M. Egard, L. Ohlsson, B.M. Borg et al., "High Transconductance Self-Aligned Gate-Last Surface Channel In0.53Ga0.47As MOSFET," *IEDM Tech. Dig.*, 303–306, 2011.

60. P.M. Solomon, A. Palevski, T.F. Fuech, and M.A. Tischler, "Low Resistance Ohmic Contacts to two-dimensional electron-gas by selective MOVPE," *IEDM Tech. Dig.*, 405–408, 1989.

61. T. Hashizume, S. Ootomo, S. Oyama, M. Konishi, and H. Hasegawa, "Chemistry and electrical properties of surfaces of GaN and GaNŌAlGaN heterostructures," *J. Vac. Sci. Tech. B*, 19(4), 1675, 2001.

62. A.V. Vertiatchikh, L.F. Eastman, W.J. Schaff, and T. Prunty, "Effect of surface passivation of AlGaN/GaN heterostructure field-effect transistor," *Electron. Lett.*, 38(8), 388, 2002.

63. R. Ventury, N.Q. Zhang, S. Keller, U.K. Mishra, "The impact of surface states on the DC and RF characteristics of AlGaN/GaN HFETs," *IEEE Trans. Electron Devices*, 48(3), 560, 2001.

64. D. Xu, K. Chu, J. Diaz et al., "0.2-μm AlGaN/GaN high electron-mobility transistors with atomic layer deposition $Al_2O_3$ passivation," *IEEE Electron Device Lett.*, 34(6), 744–746, 2013.

65. H. Wang, J.W. Chung, X. Gao, S. Guo, T. Palacios, "$Al_2O_3$ passivated InAlN/GaN HEMTs on SiC substrate with record current density and transconductance," *Phys. Stat. Sol. (C)*, 7(10), 2440–2444, 2010.

66. M.V. Fischetti, L. Wang, B. Yu et al., "Simulation of Electron Transport in High-Mobility MOSFETs: Density of States Bottleneck and Source Starvation," *IEDM Tech. Dig.*, 109–112, 2007.

67. H. Tsuchiya, A. Maenaka, T. Mori, Y. Azuma, "Role of carrier transport in source and drain electrodes of high-mobility MOSFETs," *IEEE Elec. Dev. Lett.*, 31(4), 365–367, 2010.

68. N. Moll, M.R. Hueschen, and A. Fischer-Colbrie, "Pulse-doped AlGaAs/InGaAs pseudomorphic MODFETs," *IEEE Trans. Electron Devices*, 35, 879–886, 1988.

69. K. Shinohara, A. Corrion, D. Regan et al., "220GHz fT and 400GHz fmax in 40-nm GaN DH-HEMTs with Re-grown Ohmic," *IEDM Tech. Dig.*, 672–675, 2010.

70. J. W. Chung, X. Zhao, Y.-R. Wu, J. Singh, and T. Palacios, "Effect of image charges in the drain delay of AlGaN/GaN high electron mobility transistors," *Appl. Phys. Lett.*, 92, 093452, 2008.

71. K. Shinohara, D. Regan, I. Milosavljevic et al., "Electron velocity enhancement in laterally-scaled GaN DH-HEMTs with fT of 260 GHz," *IEEE Elec. Dev. Lett.*, 32(8), 1074–1077, 2011.

72. K. Kodama, Y. Naito, H. Tokuda, and M. Kuzuhara, "Effect of image charges in the drain delay of AlGaN/GaN high electron mobility transistors," Extended Abstract, SSDM 2012, Kyoto, Japan, 862–863, Sep. 2012.

73. H. Fukui, "Optimal noise figure of microwave GaAs MESFET's," *IEEE Trans. Electron Devices*, 36(7), 1032–1037, 1979.

# 5 Group III-Nitride Microwave Monolithically Integrated Circuits

*Rüdiger Quay*

## CONTENTS

## ACRONYMS

**BT**: bipolar transistor
**CG**: common-gate
**CMCD**: current-mode class-D (amplifier)
**CMOS**: complementary metal–oxide–semiconductor
**CS**: common-source
**DCFL**: direct coupled FET logic
**DE**: drain efficiency
**DG**: dual gate
**DPD**: digital predistorsion
**DSP**: digital-signal processing
**FET**: field-effect transistor
**GaAs**: gallium arsenide
**GaN**: gallium nitride
**HBT**: hetero-bipolar transistor
**HPA**: high-power amplifier
**HR**: high-resistivity
**IC**: integrated circuit
**InP**: indium phosphide
**LINC**: linear amplification with nonlinear components
**LNA**: low-noise amplifier
**LP**: loadpull
**MAG**: maximum available gain
**MIM**: metal insulator metal
**Mix**: mixer
**MMIC**: microwave monolithically integrated circuit
**MSG**: maximum stable gain
**Osc**: oscillator
**PA**: power amplifier
**PAE**: power-added efficiency
**PCB**: printed circuit board
**s.i.**: semi-insulating
**SiGe**: silicon germanium
**Swi**: switch
**TML**: transmission line
**TRX**: transmit–receive
**VCO**: voltage-controlled oscillator
**VGA**: variable-gain amplifier

## ABSTRACT

This chapter describes the application of the outstanding device properties and results of gallium nitride (GaN)-based devices into integrated circuits (ICs). This mainly implies radio-frequency (RF) and microwave circuits, that is, amplifiers for high-power, highly linear, and high-efficiency applications. Further, drivers, robust low-noise amplifiers, mixers, and broadband amplifiers are discussed. Oscillators and digital amplifiers are also mentioned.

## 5.1  GENERAL INTEGRATED CIRCUIT CHARACTERISTICS

In general, radio-frequency (RF) and microwave integrated circuits are needed due to the radiation laws to overcome the losses of propagation of the transmission channels. This is true for both communication and sensing functions. Integrated circuits (ICs) provide low-loss transmission with very accurate impedances as bond wires and other sources of uncertainty can be avoided. As an example, Figure 5.1 gives the typical functions of RF transmit–receive (TRX) modules.

In the transmit path, high-gain or variable-gain and driver amplifiers and high-power amplifiers (HPAs) are needed. In the receive path, low-noise amplifiers (LNAs) add the most decisive contributions to signal distortion and thus require particular attention. Further oscillators, mixers, and switches are required. Thus, the main classes of ICs include the following:

- Amplifiers in their various forms: high-power amplifiers (HPAs) and driver amplifiers (DRAs)
- Low-noise amplifiers (LNA)
- Variable-gain amplifiers (VGAs)
- Oscillators (OSC)
- Linear mixers (MIX) and down- and up-converters
- RF switches (SWI)

They shall be discussed for GaN in the following.

### 5.1.1  SUBSTRATE PROPERTIES FOR III-N DEVICES AND INTEGRATED CIRCUITS

The big advantage of ICs is the possibility to integrate various functions in one circuit in a relatively small area compared to hybrid integration. As GaN does not

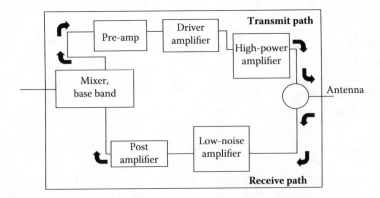

**FIGURE 5.1**  Schematic of a transmit–receive (TRX) function in communication.

yet provide a native semi-insulating GaN substrate for large areas, the choice of the substrate is more complex as availability, technical, and cost aspects have to be considered. Before we turn to the actual active circuits, we discuss the specific group III-nitride passive parameters used for ICs [1]. Gallium nitride, in various ways, is very similar to GaAs-based circuits [2]. It provides the following substrate choices:

- On semi-insulating (s.i.) silicon carbide (SiC) substrate
- On sapphire substrate
- On highly-resistive (HR) silicon substrate
- On native insulating GaN substrates

Table 5.1 gives an overview of the typical substrate properties. The loss tangent $\delta$ is related to the $Q$-factor of the material as follows:

$$\tan \delta = \frac{1}{Q} \tag{5.1}$$

Currently, s.i. is the most widely used material for GaN-based ICs. Semi-insulating SiC is available in high quality on 4-in. and 6-in. substrates from various vendors [3,4]. Sapphire is a good passive material available in every diameter with very low substrate losses; however, the material is extremely hard [5] and tricky to manufacture with respect to small viaholes. Further, the very bad thermal conductivity prevents the high-power operation of the active devices. Not exclusively, but based on the strong progress on the side of power electronics silicon substrates, GaN on high-resistivity (HR) silicon microwave monolithically integrated circuits (MMICs) has been used and implemented in MMIC libraries, for example, [6,7]. Both GaN-on-silicon devices and MMICs including the passive structures can be supported. Despite the great interest it generates, GaN-native substrates are late with respect to their availability versus time. They potentially provide semi-insulating properties as they can be easily Fe doped [8,9]. However, they are not yet available in wafer sizes

**TABLE 5.1**
**Typical Substrate Materials and Properties**

| Material | $\varepsilon_r$ | tan δ $10^{-4}$ | λ W/m K | σ Ωcm |
|---|---|---|---|---|
| s.i. GaAs | 12.9 | 6 | 46 | $10^9$ |
| s.i. SiC | 9.7 | 25–30 | 400 | $10^6$–$10^{12}$ |
| silicon (HR) | 11.7 | 40 | 138 | $10^3$ |
| sapphire | 9.4–11.5 | 1 | 38 | $10^{12}$ |
| GaN | 9.7 | 27 | 150–200 | $10^8$ |
| AlN | 8.8 | 5 | 285 | $10^{14}$ |
| Ceramics | 6.8–7.6 | 3 | 37 | $10^{11}$–$10^{14}$ |

of 4–6 in., which are required for today's volume production of large-area MMICs. Semi-insulating AlN is another material that is promising [10] as it is highly insulating and used as a passive material in many applications based on polycrystalline materials. However, crystalline AlN substrates with diameters greater than 2 in. are not available.

## 5.1.2 TRANSMISSION LINES AND PASSIVE PARAMETERS

For use in an IC, a set of passive RF components is typically used. Such components and their related functions are typically standardized in passive libraries or process design kits (PDK). This includes airbridges, power dividers [11], couplers [12], crossings, viaholes, and pad connectors. Typical passive parameters and passive elements provided included in integrated GaN on s.i. SiC and GaN on highly-resistive silicon libraries include, e.g., [13]:

- 100-V metal-insulator-metal (MIM) capacitors (with a capacitance of 112 pF/mm$^2$ [14], 180 pF/mm$^2$, or 250 pF/mm$^2$ [13,15]).
- Multiple capacitor concepts implement capacitors of 1200 pF/mm$^2$, 300 pF/mm$^2$, and 240 pF/mm$^2$ [16].
- Metal-evaporated thin film resistors (resistivity, 12 Ω/sq; power rating, 10 W/μm$^2$; maximum current rating, 1 mA/μm).
- Bulk resistors (resistivity such as 50 Ω/sq or 70 Ω/sq, and 400 Ω/sq).
- Two-Au metal levels, 3 μm thick [13] or 7 μm thick [14] with dielectric crossovers.
- Multilevel metal stack yield independent or combined interconnect metal levels of 0.7 μm, 2.0 μm, and 4.0 μm thickness to provide low loss interconnect and to maximize current handling [17].
- Slot-shaped, low-resistance through-wafer vias 30 × 75 μm$^2$ [13] or quadratic 50 × 50 μm$^2$ [14] on 100-μm substrates or even 30 × 30 μm$^2$ on 75-μm substrates [15,18].

### 5.1.2.1 Some Initial Integrated Circuit Layouts

These properties are used to build the integrated elements, as can be seen for a full wafers for multiple ICs in Figure 5.2.

Figure 5.3 shows the image of a dual-stage MMIC realized in microstrip transmission line (MSL) technology with the typical matching and direct current (DC)-supply structures. The basic composure of this amplifier is based on a dual-stage configuration with input matching networks (IMNs), output matching networks (OMNs), and interstage matching networks. DC-connections are available on both sides of the MMIC. The chip size is 4 × 3 mm$^2$ in this case for operation at X-band frequencies (8–12 GHz). As can be seen, the minimum MMIC width is determined by the width of the active transistors of the output stage with their minimum gate-to-gate pitch. Depending on the on-chip block needs, the DC-supply structures require significant area, especially, if the capacitance per area of the MIM-structures is reduced, for example, due to the high-voltage requirement of the MIM-process. However, the

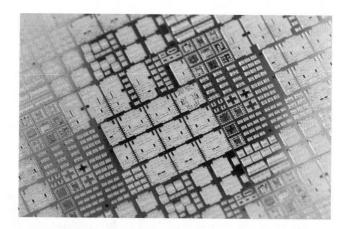

**FIGURE 5.2**   Integrated structures of GaN microwave monolithically integrated circuits (MMICs) on SiC on a full wafer.

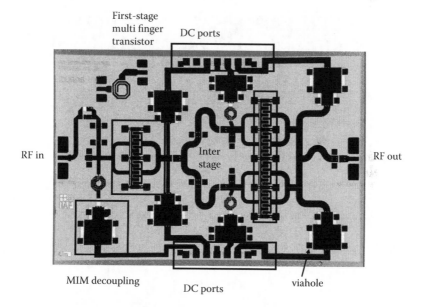

**FIGURE 5.3**   Passive structures in a dual-stage GaN MMICs on SiC in microstrip transmission line including biasing.

length of the chip, based on the transmission line (TML) parameters given above, is determined by the impedance transformation and the power combination and division required in the interstage matching network. Further care is required for the DC-line current ratings to feed the output stages, as the required significant DC-current and thus line width for maximum power operation, as seen in Figure 5.3. A very similar GaN on s.i. SiC MMIC, here in coplanar TML technology, is shown in Figure 5.4. The substrate thickness is 250–350 μm in this case, depending on the original SiC thickness of the un-thinned SiC substrates. In this case, the chip size is determined by the width of the output stage, by the need to form frontside ground lines with sufficient current capability, and by the available impedance lines suitable especially to transform the output impedance of the first stage into the low input impedance of the second stage. Again, the MIM structures and DC-lines contribute significantly to the chip area.

### 5.1.2.2 Some Transmission Lines

The basics of typical TML configurations are given in Figure 5.5. The description of some of the parameters of the different TML techniques from Figure 5.5 are given in the following. Mostly, MSLs are typically used for MMICs to 100 GHz, whereas at higher frequencies grounded coplanar transmission lines are useful due to the grounding needed for packaging in modules. Both grounded coplanar waveguide (GCPW) and MSLs require a backside process with through-via holes.

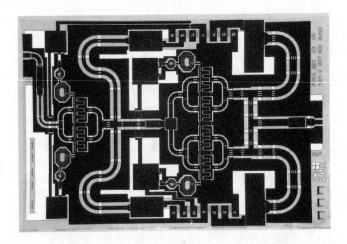

**FIGURE 5.4**   Chip image of a dual-stage coplanar X-band power amplifier MMIC, chip size $4.5 \times 3$ mm².

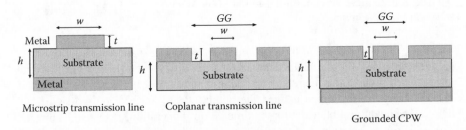

**FIGURE 5.5**   Basics of the microstrip, coplanar, and grounded-coplanar transmission lines.

### 5.1.2.3    Dimensions and Impedances of Some Typical Transmission Lines on GaN

*5.1.2.3.1    Microstrip Transmission Line*

For an MSL on GaN on s.i. SiC, we find the relative dielectric constant of $\varepsilon_r = 9.7$. Thus, for a substrate thickness of 100 μm, the characteristic impedance yields for a width of 10 μm calculated without dielectric of 260 Ω and $q = 0.55$ from a field simulation. From this we deduce impedances for the case including dielectrics.

$$\varepsilon_{eff} = 1 + q\,(\varepsilon_r - 1) \tag{5.2}$$

$$Z_0 = \frac{Z_{01}}{\sqrt{\varepsilon_{eff}}} \tag{5.3}$$

As derived from Table 5.2, the impedances of TMLs are limited by the width of the line and finally by the area consumption.

*5.1.2.3.2    Coplanar Transmission Lines*

For coplanar TMLs, similar considerations apply. Based on linecalc simulations the following TML parameters can be found, as given in Table 5.3.

---

**TABLE 5.2**

**Typical Parameters of Microstrip Transmission Lines on GaAs, GaN, SiC, and Duroid**

| Material | Thickness (μm) | Width (μm) | $Z_{01}(\Omega)$ | q | Z |
|----------|----------------|------------|------------------|------|-----|
| GaAs | 100 | 10 | 260 | 0.55 | 70 |
| GaAs | 100 | 70 | 130 | 0.64 | 50 |
| SiC | 100 | 92 | 127 | 0.64 | 50 |
| GaN | 100 | 92 | 127 | 0.64 | 50 |
| GaN | 100 | 38 | 182 | 0.66 | 70 |
| GaN | 100 | 234 | 80 | 0.7 | 30 |
| SiC | 75 | 69 | 127 | 0.63 | 50 |
| Duroid | 10 | 204 | 80 | 0.51 | 50 |
| Duroid | 500 | 1042 | 86 | 0.51 | 50 |

---

**TABLE 5.3**

**Typical Parameters of Coplanar Transmission Lines on s.i. GaAs, SiC, and GaN**

| Material | Thickness (μm) | Width (W, μm) | GG-pitch (μm) | Z (Ω) |
|----------|----------------|---------------|---------------|-------|
| GaAs | ∞ | 63 | 50 | 50 |
| SiC | ∞ | 99 | 50 | 50 |
| GaN | ∞ | 99 | 50 | 50 |
| GaN | ∞ | 20.8 | 14 | 50 |

From the basic size parameter, we see that the TML for the same impedance and thickness are a bit larger than for GaAs. For the combination GaN on SiC, the parameters of SiC are most relevant.

### 5.1.2.3.3 Transmission Line-Loss-Functions

ICs are defined in terms of size and functionality also by the losses in the passive components. For the basic propagation function γ and the characteristic impedances $Z_C$ of TMLs, read, for example, [19]:

$$\gamma = \alpha + i\beta = \sqrt{(R'+i\omega L')(G'+i\omega C')} \tag{5.4}$$

and the characteristic impedance reads

$$Z_c = \frac{R'+i\omega L'}{G'+i\omega C'} \tag{5.5}$$

A low-loss property typically holds for sapphire, s.i. SiC, and s.i. GaN substrates. The attenuation functions α of the TMLs are derived for non-resonant structures from their effective serial (here, conductive) and parallel (dielectric) losses as

$$\alpha = \alpha_{cond} + \alpha_{dielectric} \tag{5.6}$$

With the typical loss tangents and geometries of the TML on semi-insulating sapphire, SiC, and GaN, and with the conductive metals used, the Q-factors of the GaN MMICs TMLs are in the range of 30–60 for non-resonant structures [20]. The metals in this case comprise sputtered and galvanic gold or copper, where the conductivities are very similar to them on s.i. GaAs. The TML attenuation is thus very similar to TMLs on semi-insulating GaAs substrates, as can also be derived from the loss functions given in Table 5.1. For silicon, the detailed TML functions for both microstrip and coplanar TMLs are still under debate, as the HR substrates with a top of s.i. GaN layer require a detailed and process-dependent stack analysis to evaluate the Q-functions versus frequency [21-23]. For low frequencies, reasonable losses seem feasible, whereas for higher frequencies additional process adaption may be needed.

## 5.2 HIGH-POWER AMPLIFIERS

This section addresses hybrids and MMIC HPAs in GaN technology. An HPA denotes a large-signal amplifier with comparably high output power in the frequency range between 100 MHz and at least 100 GHz [24]. The classical power amplifier (PA) categories for amplifiers are generally accepted [25], however, not pursued systematically in all aspects. The typical denomination is consistent from class-A to class-J, while further types need more detailed explanation [26].

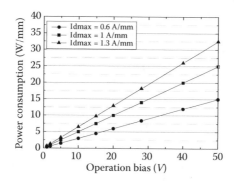

**FIGURE 5.6**  Power consumption of class-A operation as a function of operation bias for different maximum current levels.

### 5.2.1  GALLIUM NITRIDE CLASS-A, CLASS-B, AND CLASS-C MICROWAVE MONOLITHICALLY INTEGRATED CIRCUITS

Class-A operation is very desirable for any amplifier to enhance linearity and to maximize output power levels [27]. The class-A operation for GaN is very critical due to the high power density for this mode of operation. This can be estimated from the graph in Figure 5.6. It provides the dissipated power for linear operation as a function of the operation bias for class-A and maximum current typical for GaN. Most of the GaN MMIC are thus operated in class-A/B [28] with a relatively low quiescent current to minimize the quiescent power consumption. This class-A/B operation is discussed below for the various frequencies. Very powerful amplifiers up to 800 W have been presented in deep class-A/B [29].

Pure class-B operation is typically much less linear than class-A; however, it is useful for the reduction of the power consumption while maintaining maximum efficiency without additional harmonic termination [30]. Class-C operation as such is only of minor value for RF-operation, as shown e.g. in [31] it can be used in rectifier mode. Class-C operation is useful for load modulation in Doherty amplifiers [32] for the use highly-efficient linear base-station amplifiers, as shown below [33]. They are today typically designed as multiway Doherty amplifiers [34].

### 5.2.2  CLASS-D, CLASS-E, AND CLASS-F MICROWAVE MONOLITHICALLY INTEGRATED CIRCUITS

Switch-mode operation is possible for GaN field-effect transistors (FETs) due to the increased reliability and ruggedness in this mode. GaN provides extensive breakdown voltage advantages over Si and GaAs, allowing the use of these high-gain devices with a comparably high breakdown voltage while maintaining good current capability. Further, the resulting high-power density is very useful, as switches require non-distributed switching for maximum efficiency operation.

### 5.2.2.1 Class-D Amplifier

Class-D amplifiers are classically used for linear audio amplifiers in the very low frequency range from 20 Hz to 20 kHz. Two transistors are used as switches. As the switch capacitances limit the efficiency in RF- and microwave applications, the operation frequency for silicon is typically very low. GaN allows to increase the switching speed to higher frequencies [35] as this allows ample opportunities for new applications of switching at higher frequencies in the power electronics domain. Class-D operation of GaN MMICs for digital operation at frequencies of several hundreds of Megahertz has been given in literature [36,37]. For even higher frequencies beyond 500 MHz, high-efficiency digital GaN MMIC PAs for switch-mode-based mobile communication systems are given in [38]. In this case, two GaN transistors of the amplifier are used in a push–pull configuration [38]. Digital class-D operation can be achieved by feeding the amplifier from a differential rectangular bit-sequence at a given data rate, for example, 900 Mbit/s, that is, the fundamental frequency is at 450 MHz. This can reduce the need of digital to analogue conversion. The input pulse forms can be generated by using a silicon fast emitter-coupled logic IC, which gives complementary output signals. With GaN, very high efficiencies of more than 90% can be reached at frequencies beyond 450 MHz with output power levels beyond 20 W [36], drain-efficiencies of 80%, and power-added efficiencies of more than 70%. For further analysis of such GaN class-D amplifiers for more complex bandpass $\Delta$–$\Sigma$-modulated signals, see the mixed-signal Section 5.14 devoted to advanced class-S amplifiers.

### 5.2.2.2 Class-E Amplifier

The concept of the class-E amplifier with shunt capacitance was first introduced by Sokal in 1975 [39] with the extended equations [40,41]. An inductive load is presented to the transistor. This requires a very high voltage swing capability of more than 3.5 of the operation bias due to the resulting inductive peaking. The class-E amplifier is designed to provide a minimum overlap of current and voltage over time to minimize the power loss during RF-switching. The specific characteristics of the ideal class-E amplifier combines zero-voltage switching (ZVS) with zero-voltage derivative switching (ZVDS), which forms clear conditions to achieve class-E operation. The ideal on- and off-switching are defined to occur up to a maximum frequency. The saturated amplifier as a derivation of the class-E amplifier is described in the work of Jee et al. [42] for operation beyond the maximum frequency. For this, the maximum intrinsic current and the capacitance scale in the same with respect to varying device periphery, thus the ratio is independent of the device size. The ratio only depends on the device technology and the scaling of the parasitic environment, most importantly on the effective output capacitance [43]. Initial reports on GaN class-E MMICs are found in the work of Gao et al. [44] with both single- and dual-stage devices. As in many reports, the efficiency is significantly enhanced as compared to class-A/B operation. Many examples of GaN hybrid class-E PAs have been presented. Very high-efficiency class-E amplifiers have been presented with output power levels as high as 70 W in the work by Calvillo-Cortes et al. [45], and efficiencies as high as 84% at 2 GHz and 10 W of output power [46]. Design considerations for the design of hybrid GaN class-E amplifiers are given in [17]. The principal trade-offs are based on the fact that the maximum frequency for class-E operation will drop as a function of drain bias; as the capacitance to be matched for class-E operation for a given device size will drop. Eventually, the given parasitic output capacitance will limit the frequency to operate in class-E for a given device size and output voltage.

The class-E amplifier can also be used as a core amplifier in more advanced linear concepts, for example, the outphasing concept in the work of van der Heijden [47] in combination with high-voltage CMOS (complementary metal–oxide–semiconductor) IC switch-drivers [48]. Two core amplifiers are used to switch two phases (one is the converted amplitude information) to form a linear amplifier in Chiriex [49,50] configuration suitable for efficient back-off operation [47]. High-efficiency class-E GaN-integrated MMIC HPA even at higher X-band frequencies are given by Moon et al. [51]. The devices are biased at pinched-off conditions in this case. The switching at 8.5–11.5 GHz requires a cut-off frequency of 58 GHz, that is, an oversampling of a factor of five, to enable the efficient switch-mode operation to amplify the harmonics.

### 5.2.2.3   Class-F and Inverse Class-F Amplifier

The concept of class-F and inverse class-F was proposed by Raab [52]. Other than class-E, it requires explicit termination of the first harmonics.

For GaN MMICs, class-F operation is demonstrated at 4 GHz by Zomorrodian [43]. In class-F, a square wave current and a sinusodial voltage are created with minimum overlapp, which leads to a maximum efficiency. Very high efficiencies of more than 80% have been reached for class-F [53]. For class-F, the following load conditions hold for the impedances of the first three harmonics:

$$Z_L(f_0) = Z_{opt} \tag{5.7}$$

$$Z_L(2f_0) = 0 \tag{5.8}$$

$$Z_L(3f_0) = \infty \tag{5.9}$$

which are inverted between second and third harmonics for class-F.

$$Z_L(f_0) = Z_{opt} \tag{5.10}$$

$$Z_L(2f_0) = \infty \tag{5.11}$$

$$Z_L(3f_0) = 0 \tag{5.12}$$

GaN class-F amplifiers with higher power levels are reported [54] with 12 W of output power at 3.5 GHz. Similar results are obtained by Chin and Clark [55].

### 5.2.3   GaN Continuous Mode MMICs

The efficiency continuum will allow the building of MMIC with broadband efficiencies. The IC concept allows very precise definition of the harmonics required even for relatively low frequencies below 6 GHz, where the use of GaN MMIC is not always considered. In combination with some hybrid decoupling capacitors and inductors, this allows for a new class of very broadband high-power MMICs.

The continuous modes for harmonic operation for broadband circuits were first suggested by Wright et al. [56] and Cripps et al. [57]. Various demonstrations using GaN are given on hybrid and MMIC level to achieve a very broad bandwidth with very high efficiencies [58-60]. MMICs in continuous modes on GaAs with very good power-added efficiency (PAE) values have been shown by Powell et al. [61]. The Class-B to Class-J continuum of modes offers Class-B levels of efficiency over a continuum of impedance matching conditions, making wider bandwidth designs more feasible. The continuous modes are very attractive for high-efficiency operation for GaN due to the high breakdown voltage required for the realization of the broadband harmonic matching. Integrated design with class-J modes is given [20] for an integrated 0.5 W GaN MMIC class-J PA along with an analysis of the power-efficiency design paradigm. Derivations of optimum load impedances, output power, and efficiency are presented to demonstrate their dependence on the quality factor of the output-matching network.

## 5.3   MICROWAVE MONOLITHICALLY INTEGRATED CIRCUIT EXAMPLES FROM C-BAND TO K-BAND FREQUENCIES

### 5.3.1   S-Band and C-Band MMICs

ICs are very attractive to reduce the losses during packaging and to improve the accuracy of the matching of bond wires of wider power cells and power bars [62]. However, low frequencies below 3 GHz cause MMIC integration including passive structures to become large and thus too costly. Area-saving strategies are thus required. A compact 70 W PA MMIC utilizing S-band GaN on SiC HEMT process is given in [63]. At an operation bias of $V_{DS} = 40$ V, the MMIC is capable of output powers of greater than $P_{sat} = 70$ W at a frequency of 3.5 GHz. The MMIC integration can be useful to obtain a compact chip suitable for transmit-array applications under the constraint of the wave length. With a compact die area of $4.1 \times 3.1$ mm$^2$, an output power density of 5.6 W/mm$^2$ per die area for a single fully monolithic S-band HPA is demonstrated. At C-band, an AlGaN/GaN MMIC HPA for synthetic aperture radar applications is given in [64]. The HPA delivers up to 16 W of output power with PAE over 38% at 6 dB gain compression within a 900-MHz bandwidth around 5.75 GHz. The chip size amounts to $4.5 \times 3.5$ mm$^2$. Up to 100 W of output power has been reported with PAE levels of $\geq$60% at 4 GHz using GaN hybrid powerbars [65]. Second and third harmonic tuning on hybrid substrates are integrated in the package to maximize the PAE for space operation. On-chip second harmonic integration is given in [66] for the same power level and efficiency in continuous wave

(CW) operation. The example shows the area consumption in C-band to achieve this on-chip matching: it amounts to 25%–30% of the chip area and is thus considerable.

### 5.3.1.1 Discussion

The C-frequency band (4–8 GHz) is a very interesting band for GaN MMICs, as the high efficiencies are not obtained so easily as at L-band; however, still very high power levels are required to replace, for example, traveling wave tubes. Further MMIC integration helps with the harmonic termination. A C-band 50-W high-power MMIC for C-band radar applications is given by Jeong et al. [67]. The amplifier demonstrated a saturation output power of 47 dBm and higher than 35% PAE with an output power density of 3.2 W/mm² [67]. Even higher PAE levels of MMICs at C-band frequencies at 6 GHz have been presented in [68] with a power density of 5.5 W/mm in pulsed operation.

Figure 5.7 gives an image of a single-stage C-band design [68]. The chip size amounts to $5 \times 3$ mm². In this case, a process with a gate length of 0.25 µm is used to achieve good maximum stable gain (MSG)/maximum available gain (MAG) values for very wide power cells around 6 GHz to improve the efficiencies. The MMIC with an output periphery of 6.4 mm delivers more than 40% PAE and a maximum output power of 45.5 dBm at $V_{DS} = 40$ V. For wideband operation power levels of more than 25 W from a few megahertz to 6 GHz are described in the work of Pengelly [69] based on nonlinear distributed power amplifiers (NDPA). The NDPA enables this broadband operation from 20 MHz to 6 GHz with more than 20 W of output power with a PAE > 30%. This efficiency level can only be obtained using the IC approach to avoid the phase uncertaincy induced through bond wires for the hybrid approach and to obtain stable operation.

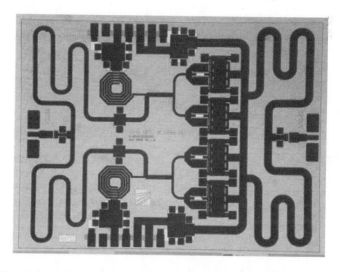

**FIGURE 5.7** Image of a single-stage C-band high-power amplifier. (From Kühn et al., *International Journal of Microwave and Wireless Technologies*, 2(1), 115–120, 2010 [68].)

### 5.3.2   X-Band Frequency MMICs

#### 5.3.2.1   State-of-the-Art

X-band frequencies between 8 and 12 GHz provide one of the main applications for which high-power GaN amplifier MMICs are developed. This main application yields transmitters for solid-state active-electronically steered aperture (AESA) radars. This is mainly due to the inability of GaAs MMICs of the same type to provide efficiently output power levels beyond 10 W per IC, as the power combining [70] has a great impact on the MMIC and resulting system efficiency for mobile radar. Further, the TR-module architecture can be simplified with a reduction of less MMICs per module, and the module size can be potentially reduced in the view of TR-size limitation due to the array-pattern requirements.

Initial reports on GaN MMICs suitable for airborne radar can be found in [71,72]. High power beyond 10 W is easily achievable. X-band high-power MSL MMICs with 20 W GaN HPAs for next generation X-band T/R-modules can be found in studies by van Raay et al. [73,74]. The typical power cell size for X-band is about 1 mm with the combination of four or eight transistors in the final stage. Typically, dual- or tripple-stage MMICs are realized, which provide a gain of more than 10 dB per stage. Similarly, a 43-W, X-Band GaN HEMTs MMIC HPA with 52% PAE can be found in th ework by Piotrowicz et al. [75]. The power cell devices are based on oversized $16 \times 100$-μm HEMTs. This power device has a maximum available gain of 11.8 dB at 10 GHz; however, it is not not suitable for operation at higher bandwidth of 12 GHz. The electroplated gold layer of 6 μm thickness was used for the combiner networks. Parallel RC-networks in series at the input of each transistor enhance the loop-stability [76] and prevents parametric oscillations [75].

Balanced versions of an X-band PA MMICs can be found in [77]. Microstrip MMIC directional 3-dB couplers with low impedance levels on s.i. SiC are designed for a center frequency of 10 GHz and show a coupling factor of 3.5 dB and a low net insertion loss of 0.3 dB.

High-efficiency X-band Class-E GaN MMIC HPAs are given in the work by Moon et al. [51]. The MMICs were fabricated in a microstrip layout with 50-μm-thick SiC substrates and source via holes. The maximum PAE achieved is 60% with 5 W and 50% with up to 10 W of output power at 8.5–11.5 GHz.

#### 5.3.2.2   Discussion

With a targeted output power of ≥20 W, four GaN HEMTs are typically used in parallel. For an operation bias of 28–40 V the typical power density of a power cell with a gate width of 1 mm amounts to 5–8 W or W/mm. The output impedance level for 40-V operation is in the range of 40 Ωmm. Figure 5.8 shows the image of a dual-stage PA in microstrip TML. The chip size is $4.5 \times 3$ mm$^2$.

For maximizing the efficiency of the driver stage relative to the final stage, two options can be chosen from: (1) the total cell size of this stage can be lowered (2) the impedance of the load line can be increased. A typical ratio of the peripheries for GaN-stage is about 2:1. A cell size reduction for the first stage is preferred as an increase of the impedance level of the driver output results in a more critical matching transformation to the low-value real part of the PA stage input impedance [78].

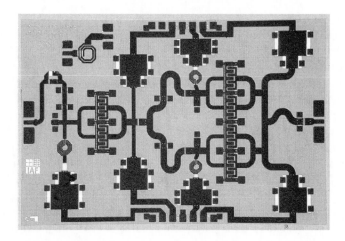

**FIGURE 5.8**  Image of an X-band dual-stage power amplifier in microstrip transmission line technology. (From Kühn et al., Design of Highly-Efficient GaN X-Band-Power-Amplifier MMICs, IEEE MTT-S International Microwave Symposium Digest, 661–664, 2009 [78].)

The impact of harmonic termination on X-band MMIC is discussed in [79]. For small bandwidth below 10%, a positive impact of the harmonic matching at input and output on the PAE can be observed for small power levels and single transistors. For a higher number of FET to be combined, this effect needs to be traded for the additional losses of the harmonic structures. Further analysis is given by Canning et al. [59]. Input second harmonic termination can greatly increase the efficiency, which was for GaAs proven by active waveform analysis by Canning et al. [80].

### 5.3.3  Ku- and K-band Frequency MMICs

Ku- (12–18 GHz) and K-band (18–26 GHz) MMICs are very suitable for tube replacement for SATCOM frequencies, and GaN MMICs have been suggested early for this application [81]. Very high efficiencies at high frequencies are required here, which are hard to be obtained by any semiconductor technology in comparison to advanced waveguide tubes. However, as the system efficiency is to be maximized, especially distributed antennae systems can be realized to achieve a better overall performance, as the multiple MMIC-based radiating elements can be put closer to the actual antennae outputs, thus feeder losses can be reduced, and additional redundancy be created.

The K-frequency band is served with GaN processes in the gate length range of 0.25–0.15 μm. It is again an interesting band to design in, as this band does not provide other typical mainstream applications, however, the requirements are harsh. Satellite services around 20 GHz require maximum efficiencies for the spaceborn components, while the ruggedness requirements, for example, voltage standing wave ratio (VSWR), are also very high.

#### 5.3.3.1  State-of-the-Art

Using conventional MMIC design, an AlGaN/GaN HEMT PA MMIC at K-Band is given in the work of Friesicke et al. [82], investigated for maximum PAE. The measured amplifier has a maximum PAE of 41% and an associated output power of 31.4 dBm when operated at a drain voltage of 20 V. At 35 V drain voltage, the amplifier exhibits a PAE of 34% and an associated output power of 34 dBm. A linear PA at K-band with 4 W of output power using GaN HEMTs with a gate length of 250 nm is given in [83]. The maximum PAE is 34%, which is lower for the low-power version, however, is still good. Third-order (OIP3) is 40 dBm at 1 MHz off-set in a two-tone test. As K-band amplifiers are typically used for linear operation, the back-off efficiencies are of critical importance. In the work of Campbell et al. [84], the Doherty concept with load modulation is used at 10 times the frequency of typical Doherty operation. A K-band Doherty amplifier MMIC utilizing 0.15 μm GaN HEMTs with 5 W of output power at 22- to 24-GHz bandwidth with a maximum PAE of 48% is described. The linear gain is 15 dB. The MMIC die dimensions are $3.4 \times 2.0$ mm$^2$. With this demonstration concept known from the microwave frequencies, the linear efficiency is found to be as high as 27% in linear 256 QAM operation at 22 GHz, which is promising.

### 5.3.4  KA-(26–40 GHz) AND Q-BAND (33–50 GHz) FREQUENCY MICROWAVE MONOLITHICALLY INTEGRATED CIRCUITS

At millimeter-wave frequencies, GaN can provide much higher power levels than any other semiconductor technology. For higher frequencies, GaN amplifiers are attractive for the frequency range of 26–50 GHz with gate lengths of 0.2–0.1 μm [85,86].

Power densities on FET level up to 10 W/mm at 30 GHz were reported, which, however, can only partly be used for MMICs. Not all available results are or will be published as the frequency range is very suitable for military applications, for example, for radar and battle field communication. Initial results of coplanar Ka-band AlGaN/GaN HEMT high-power and driver amplifier MMICs are presented [87]. Output powers of more than 2 W at 27 GHz have been achieved. A 5-W GaN MMIC for millimeter-wave applications with a reasonable PAE of 20% are given by Boutros et al. [88].

PAE is critical to the whole design for thermal and reliability reasons. Design methodologies of 5 W power MMIC around 30 GHz are given by Cheron et al. [18]. The final two-stage PA MMIC was designed to operate from 29 to 31 GHz. The work further reports the design and performances of three PA MMICs providing output powers higher than 1, 2, and 4 W with 41%, 33%, and 28% of PAE, respectively. Both higher efficiencies and higher power levels can be found in the work of Campbell et al. [89] with more 10 W of output power, both realized in MSL technology. In this case, the MMIC is given in balanced configuration with 25% of PA with a chip dimension of $3.24 \times 3.6$ mm$^2$. Even higher output powers have been obtained by Northrop Grumman [90]. Up to 40 W of output power is achieved at 27 GHz using transistors with a periphery of $8 \times 82.5$-μm-wide gate fingers with a gate length of 0.2 μm. A minimum of 30% PAE is achieved. The power density of the process is 4 W/mm tuned for maximum power at $V_{DS} = 28$ V, and 3.4 W/mm in the design for maximum PAE. The chosen output periphery is 10.67 mm.

#### 5.3.4.1  Discussion

Figure 5.9 gives the image of a Ka-band high-power MMIC amplifier in coplanar technology. The substrate thickness for coplanar circuits is typically 250–300 μm based on the original substrate thickness of s.i. SiC on 3-in. and 4-in. wafers. The device initially provides an output power of 34 dBm with a gain of 13 dB at 27 GHz. PAE was still considered a great challenge [87]. To that end, substantial improvements were required and achieved after the initial developments. In the work of Cheron et al. [18], a 1-W MMIC was achieved with a PAE $\geq$40% at 32 GHz based on an improved technology. Using the same technology, a dual-stage MMIC was realized with more than 4 W of output power with a PAE of more than 27%. The typical cell size is still $8 \times 60$ µm for the final stages with four devices in parallel and approximately a 2:1 driver ratio for each stage.

Figure 5.10 gives the image of a 9-W PA at 30 GHz [91]. The device is based on a 100-nm gate technology and achieves a maximum efficiency of $\geq$30% at 30 GHz.

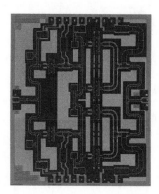

**FIGURE 5.9** Image of a Ka-band power amplifier in coplanar transmission line technology. Chip size is 2.75 mm $\times$ 3.25 mm. (From van Heijningen, M. et al., Ka-Band AlGaN/GaN HEMT High Power and Driver Amplifier MMICs, Proceedings of 15th European Gallium Arsenide and other Compound Semiconductors Application Symposium, Paris, France, pp. 237–240, 2006 [87].)

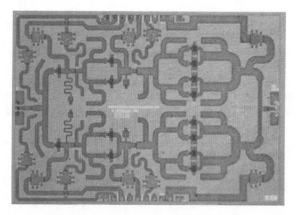

**FIGURE 5.10** Image of a Ka-band power amplifier in microstrip transmission line technology with a chip size of 5 mm $\times$ 3 mm. (From Cheron, J. et al., High-gain Over 30% PAE Power Amplifier MMICs in 100 nm GaN Technology at Ka-Band Frequencies, Proceedings of the 10th European Microwave Integrated Circuits Conference (EuMIC), Paris, France, pp 3, 2015 [91].)

The MMIC is a three-stage device, which is balanced from a smaller dual-stage MMIC device. Eight $8 \times 60$-μm devices are used in the final stage leading to a periphery of 3.84 mm. The linear gain is $\geq 18$ dB. This device is very suitable to achieve very high linear power levels for point-to-point links in backhaul data communication.

## 5.4 BROADBAND AMPLIFIER MICROWAVE MONOLITHICALLY INTEGRATED CIRCUITS

The concepts to realize broadband amplifiers with octave (i.e., the upper band edge frequency is twice the lower band edge frequency), or even decade (i.e., an order of magnitude) bandwidths is a specific requirement we observe from the applications. The bandwidth ratio (BWR) is typically defined as

$$BWR = \frac{f_{upperbandedge}}{f_{lowerbandedge}} \qquad (5.13)$$

On the basis of fundamental laws of nature, such as the Kramers–Kronig relation, the specific matching of a broadband amplifier cannot be perfect for all frequencies [92]. This formulates a fundamental challenge for the MMIC designer. Compromises have to be found for matching with respect to either output power, efficiency, gain, PAE, sensitivity, and survivability [93]. Both good matching and the provision of high power and efficiency providing low-frequency dependence and low-gain ripple yield the following circuit concepts suitable for broadband operation, as reported for GaN [93]:

- Reactively-matched PAs [94]
- Feedback amplifiers [95], with both Cascode and Darlington concepts
- Distributed Amplifiers (DA) and NDPA such as traveling wave amplifiers (TWA) [94]
- Combinations of the above [96]

The Bode–Fano criterion [97] is one possible principal limitation, which is often cited; however, it is by far not the only limit. The limitations of broadband GaN PAs and the trade-offs include considerations regarding availability of gain at the upper band edge frequency of the PA, low-frequency flatness, maximum bandwidth, and the variation of power, gain, and efficiency over bandwidth. The next paragraph gives an overview on the potential and the measures taken accordingly.

### 5.4.1 STATE-OF-THE-ART

An overview of the design techniques for multi-octave PAs is given by Kobayashi et al. [98]. The principal choices for broadband PAs are discussed along with a discussion of broadband LNAs. Cascodes and dual-gate transistor provide more gain and bandwidth, whereas efficiency and stability are compromised as compared to the common-source solution [99].

The development of broadband amplifier based on GaN HEMTs from 30 to 4000 MHz (i.e., more than 2 decades) is described in literature [100]. Feedback and distributed amplifier concepts are used. A feedback concept provides 42 dB gain with ± 1.75 dB flatness and 32 dBm of output power. C- to Ku-band ultra-broadband GaN MMIC amplifier with 20 W output power is given in the work of Masuda et al. [101] for the bandwidth of 6–18 GHz with a power gain of 9.6 dB. A relative bandwidth of more than 115% is reached. The circuit dimension is $4.8 \times 4$ mm$^2$. A reactive circuit concept is used for the range 6–18 GHz, as it is argued that the output power of a distributed amplifier is lower in this case than for the reactive matching of the amplifiers, since the gate periphery of the distributed amplifier cannot be large enough to provide the same power based on the process used.

A decade bandwidth GaN HEMT PA MMICs from 2 to 20 GHz using various field-plate technologies are given by Komiak et al. [102]. The MMICs with dual-field plate (DFP) achieved a $P_{3\,dB}$ of 15.4-W average, with 19.8% average, and a gain 8.6-dB average from 2 to 20 GHz. MMICs with no field-plated FETs achieved an improved saturated power $P_{3\,dB}$ of a 16-W average and 25.9% average PAE, and 9.7 dB of average power gain from 2 to 20 GHz. This is based on the improved (i.e., reduced) feedback capacitances and the overall reduced output capacitances for the devices without field plates to be matched over such a wide bandwidth. GaN-based amplifiers for wideband applications from 2 to 18 GHz are discussed in the work of Schuh et al. [103]. Output power levels of 38 W for hybrid amplifiers at lower frequencies from 2 GHz up to 6 GHz, and about 15 W for the MMIC PAs at higher frequencies in the frequency range 6–18 GHz are measured. These are examples of amplifiers for classical electronic counter measures, which require this combination of power and bandwidth. A three-stage resistively-matched 6–18 GHz high-gain HPA is given by Mouginot et al. [104]. At 18 GHz, the MMIC achieves 10 W of output power in CW-mode with 20 dB linear gain and a PAE of 20%. The HPA provided 6–10 W of output power over the full range of 6–18 GHz with a minimum small-signal gain of 18 dB. An ultra-wideband (UWB) GaN MMIC chip sets and an HPA module for multifunctional AESA, again based on reactive matching applications, are discussed in the work of Schmid et al. [105]. The potential of frequency-selective ultrawide-band HPA transmit modules for multifunctional (MFC) active electronically scanned antenna radar combining electronic warfare and communication applications from 6 to 18 GHz is given. The AESA modules must comply with a lateral grid of half the free-space wavelength at the highest RF frequency. To date, these high power-gain values can only be achieved with a reactively matched multistage MMICs. Measurements at the module level indicate 18.5 W of typical output power in both pulsed and CW operation combining two MMICs. A copper housing serves for an optimized heat flow away from the MMICs to the active cooling unit.

Using a more innovative InAlGaN device technology, MMICs in MSL technology for wideband applications are given by Schuh et al. [106], again based on reactive matching. InAlGaN/GaN HEMTs provide potentially more power density especially for wideband applications, as the impedance to be matched at the output are more suitable for wideband matching compared to AlGaN/GaN FETs. The InAlN- or

InAlGaN technology allows to adjust the impedances of the broadband amplifiers in a ratio-operating voltage and maximum drain current [107]. Further, very high efficiencies can be achieved, at the same time, the shape of the transconductance function versus bias can be tailored toward improved linearity.

### 5.4.2 DISCUSSION

On the basis of the previous examples, the main limitation of broadband PAs are to be discussed. The typical limitations include

1. The often cited Bode–Fano criterion [17,97] for the limitation of the band-width with respect the matching of a reactive load
2. Actually more general: the Kramers–Kronig relation [92]; for the relation of bandwidth (BW) with ideal matching for the desired target
3. The impedance levels for transformation to be matched
4. The available gain in compression at the maximum frequency of operation, that is, the upper band edge
   The Bode–Fano criterion is often cited to limit the bandwidth of GaN wide-band MMIC amplifiers [16,108]. The equation reads

$$\int_0^\infty \ln\left|\frac{1}{\Gamma(\omega)}\right| d\omega \leq \left|\frac{\pi}{R_{\text{Load}}C_{\text{out}}}\right| = \frac{\pi}{\tau} \tag{5.14}$$

In the case of an FET to be matched, the load resistance $R_L$ and the effective output capacitance $C_{\text{out}} = C_{\text{ds}} + C_{\text{gd}}$ have to be low. Equation 5.14 states in other words that, for a given complex load, there exists a fundamental limit of achievable bandwidth on an amplifier. However, this limit is not achieved for GaN in all cases due to multiple other effects [108]. First of all, the Kramers–Kronig relation is a principal limitation based on the casuality [92]. No perfect reactive matching can be achieved over unlimited bandwidth as otherwise the casuality is violated. In the work of Campbell et al. [17], it is argued that the impedance transformation for a matching situation of 20 dB will reduce the bandwidth of a reactively matched PA. However, the perfect matching considered for GaN may not be required as the load circles of GaN may be relatively broad, so that the loss of power, PAE, and gain even in a mismatch situation may not be as critical. The impedance-level transformation deduced from matching of both input- and output-matching and especially from interstage-matching of multistage amplifiers for high bandwidth is critical. Another limit is the gain achieved for this wideband matching [96]. This critical feature is found to be one of the main limitations for wideband matching in GaN MMIC beyond an octave bandwidth. It implies two considerations: first of all the product of gain and bandwidth of any technology is limited. This results in a trade-off situation of the maximum allowable FET size of the power cell with respect to gain (MSG or MAG), which thus defines the maximum absolute output power, and gain in power compression is critical. GaN HEMTs are considered a material with a soft power compression [109]. This compression is more complex than in GaAs or silicon due to heat considerations of the high-power density, related thermal constraints residual dispersion, and the class-A-B operation typically applied. Independent from the impedance transformation and the matching, the available power gain and related power density and PAE in power compression at the upper band edge are decisive [105]. This power density at the upper band edge is a critical limit as is defined by the ration of current and operation bias to define an impedance defined by the power requirement per power cell. As the PAE is decreasing with frequency, this also defines the maximum heat flow, for which the packaging has to be optimized.

### 5.4.3 Design Procedure

This combination of limited gain, that is, the power gain limitation at a given compression level and the associated output power, typically leads to the following design procedure for reactive matching: a nearly ideal matching at the maximum frequency of operation is produced to maximize the power of the overall circuit. In the trade-off, this maximum output power at $f_{\text{upper band edge}}$ is paid for by some critical matching situation, which occurs somewhere within the wide bandwidth $f_{\text{upper band edge}} - f_{\text{lower band edge}}$.

### 5.4.4 Feedback Amplifier Microwave Monolithically Integrated Circuits

As a simple, however, effective means to reach broadband operation, feedback is used. The technique is useful for GaN to overcome the gain shortage at the upper band edge, for example, at Ku-band at 18 GHz [95]. A feedback loop can be used to improve gain flatness and especially the absolute gain at 18 GHz, as suggested by Schmid et al. [110]. In the example presented by Schuh et al. [95], two-tone measurements show good linearity of the particular feedback amplifier. For up to 26 dBm of output power, the IM3 value is better than 30 dBc. The image of a feedback amplifier realized with GaN HEMTs is given in Figure 5.11. Both parallel and serial feedbacks are applied. In this case, only two viaholes are applied to realize the parallel feedback for the amplifier in each of the three stages.

#### 5.4.4.1 Discussion

Feedback is a useful technique to enhance the bandwidth, especially for more than one octave bandwidth [111] and the ability to operate down to baseband frequencies. The low-frequency operation of such devices is typically limited by the finite on-chip feedback in passive MMIC processes. Darlington feedback produces additional noise for the broadband LNA.

Very wideband LNAs have been realized using GaN [112]. BWRs of 25:1 have been achieved with 13 dB of gain for a bandwidth of 1–25 GHz using modified resistive-feedback topology. To obtain such a wide bandwidth, several bandwidth-enhancement techniques are used. An inductor connected to the source of the input transistor ensures good input matching over the bandwidth. A shunt feedback loop and the inductive source-degeneration minimize the inductor values required for

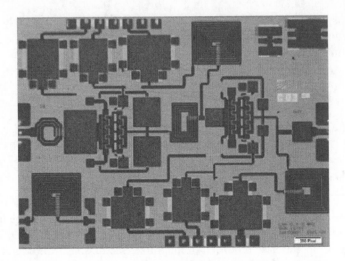

**FIGURE 5.11** Schematic of a dual-stage broadband low-noise amplifier using feedback. Chip size is 3 mm × 2.5 mm. (From Maroldt, S. et al., QFN-packaged Highly-linear Cascode GaN LNA MMIC from 0.5 to 3 GHz, 8th European Microwave Integrated Circuits Conference (EUMIC), Nuremberg, Germany, pp. 428–431, 2013 [113].)

broadband matching. The LNA MMIC amplifier is comparable in performance to distributed amplifiers, however, with significantly reduced power consumption and smaller area, as described in Section 5.7.

### 5.4.5 GALLIUM NITRIDE DISTRIBUTED OR TRAVELING WAVE CONCEPTS

#### 5.4.5.1 State-of-the-Art

Distributed amplifier can provide more than an octave bandwidth. High-efficiency GaN PA using a non-uniform distributed (NDPA) topology is proposed by Gassmann et al. [114]. The NDPA topology is useful, as it improves the matching conditions of various FETs. A comparison of GaN to MMIC amplifiers in GaAs technology is given by Meharry et al. [115]. The GaAs amplifiers are found to be more complex, whereas the ration of the GaAs to the GaN gate periphery is about 14.4 mm to 2 mm, that is, an improvement for GaN of 7:1.

To that end, over 10 W C to Ku-band GaN MMIC non-uniform distributed PAs with additional broadband couplers are described by Masuda et al. [101]. The fabricated MMIC PA with 0.25 μm GaN HEMTs with an $f_T$ of 21 GHz delivered an output power of more than 10 W with average PAE of 18% over 6–18 GHz. Lange couplers are used to combine two NDPAs to provide the maximum power.

#### 5.4.5.2 Discussion

Some of the considerations discussed earlier are based on a parallel combining of devices that are combined in phase. However, to overcome the bandwidth limitation of purely reactive matching there is an alternative making use of the phase relations

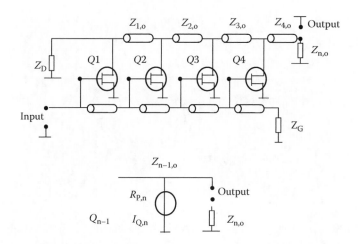

**FIGURE 5.12** Principle of the traveling-wave amplifier in a wave representation (top) and baseline matching approach (bottom).

[116,117]. Figure 5.12 gives the principle of a TWA. The amplifier consists of multiple devices, combined with respect to the phase correlation. The input signal is sent through multiple parallel paths, which are designed to interfere constructively at the output TML. The input and output capacitances are absorbed in the TMLs and are thus not limiting the bandwidth.

The principle leads to a combination of the individual output powers of each path. This concept is especially suitable for broadband amplifiers; as in conventional amplifiers, one cannot increase the gain-bandwidth product by paralleling FETs, as the resulting increase in transconductance $g_m$ is compensated for by the corresponding increase of both input and output capacitances. An example of one of the first TWAs realized in GaN technology is shown in Figure 5.13. In this case, the devices are still identical in device size and thus are the impedances per device.

The TWA depicted in Figure 5.13 is based on 0.15-μm gate length and delivers more than 9 dB of gain for the frequency range 2–20 GHz with an output power of more than 1 W [77]. This initial design does not yet reflect on the impedance issue observed for GaN devices due to their very high impedances.

The output power of a TWA is limited by the load impedance and is, ideally without bandwidth limitation, proportional to that found, e.g., by Campbell et al. [17]:

$$P \approx \frac{(V_{DC} - V_{knee})^2}{R_L} \tag{5.15}$$

This explains, for a given load impedance, why GaN provides such an increase in broadband output power, as the DC-voltage can be significantly increased, whereas the impact of the knee voltage needs to be minimized. The current is composed as given in Figure 5.12, bottom. As described by Campbell et al. [17], the impedance

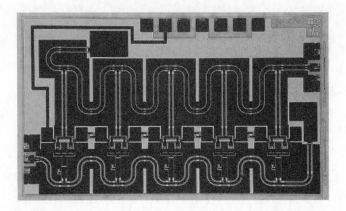

**FIGURE 5.13** Example of the first GaN traveling-wave amplifier realized in GaN coplanar transmission line technology with a low gate length of 0.15 μm, chip size 2.5 × 1.5 mm². (From Kühn et al., Proceedings of the EuMIC 2008 European Microwave Integrated Circuit Conference, 95–98, 2008 [77].)

of the $n^{th}$ transistors in the TWA required for the TML in Figure 5.12 actually can be expressed as

$$Y_{0,n} = \frac{1}{Z_{0,n}} = \frac{1}{R_P (\text{Ohm} \times \text{mm})} \cdot \sum_{\hat{i}=1}^{n} W_{Q,i} \qquad (5.16)$$

The resulting impedances $Z_{0,n}$ of the TMLs are very high for the first transistors to be realized, as the last transistor $n$ has to yield a minimum value of the final load impedance $Z_n,0$. With typical values for the resistor $R_P$, it is more efficient to choose bigger transistors for the first $n$. Various NDPAs have thus been realized in GaN as the TWA principle in a non-uniform way [103,118].

A design with a non-uniform distribution is depicted in Figure 5.14 [103]. A gate width of 6 × 125 μm is used for the first HEMT, and each of the following nine transistors have a gate width of 6 × 50 μm. The design goal was to find matching lines at gate and drain as short as possible to obtain small chip sizes. The single-stage NDPA yields an output power of >8 W between 6 and 17 GHz in CW and more than 15 W of output power in pulsed operation. The chip size is 5.5 × 3 mm².

### 5.4.5.3 Wideband GaN for Millimeter-Wave Broadband Operation

For wider bandwidth and higher frequencies an 8–42 GHz GaN non-uniform distributed PA MMICs in MSL technology are discussed by Dennler et al. [119]. A dual-stage NDPA is presented. The goal for the dual-stage NDPA design and investigation is to significantly increase the gain while maintaining the saturated power and large bandwidth. This was achieved by properly choosing the active device geometries of the transistors with a gate length of 100 nm and a low interstage impedance of 32 Ω The gate widths are $W_g = 4 × 90$ μm and $W_g = 4 × 45$ μm, which are used for the five FETs Q1 and Q2–5 in each stage. An extension to higher power levels using a similar non-uniform distributed 6–37 GHz PA using an MMIC with dual-gate FET driver

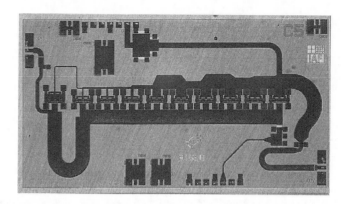

**FIGURE 5.14** Example of a GaN nonlinear distributed power amplifier realized in GaN microstrip transmission line technology with a gate length of 0.25 μm, chip size 5.5 × 3 mm². Courtesy Airbus Defense and Space, Ulm. (From Schuh, P. et al., *International Journal of Microwave and Wireless Technologies*, 2(1), 135–141, 2010 [103].)

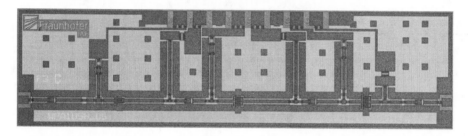

**FIGURE 5.15** Image of the U-band broadband power amplifier for the frequency range 36–62 GHz, chip size 3.5 × 1 mm. (From Schwantuschke, D. et al., A U-band Broadband Power Amplifier MMIC in 100 nm AlGaN/GaN HEMT Technology, Proceedings of the 7th European Microwave Integrated Circuits Conference, EuMIC, Amsterdam, the Netherlands, 703–706, 2012 [121].)

stage is given by Dennler et al. [120]. The MMIC with a dual-stage topology uses dual-gate HEMTs in the driver stage to boost the gain of the overall amplifier. The measured $S_{21}$ is 17 dB with a gain flatness of ±1 dB. This is a significant increase of 3 dB as compared to a situation when using a driver stage using standard common-source HEMTs. An output power well beyond 1 W over the entire frequency range is obtained. For even higher frequencies, a U-band (40–60 GHz) broadband PA MMIC is described by Schwantuschke et al. [121]. The matching is resistive for the BWR of 1.7:1. This corresponds to a very high small-signal bandwidth of over 48%. Output power levels beyond 24 dBm can be reached at these frequencies for the full bandwidth. The corresponding chip image is depicted in Figure 5.15. A small-signal gain of over 20 dB was measured in the frequency range from 38 to 62 GHz for the designed amplifier. As compared to the current-gain cutoff frequency $f_T$, the bandwidth of the amplifier is 30% of $f_T$ and the gain-bandwidth-product is approximately five times $f_T$.

An input and output matching of better than −5 dB were determined. The measured saturated output power at 58 GHz is 25.8 dBm (380 mW). The power density of the output stage is over 1 W/mm for an operation bias of 12.5 V at this frequency. These amplifiers suggest the enormous potential in power density and absolute output power up to at least 100 GHz with scaled gate lengths.

## 5.5   RADIOFREQUENCY-SMALL-SIGNAL AND LOW-NOISE AMPLIFIER MMICS

Small-signal amplifiers on GaN MMIC are also attractive and have been realized in greater numbers to achieve high-linear gain, or high-linear gain with very little distortion [122]. At the receiving end of an RF-system, the signal from the antenna has to be amplified because of its low power level derived from the radiation laws at some distance from the transmitter. The signal-to-noise ratio (SNR) is defined as

$$SNR = \frac{Signal}{Noise} \tag{5.17}$$

This is, independent from the gain consideration, a fully independent consideration to minimize noise on the low end. Further, linearity needs to be improved at the high end of the dynamic range. GaN can provide a very attractive combination of the three properties. In addition, the overdrive capability and survivability levels are exceptionally high for GaN, which has been demonstrated at least up to millimeter-wave frequencies [123]. GaN gives good sensitivity at low power input while providing reasonable gain and very good input linearity at high power levels.

### 5.5.1   STATE-OF-THE ART

Initial reports on GaN LNA MMIC at X-band frequencies are given in literature [124,125]. The aspect of high-dynamic range at the upper edge of the input power range of GaN HEMT C-band LNAs is discussed by Xu et al. [125]. For input power levels larger than 20 dBm, the device gate diode is partially forward biased during each RF-cycle. This level is typically at least 6–10 dB higher than what can be done today with GaAs pseudomorphic High Electron Mobility Transistor (PHEMT). With the increasing of the drive level to 31 dBm for a 2 × 100-μm device, the forward gate current increases, resulting in device burn-out when the diode reaches a forward current level of 400 mA/mm. This is typically far beyond the safe-operating-area (SOA) of a typical Schottky diode. As an example, T/R-modules for X-band front-end based on dual-stage GaN LNA MMICs are given by Schuh et al. [126]. For decoupling purposes, but also to increase the robustness, a resistor is used in the gate bias network. For an input power of 40 dBm, the LNA yields a gate current of nearly 4 mA equivalent to 8 mA/mm gate width in forward operation. In this initial comparison for frequencies up to 10.5 GHz, the noise figure (NF) of this GaN LNA is only about 0.5 dB worse than the NFs obtained for the GaAs LNAs used normally. The GaN devices, however, potentially, can survive the imminent input power without the protection of a circulator. As this component is costly, bulky, and typically causes additional losses for a GaAs receiver of the same size, on the system level, GaN and GaAs LNAs are at least comparable at this point.

Several dual-stage LNA designs have been realized with additional gate resistors in the work of Schuh et al. [103] for X-band frequencies up to 12 GHz. The measured NF of the LNA is in the range of 2 dB between 6 and 12 GHz. The resistor was chosen to ensure a safe limit in gate current dependent on the device gate width ($I_G = 10$ mA/mm) and simultaneously not exceeding the breakdown voltage of the gate diode (in this case breakdown voltage [BV] = 80 V). Robustness tests have been performed with input power levels of up to 10 W, which, however, do not lead to any destruction at the LNA. At this input power level, the LNA is in compression with a compression level of −30 dB. The saturated output power of the LNA is about 27 dBm, which demonstrates, on the one hand, a limiting capability of the LNA; however, the output and the related power consumption are still substantial for a classical receiver function. This is important, as receiver functions are typically operational for most of the duty cycle. This is a design challenge with respect to the level setting in the receiver chain and with respect to the power consumption. Further, GaN devices for switching and low-noise applications are presented in the work of Pengelly et al. [127]. An LNA S-band MMIC with a dual-stage design demonstrates a NF of less than 2 dB over the wide bandwidth from 2.5 to 4.5 GHz. The linear gain amounts to about 30 dB. The linearity output third-order intercept (TOI) point is greater than 38 dBm when operating at an operation voltage of $V_{DS} = 24$ V. This voltage is compatible with a typical voltage rail in a GaN module; however, it leads to a substantial power consumption in the receiver path. Robust X-band low-noise MMICs including additional limiting amplifier functions are described by Schuh and Reber [128]. The reduction of output power and the DC-power-consumption under drive is the main issue of the LNA in this work. Two different robust LNAs with different limiter functionalities are given. The first version with 1.6-dB NF survives 4 W of input power, and with the included current limiter, the maximum output power is only 17 dBm, which significantly improves the usefulness for receiver chains. This is required to be integrated in the LNA as otherwise further limiting functions would be required to control the reception chain in high overdrive. The other version with input limiter survives more than 10 W of input power, while the NF is 2 dB. Further, broadband sub-decibel NF GaN MMIC LNAs with multi-octave bandwidth from 0.2 to 8 GHz with output power level up to 2 W are presented by Kobayashi et al. [98]. At low bias ($V_{DS} = 12$ V, 200 mA), the amplifier achieves an NF as low as 0.5 dB, which is very low as reported for a multi-octave MMIC amplifier in the S- and C-band frequency range. The linearity of the amplifier under power-hungry class-A operation is very high with an average output OIP3 between 43.2 and 46.5 dBm across the bandwidth of 2–6 GHz. The DC-power consumption under small-signal operation is as high as 6 W in this case. The exceptional linearity is traded with the high DC-power consumption, which is critical in many applications, as a receiver usually considered a low-power function [110]. Even wider band dual-gate GaN HEMT LNAs for front-end receivers between 1 and 12 GHz are reported by Aust et al. [129]. The amplifier uses dual-gate devices including current feedback and drain bias network to attain wideband performance in terms of both lower noise and higher gain. The use of dual-gate device can thus increase the gain and bandwidth product for low-noise operation.

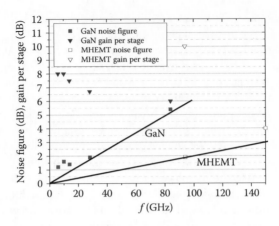

**FIGURE 5.16** Noise figure and associated gain of GaN devices over frequency.

The current level is chosen at 125 mA/mm at a drain voltage of $V_{DS} = 10$ V to achieve a gain level of 12.5–18 dB, and an NF of 1.3–2.5 dB. In comparison to an InAlAs/InGaAs HEMT MMICs, the GaN device provides similar noise and gain levels to 8 GHz on chip level. However, the input protection of the GaN LNA can be avoided; thus, on a system level the GaN LNA may provide advantages.

For even higher frequencies, a highly linear 84 GHz LNA MMIC in AlGaN/GaN HEMT technology is described by Kallfass et al. [130]. In this case, a technology with a gate length of 100 nm is used. The LNA achieves over 25 dB of gain and an NF of 5.6 dB at 84 GHz. A similar device is co-integrated to a full 77-GHz heterodyne receiver by Kallfass et al. [131]. As no additional device protection is required, the NF is to be considered quite good for this frequency. In a more general comparison, the NFs and associated gain of GaN transistors over frequency are given in Figure 5.16 in comparison with very-low-noise InAlAs/InGaAs HEMTs for the same frequency range.

The comparison over frequency [132,133] yields better gain- and noise-performances for the GaAs devices; however, the GaN devices still serve the frequencies with longer gate lengths. Thus, further improvement of GaN HEMTs can be expected while the comparison should be performed on system level, that is, for the actual operation with all functions and trade-offs such as power consumption included.

### 5.5.2 Discussion

The overdrive ruggedness of GaN MMICs has further been investigated in various works. The survivability of AlGaN/GaN HEMT is reported by Chen et al. [134]. Two catastrophic failure mechanisms are identified for the LNA operation. At low quiescent drain-source voltages, that is, $V_{DS}$ (≤10 V), the forward turn-on current of the gate diode may exceed the given burn-out limit, resulting in a sudden device failure. Increasing the quiescent drain-source voltage $V_{DS}$ increases the peak drain-gate voltage and changes the failure mechanism to a gate-drain reverse breakdown. This parameter can be influenced by the setting of the gate-drain contact spacing of the FETs used. A more detailed analysis of the survivability of GaN LNAs is further given by Rudolph et al. [135]. It is shown that the gate DC-current, which occurs due

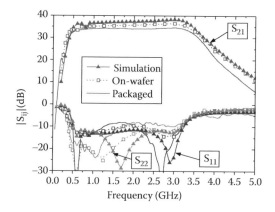

**FIGURE 5.17**  Simulation and measured S-parameter results of a dual-stage GaN wideband low-noise amplifier for the frequency range 0.5–3 GHz. (From Maroldt, S. et al., QFN-packaged highly-linear cascode GaN LNA MMIC from 0.5 to 3 GHz, 8th European Microwave Integrated Circuits Conference, Nuremberg, Germany, 428–431, 2013 [113].)

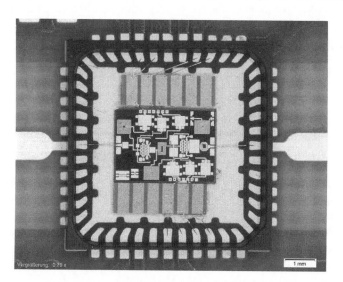

**FIGURE 5.18**  Broadband low-noise amplifier integrated in a QFN-package.

to self-biasing, is the most critical factor regarding LNA survivability. A series resistance in the gate DC-feed can reduce this gate current through a feedback mechanism, and may be used to improve LNA ruggedness at the expense of additional NF.

Figure 5.17 shows the image of the low-noise results of the wideband dual-gate dual-stage LNA for the frequency range of 1–3 GHz from Figure 5.11 [113]. A gain level of more than 30 dB is achieved with a matching of better than −10 dB is achieved between 0.5 GHz–3 GHz for both an on-wafer and packaged situations. The important integration scheme is given in Figure 5.18.

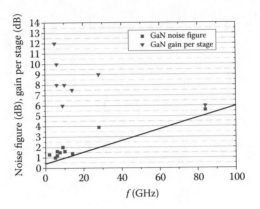

**FIGURE 5.19**  Noise figure of GaN low-noise amplifier MMICs and gain per stage over frequency.

Figure 5.19 shows the NF of MMICs and gain per stage as a function of frequency. For multistage devices, the overall gain is divided by the number of stages. The summary yields a slope of the NF with about 6 dB/100 GHz in GaN.

## 5.6  MIXER MICROWAVE MONOLITHICALLY INTEGRATED CIRCUITS

So far, GaN HEMTs have been widely used in PAs and LNAs MMICs. This is based on the capability to handle both large voltages and high currents at the same time with high speed. Besides this advantage, GaN technology also demonstrates a good combination of noise, linearity, and robust performance for mixer and converter applications. Resistive GaN FET mixers are inherently highly linear and with very low (or zero) DC-power dissipation. This principle relies on the use of the channel resistance as a mixing element [136]. Prominent criteria for good mixer transistors are

- To achieve instantaneous switching on and off to a theoretical minimum conversion loss of 2/φ or −3.9 dB loss.
- To have low-noise capabilities
- To handle the high local oscillator (LO) power

### 5.6.1  State-of-the-Art

Initial resistive GaN mixers were presented by Kaper et al. and Lan et al. [136,137]. From the beginning of the development, the excellent linearity of the mixers was suggested, whereas conversion loss and gain initially formed an issue of improvement. Mixers with very low intermodulation distortion were observed in the study of Kaper et al. [137]. A conservative estimate of the input TOI point based on the slope of the third intermediate frequency (IF) harmonic yields the minimum value of 40 dBm with a corresponding 1 dB compression point of 30 dBm.

C-band to Ka-frequency band resistive mixer MMICs are presented by Do et al. [137]. The MMICs are based on both 0.25- and 0.15-µm gate lengths. The article discusses the need to modify the receiver architecture to adjust the power level settings. This is required as any GaN LNA, if not leveled with limiting functions, will put out significant amounts of output power. Three breadboards of RF-front-end receivers were manufactured to evaluate the robustness of such technology. The measured conversion gains are +21, +20, and +14 dB, and the $P_{-1\,dB}$ are 10, 11, and 3 dBm for the C-, Ku-, and Ka-band, respectively. Resistive wideband mixer with a 3:1 BWR from 6 to 18 GHz are reported by Di Giacomo et al. [139]. The frequency range is 6–18 GHz for local oscillator (LO) and RF, and DC –6 GHz for the IF. The conversion loss is –16 dB for this frequency range. Wideband mixer MMICs dedicated to compact, wideband, and high spurious-free high dynamic range (SFDR) receivers are reported in the work by Mallet-Guy et al. [140]. As a topology, double-balanced topology is used. Over the RF-frequency band of 6–18 GHz and the IF frequency of 1.4–2.4 GHz, the mixer exhibits a conversion loss of –12 dB.

To explore the possibilities and limitations of a GaN mixer, a double-balanced image-reject FET quad-ring topology is explored by van Heijningen et al. [141]. Double-balanced image-reject mixer MMICs in 0.25 µm AlGaN/GaN technology are presented. The image rejection supports wideband operation without additional filtering. This design features an integrated LO-amplifier and active IF-balun. This is useful as GaN can provide easily the high LO-power levels required for conversion with very high dynamic range. The measured conversion loss is less than 8 dB for the frequency range from 6 to 12 GHz, at LO-power levels of 0 dBm.

For even higher frequencies, high-linearity active GaN-HEMT down-converter MMICs for E-band (71–84 GHz) radar applications are given by Kallfass et al. [142]. The proposed stand-alone mixer employs single-ended unbalanced fundamental transconductance mixer topology resulting in a conversion gain of –7.5 dB at 77 GHz. The input-referred 1-dB compression point is measured at 13 dBm, which is a very high mixer linearity obtained at this frequency. Combining two mixers, again in a balanced topology, could further enhance linearity by a factor of 2.

## 5.6.2 DISCUSSION OF GaN MIXERS

Figure 5.20 gives the image of a resistive C-band GaN mixer MMIC based on a MMIC technology with a gate length of 0.25 µm [138]. The mixers yields again a balanced image rejection topology using two cold GaN HEMT mixing cells. The required 90-degree combiner for the IF is not integrated on the MMIC mixer to save GaN wafer area, regarding the low IF frequency below 6 GHz. In addition to the image frequency cancellation, this architecture naturally provides a good LO to IF isolation, thanks to the Wilkinson power divider at the RF input, and 90-degree Lange coupler at the LO input. Both GaN transistors are nominally biased close to the pinch-off voltage, and the nominal LO-input power is set to 10 dBm. This LO drive can be tuned to improve the mixer linearity. DC-consumption is expected to be 0 mW because only gate biasing is required in this structure. The MMIC demonstrates very good linearity. At mixer level, the measured conversion loss are –10 dB for the C-band. The input $P_{-1\,dB}$ compression points are –1 dBm. At the receiver level,

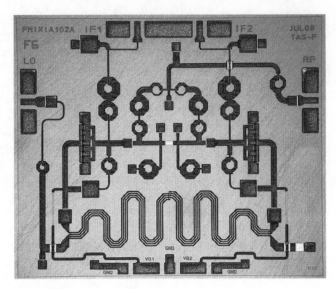

**FIGURE 5.20**   Image of a GaN resistive mixer for C-band applications, chip size 2.5 × 3 mm², courtesy Thales-Alenia Space, Toulouse. (From Do et al., AlGaN/GaN Mixer MMICs, and RF Front-end Receivers for C-, Ku-, and Ka-band Space Applications, Proceedings of the 5th European Microwave Integrated Circuits Conference, EuMIC, London, UK, 57–60, 2010 [138].)

the conversion gains are +21 dB with $P_{-1\,\mathrm{dB}}$ of +10 dBm. This example demonstrates the very good combination of linearity and robustness at high-power drive.

## 5.7   GALLIUM NITRIDE RF-SWITCHES

Similar to mixers and LNAs, GaN-based low-loss switches are very promising for linear operation at comparably high power levels. GaN provides a high-power capability through the combination of high maximum current $I_{\max}$ and high breakdown voltage $V_{\mathrm{break}}$. This allows switched power transmission in the multi-10-W range, as indicated in various publications [143,144]. This provides the opportunity to engineer the off-capacitance $C_{\mathrm{off}}$ in a trade-off with $R_{\mathrm{on}}$, that is, trade isolation requirements for the loss function over frequency. Further, GaN on s.i. SiC substrate provides a low thermal resistance and a high maximum temperature capability, which is favorable in comparison to competing semiconductor-based GaAs PHEMTs [145] and silicon-based PIN-diode technologies [146].

### 5.7.1   STATE-OF-THE-ART

The design targets for such switches are

- Linearity in terms of compression over transmitted power
- Bandwidth
- Isolation in both switching states
- Low insertion losses

Initial reports of GaN switches are given by Kaper et al. [137]. The switch configuration in this case is series transistors with a total gate width of 400 μm. A linear power defined by the compression point $P_{-1\,dB}$ of 1 W can be be transmitted at frequencies of 2–3 GHz. Various MMIC examples are given in the literature for switches: Ref. [147] yields wideband single-pole double-throw (SPDT) switch MMICs for the frequency range up to 18 GHz. The circuits are designed to cover frequency ranges of DC-6 GHz, DC-12 GHz, and DC-18 GHz with input power handling optimized over the specified bandwidth. Measured maximum small-signal insertion loss is 0.7, 1.0, and 1.5 dB, respectively, for the 6 GHz, 12 GHz, and 18 GHz bands. This level is superior to any other semiconductor technology. Measured continuous-wave power data demonstrates typical input RF-power handling of 40 W, 15 W, and 10 W. The article further describes the trade-off of linear losses represented by $R_{on}$ versus the breakdown voltages of the switch transistors. Because of the high breakdown voltage and the high electron mobility of GaN HEMT, the switch has high power handling capability and low insertion loss due to the low-sheet resistance of the GaN HEMTs. Further, it does not require DC power consumption, unlike the PIN diode switches [148].

The development of more complex high-power SP4T RF-switch in GaN HEMT technology on silicon is described by Yu et al. [149]. The transmit paths were optimized for insertion loss ranging at 1.4 dB at 1.5 GHz. An isolation level of 25 dB up to 2.5 GHz is obtained. The measured $P_{-0.1\,dB}$ compression point at 1 GHz was measured to be 43 dBm. Further reports for SPDT switches for Ka-band frequencies are given by Kaleem et al. [150]. In the transmit-state, large-signal measurements at 19.5 GHz with a control voltage of −5 V yield an input compression point $P_{-1\,dB} =$ 29 dBm and an input TOI of 45 dBm. With control voltage set to −20 V, the insertion loss decreases to ≤ 0.8 dB, and the input $P_{-1\,dB}$ increases to 40 dBm. In the isolation state, LS-measurements at 19.5 GHz with control voltage of 0 V gave an input $P_{-1\,dB}$ at 36 dBm. Regarding the reduction of the considerable control voltages, the study by Campbell [151] gives a gate bias circuit that allows high-power RF-switches to operate at rated power with a reduced control voltage. Results demonstrate up to an order of magnitude increase in power handling for the modified design when operated at a −10 V control in comparison to −40 V.

### 5.7.2 DISCUSSION

Figure 5.21 gives the image of a wideband GaN SPDT switch to 18 GHz realized in MSL technology based on a process with 0.25 μm gate length.

This results in the trade-off of isolation and insertion losses discussed by Campbell and Dumka [147]. In the example in Figure 5.21, the bandwidth requirement to 18 GHz is traded for the insertion losses imposed for small-signal and especially for large-signal operation. The isolation of both paths ($S_{12}$ for path 2 and $S_{23}$ for path 3) shall be better than 20 dB at all frequencies, especially at 18 GHz.

$$S_{\text{insertion}}[dB] = S_{11}[dB] + S_{31}[dB] + S_{33}[dB] \qquad (5.18)$$

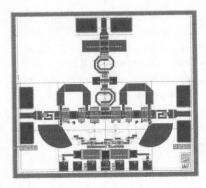

**FIGURE 5.21**    Image of a GaN single-pole double-throw switch to 18 GHz with a chip size of $2 \times 1.75$ mm$^2$.

Because of their direct impact on the insertion loss in the transmit state, the matching parameters $S_{11}$ and $S_{33}$ are optimized to be better than _15 dB over the full bandwidth.

As the $S_{31}$ compresses under LS-drive for various power levels, in this case up to 10 W, a $P_{-0.1\,dB}$ better than 1 W is reached at all frequencies; further, a minimum compression is required for the full bandwidth up to 18 GHz for a power level of 10 W. Similar results have been obtained by Campbell and Dumka [147]. The maximum transmission power transmission in the off-state amounts to

$$P_{\text{Max},off}(W) = \frac{(V_{\text{breakdown}} - V_{pi})^2}{2 \cdot Z_0} \tag{5.19}$$

In this case, the pinch-off voltage is $V_{pi} = -4.0$ V for $V_{DS} = 0$ V. For $Z_0 = 50\ \Omega$, Figure 5.22 gives the transmission power in the off-state, if the gate voltage can be safely centered between the breakdown voltage $V_{BD}$ and the pinch-off voltage $V_{pi}$ of the FET technology.

From Figure 5.22, it can be deduced that there is a potential for multi-10-W transmission for GaN switches up to 100 W, as mentioned by Hangai et al. [152].

## 5.8   MILLIMETER-WAVE AND SUB-MILLIMETER-WAVE INTEGRATED CIRCUITS

Classical RF and microwave circuits based on GaN can be taken at least to W-band frequencies [153] and even beyond [154] to at least 180 GHz due to the strong advancements of device technologies with very low gate lengths of 100 nm and below [155]. The main advantage of GaN-based millimeter-wave devices results from the typical limitation of Si, InP, and GaAs devices with respect to maximum breakdown voltages, which normally amount to a few Volts only [156,157], and thus are limited to less than 500 mW and less in output power at these frequencies.

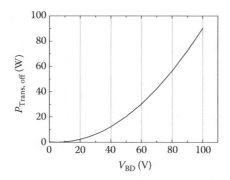

**FIGURE 5.22**    Maximum transmission power, $P_{Trans,off}$, as a function of breakdown voltage VBD.

### 5.8.1  GaN Microwave Monolithically Integrated Circuit Operation at W-Band (75–110 GHz) Frequencies

Watt-level MMIC power operation at W-band frequencies is presented by various authors [153,158,159]. This requires the scaling of the GaN device technologies to gate length of 150 nm and below, which have to be reliable at the same time. This is a device design challenge, as the field-plates used for lower frequencies have to be replaced by T-gates still suitable for high-voltage operation. Initial W-band GaN MMICs in microstrip TML technology operating at 88 GHz with 842 mW output power are given by Micovic et al. [158]. The three-stage GaN MMIC chip yields a periphery of 600-μm-wide FETs in the output stage. The output stage consists of four FETs with a periphery of $4 \times 37.5$ μm each, equivalent to the overall periphery mentioned. The backside of the substrate is thinned to 50 μm in this case. Integrated in a split block module with WR-1O waveguide connectors for power characterization, the module provides the quoted CW-output power with an associated PAE of 14.8%. The linear gain is better than 15 dB with about 10 dB of power gain under compression. Further advancing the art, 92–96 GHz GaN PAs with increasing gate peripheries to 1200 μm are reported by Micovic et al. [159]. The eight-way PA at 95 GHz provides a peak output power $P_{out}$ of 2.1 W with an associated PAE of 19% when operated at $V_{DS}$ = 14 V. This MMIC provides an output power $P_{out}$ of more than 1.5 W over the frequency range from 92 to 96 GHz with an associated PAE over 17.8%. The power is proven to scale linearly from a gate width of 150 to 1200 μm.

Based on T-gate GaN HEMT process with a gate length of 150 nm, further microstrip MMICs have been reported [152]. Again, the substrate thickness is 50 μm. The single-power cells provide an output power exceeding more than 300 mW and a peak PAE of 37%. The four-stage MMICs are composed of 15 FETs with an overall periphery of 2.4 mm with a 2:1 driving ratio at each stage. The power cells thus yield a periphery of 0.16 mm. The resulting chip size is $2.5 \times 1.6$ mm². The devices are operated at $V_{DS}$ = 15 V at a drain current density of 250 mA/mm. A small-signal peak gain of 15 dB is reported at 91 GHz. Eight devices provide an output power of more than 1.7 W based on power combining of two identical devices with Wilkinson combiners.

GaN on silicon MMICs for W-band operation in coplanar TML technology have been reported by Yoshida et al. [160]. The three-stage PA exhibit an output of over 12 dBm with 5-dB gain in the frequency range of 75–81 GHz.

## 5.8.2   E-Band MMIC Operation

GaN MMIC examples at E-band frequencies (60–90 GHz) for high-capacity data transmission point-to-point links are given by Micovic et al. [161]. These reports provide three-stage GaN MMIC PAs that produce 500 mW of saturated output power in CW-mode and have >12 dB of power gain. Two 150-μm-wide devices on the output stage were used, thus achieving a power density of up to 2.1 W/mm at this frequency. In the comparison given by Micovic et al. [161], the output power density of FETs in similar GaAs PHEMT MMIC is 0.28 W/mm for the same frequency for a similar output periphery of 600 μm. This manifests an improvement of nearly a factor of eight in power density, which also results in a similar improvement in the absolute output power.

Similar frequencies are addressed by Raytheon [162]. E-band PA MMICs have been designed and fabricated that demonstrate output powers of 3 W within the 71–76 GHz band and 2 W in the 81–86 GHz band, again, in MSL technology. The gain amounts to 15 dB for a 3-dB bandwidth of 7.5 GHz. For maximum efficiency biasing the MMIC produce levels of the PAEs exceeding 25% with output powers exceeding 1 W. For the maximum power of 3 W, the devices yield PAE levels around 14%. Earlier examples for GaN on SiC MMICs showed 1.3 W of output power at 75 GHz [163]. It is based on four power cells at the output with a gate width of $4 \times 50$ μm, which is equivalent to 1.6 W/mm of output power density. The efficiency is 7% for a common source three-stage MMIC with a remarkably high operation voltage of 35 V and 480 mA of quiescent current equivalent to 17 W of DC-power. The MMIC is realized in GCPW technology. The chip size is $3.0 \times 2.4$ mm$^2$.

## 5.8.3   Forms of Millimeter-Wave Power Combining

The W-band frequencies can also be addressed with distributed forms of power combining using GaN MMICs. Using an on-chip traveling-wave power combiner circuit, the W-band MMIC [164] achieves power levels of greater than 1 W in CW-operation for the bandwidth from 80 GHz to 100 GHz. The peak power of this MMIC is 2 W CW at 84 GHz with a small-signal gain of 8 dB from 75 GHz to 102 GHz. In pulsed operation, even power levels of 3.2 W at 86 GHz and 10% duty cycle are achievable. The power cell at the output is $4 \times 37$ μm.

Spatial power combining of such high-performing MMICs enables even higher power levels [165]. An output power of 5.2 W was achieved by combining 12 GaN MMICs with a rated output power of 400 mW operated at $V_{DS} = 12$ V in a low-loss radial-line combiner network. Output power levels of 5.2 W are reached at 95 GHz with 7.4% PAE and power greater than 3 W over the bandwidth from 94 GHz to 98.5 GHz. The 12-way combiner demonstrates an overall combining efficiency of 87.5%. The

12-way device is driven by a four-way driver derived from the same MMICs, which is power combined in a planar fashion. The overall gain amounts to a maximum of 29 dB.

### 5.8.4 Discussion of Millimeter-Wave Operation

PAs require specific matching conditions for the input and output to maximize gain, output power, and PAE. Figure 5.23 gives the image of a single-stage common source PA with a power cell of $4 \times 45$ μm.

GaN is considered a high-impedance material at the output for power cells of, for example, $4 \times 45$ μm, which is equivalent to a real impedance of 100 Ω for $V_{DS} = 12$ V and $I_d$, max = 1.3 A/mm at low frequencies. For any millimeter-wave operation, however, this relatively high impedance is transformed to very low values based on their parasitics through the transformation. An overview of the impedances for the various bias settings and device sizes for a millimeter-wave technology can, for example, be viewed in the work of Schwantuschke et al. [166].

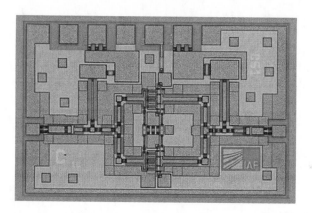

**FIGURE 5.23**   Chip image of a single-stage V-band power amplifier MMIC in GCPW-technology, chip size $1.5 \times 1$ mm².

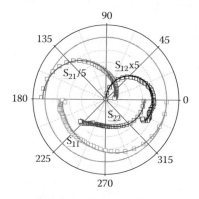

**FIGURE 5.24**   S-parameters si j from near-DC to 100 GHz of the power cell $4 \times 45$ μm as used for the design in the single-stage GaN MMIC taken at $V_{DS} = 10$ V.

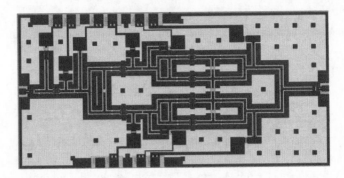

**FIGURE 5.25** Image of a multistage E-band high-power amplifier 71–76 GHz in grounded coplanar waveguide technology with a substrate thickness of 75 μm, chip size 3.75 × 2 mm². (From Schwantuschke, D. et al., 2014 IEEE MTT-S International Microwave Symposium Digest (IMS), Tampa, FL, pp 3, 2014 [167].)

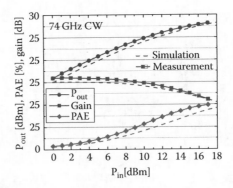

**FIGURE 5.26** Output power, gain, and power-added efficiency versus input power in the multistage E-band high-power amplifier at 74 GHz. (From Schwantuschke, D. et al., Q- and E-band Amplifier MMICs for Satellite Communication, 2014 IEEE MTT-S International Microwave Symposium Digest (IMS), Tampa, FL, pp 3, 2014 [167].)

As an example, Figure 5.24 gives the S-parameters to 100 GHz of a power cell of size 4 × 45 μm. We see that the transformation is significant for $S_{11}$ and $S_{22}$ from DC to 100 GHz even for optimized GCPW power cell layouts. Around 100 GHz, the input impedance reaches nearly ohmic values with very low-impedance levels.

For higher power operation, Figure 5.25 gives the image of a dual-stage PA for E-band frequencies [167], again in GCPW technology). The power cells are based in common-source devices with a current-gain cut-off frequencies of 80 GHz. For the backside, the SiC is thinned to 75 μm (3-mil) and 30 × 30-μm² through-wafer via holes are applied. This forms GCPW technology with a ground-to-ground spacing of 50 μm to suppress undesired substrate modes when packaging these devices.

Figure 5.26 gives power measurements of the E-band common source amplifier. Up to 28 dBm of output power is reached at 74 GHz in the three-stage design with 15 dB linear gain.

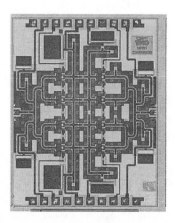

**FIGURE 5.27** Chip image of a dual-stage W-band power amplifier microwave monolithically integrated circuit, size 2.5 × 2 mm². (From van Heijningen, M. et al., W-band Power Amplifier MMIC with 400 mW Output Power in 0.1 μm AlGaN/GaN Technology, Proceedings of the 7th European Microwave Integrated Circuits Conference (EuMIC), Amsterdam, the Netherlands, pp. 135–138, 2012 [168].)

For even higher frequencies and then based on dual-gate devices, Figure 5.27 gives the image of a dual-stage dual-gate PA for W-band operation [168].

The typical cell size of the dual-gate power cell is again 4 × 45 μm in this case for operation at 88–94 GHz. The dual-gate devices provide an even lower output impedance due to the transformation of the high-intrinsic impedance through the parasitics [156]. As explained by van Heijningen et al. [168], the optimum load determined by simulation is (−0.66 + j 0.57) around 90 GHz for a dual-gate device. The achievable matching to the optimum load across the 10% bandwidth is around 10 dB. The loss of the matching network is around 1.4 dB for four power cells in parallel. The linear gain around 90 GHz is 15 dB. The saturated power density of the cascodes at an operation bias of 20 V is 540 mW/mm at 94 GHz. It has been reported by many authors that, for GaN, the classical 2:1 driver ratio may be changed to 1:1 for the interstage matching as the loss can be reduced due to that ratio. Appropriate waveform design is used in this case for the first stage to reduce the power consumption of the driver stage for the ratio of 1:1.

## 5.9 GaN MICROWAVE MONOLITHICALLY INTEGRATED CIRCUIT OSCILLATORS

GaN FETs provide very powerful high-power amplifiers. However, this feature can also be used for power sources [137,163,169]. It has to be noted that, in general, FETs provide a higher phase-noise than bipolar devices. Thus, the phase noise performance required in applications such as radar or communication systems is usually inferior for free-running high-electron mobility transistor (HEMT)-based oscillators, which are known for their moderate phase noise performance for any other material system. However, especially a scaled GaN HEMT technology offers very robust oscillators with high output-power density [169].

### 5.9.1  State-of-the-Art

Initial reports of GaN oscillator MMICs are found in the literature [137] with very high-output power levels of 35 dBm at X-band frequencies around 10 GHz. Various demonstrations can be found since [169]. A Q-band low phase noise monolithic AlGaN/GaN HEMT voltage-controlled oscillator (VCO) is given by Lan et al. [170]. The GaN VCO delivered an output power as high as +25 dBm with a phase noise level of 92 dBc/Hz at 100 kHz offset from the carrier, and 120 dBc/Hz at 1 MHz offset. At millimeter-wave frequencies a V-band (50–75 GHz) monolithic AlGaN/GaN VCO is reported [171]. The VCO delivered an output power of +11 dBm at 53 GHz with an estimated phase noise of 97 dBc/Hz at 1 MHz offset. No buffer amplifier was used. A 67-GHz GaN VCO MMIC with high output power is reported [172]. The oscillation frequency of the fixed-frequency oscillator is 65.6 GHz, while the VCO can be tuned from 65.6 to 68.8 GHz. This leads to a relative bandwidth of 5%. The measured phase noise of the VCO is −83 dBc/Hz at 1 MHz offset. Again, no buffer amplifier is used; however, the oscillator achieves output power levels of almost 20 dBm. A similar 92-GHz GaN HEMT VCO MMIC is given by Weber et al. [173]. The VCO can be tuned between 85.6 and 92.7 GHz, which is a relative tuning bandwidth of 8%. The achieved maximum output power is 10.6 dBm. The phase noise of the VCO varies from −80.2 to −90.2 dBc/Hz at 1 MHz offset from the carrier over the full-tuning voltage range. Very high output power levels of 85 mW have been reported at 70–75 GHz. The MMIC uses a buffer amplifier to achieve this power level. The measured phase noise was only −65.2 dBc/Hz at a 1-MHz offset. The DC power consumption was 1.47 W, which is quite substantial. GaN oscillators can thus be used to obtain high-output power levels with high power densities and high phase noise [137,169]. Figure 5.28 gives the chip image of the W-band oscillator [173].

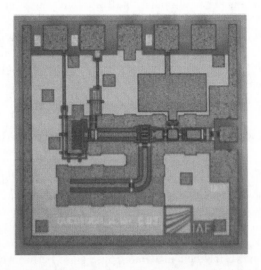

**FIGURE 5.28** Chip image of the W-band oscillator, chip size 1 × 1 mm². (From Weber et al., IEEE MTT-S International Microwave Symposium, Tampa, Florida, pp 4, 2014 [173].)

The chip size is $1 \times 1$ mm$^2$ with a lot of empty area. No buffer amplifier is used; instead, an attenuator is used to reduce the impact of the non-optimal loading. The phase noise level at 1-MHz off-set amounts to −90 dBc/Hz.

## 5.10  OTHER FUNCTIONS REALIZED WITH GaN MICROWAVE MONOLITHICALLY INTEGRATED CIRCUITS

Refering to the diagram in Figure 5.1, further functions are required. This is especially true for phased-array applications where both magnitude and phase are to be controlled in an array [174]. GaN provides the potential of extremely linear operation even for high power levels. In the study of Diebold et al. [175], a GaN millimeter-wave VGA MMICs are given. They have a center frequency of 94 GHz. Phase variation techniques are applied, which yield a phase variation over dynamic range from −17 dBm to 7 dBm with a phase variation of only 12.6°. The TMLs connecting the common-source (CS) and common-gate (CG) stage in a cascode device have a strong influence on the phase stability of the VGA. The gain-variation per phase-variation value is

$$\text{GPV} = \frac{\Delta G}{\Delta \text{phase}} \tag{5.20}$$

achieved with GaN ranges from 1.29 to 1.92 dB/deg. This corresponds to a gain tuning range from 16.7 to 30.3 dB with an associated parasitic phase variation ranging from 10.8 to 23.5 deg. For the related phase, control phase shifters have been realized using GaN MMICs [176]. The design on GaN was performed with the particular emphasis to increase the bandwidth. Further, the power-handling capability of the phase shifter circuits based on a 0.5-μm process was shown to be very high. The power handling was well beyond 10 W, and the TOI point was determined to be near 50 dBm in the frequency range 7 GHz–13 GHz. The phase variation is only about 1 degree over this band. This shows the enormous potential of GaN MMIC phase shifters for phased-arrays.

## 5.11  GaN MIXED-SIGNAL INTEGRATED CIRCUITS AND INTEGRATED DIGITAL-POWER AMPLIFIERS

### 5.11.1  MIXED-SIGNAL CIRCUITS

Apart from purely analogue designs, switch-mode and mixed-signal circuits can also be of particular interest for GaN ICs based on the robustness of the switches. GaN further provides very high currents and high voltages for high-power switch-mode operation. This combination of high-current levels of ≥1 A/mm and high break-down voltages cannot be achieved by GaAs, so that there is typically a reserve in breakdown, which can be used. Mixed-signal operation provides a combination of switching speed [177] and low-power consumption [178]. GaN transistors with an exceptional Johnson Figure of Merit (JFOM) defined as cutoff frequency $f_T \times V_{break}$ products of more than 5 THz Volt have been achieved, which is a substantial achievement over other technologies such as silicon and GaAs [178]. Small break-down voltage of other devices may restrict the dynamic range of the circuit and may represent a severe limitation [177].

So far, GaN was considered a technology with a few transistors only. However, in the study by Albrecht et al. [178], direct-coupled FET logic inverters and 501-stage ring oscillators are fabricated using highly-scaled GaN HEMTs with gate lengths of 20 nm and 40 nm [179]. The circuits yield up to 1006 FETs in the same circuit. This approach yields the ability to trade the enormous robustness of GaN FETs for speed and vice versa for switching power. This E/D-logic (enhancement/depletion) capabil-ity has previously been unavailable in nitride semiconductors and can now be used for advanced mixed-signal functionality to simplify circuit design [178].

### 5.11.2  DIGITAL-POWER AMPLIFIER

GaN can further be used on the microwave high-power side. In this case, the AlGaN/GaN-HEMTs are driven in a highly-efficient digital switch-mode operation. As an example, Figure 5.29 depicts a typical differential driver core chip MMIC with the functions assigned, which can be used either in class-D or further down in class-S operation.

### 5.11.3  CHIP EXAMPLE

The IC in Figure 5.29 consists of a circuit in differential mode. The two input signals are inserted in differential mode. The actual circuits consist of an inverter each, which can be driven in a current mode or a voltage mode. Figure 5.30 gives the concept of both operation modes. The current-mode (CM-class-D) requires a simple differential input drive. Two choke inductances are used to supply the transistors, and the load voltages are linear in the supply voltage. The chokes isolate the DC from the RF-current. ZVS will occur in this configuration as there is no voltage across the transistor when switching the current. The output capacitance becomes part of the output becomes, and the related loss problem is minimized. The output will be bal-anced, for example, by using a balun [180]. The voltage mode (VM-class-D) requires level shifters for the input for GaN. As compared to the CMCD, the VM creates switching loss due to the additional parasitic $C_{out}$ of the transistor, which is charged for every circle.

In Figure 5.29, the output-stage consists of two GaN 1.2-mm devices, which have the capability to be driven to a maximum current of 1.2 A each. The FET technol-ogy used yields a current-gain cutoff frequency of 30 GHz for the gate length of 0.25 μm while maintaining a breakdown voltage of $\geq$100 V. No other technology than GaN can provide the combination of current drive, breakdown voltage, and switch-ing speed. The actual operation mode between class-D and class-S is determined exclusively by the external circuitry, that is, the filtering, and by the input signal format. A square-wave signal will lead to a class-D operation. As an example, for the 2-GHz operation a bitrate of 4 Gbit/s is required. Class-S operation will be achieved if the input signal is bandpass $\Delta$–$\Sigma$ modulated. For this mode, an oversampling of 4:1 is required, that is, for 2 GHz operation a bitrate of 8 Gbit/s is required.

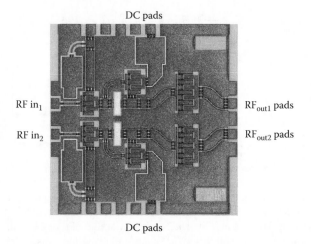

**FIGURE 5.29** Chip image of a GaN class-D integrated circuit, chip size $1.5 \times 1.75$ mm$^2$. (From Maroldt, S. et al., An Integrated 12 Gbps Switch-mode Driver MMIC with 5 V(PP) for Digital Transmitters in 100 nm GaN Technology, Proceedings of the 7th European Microwave Integrated Circuits Conference (EuMIC), Amsterdam, the Netherlands, pp. 115–118, 2012 [181].)

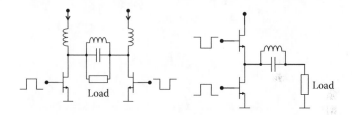

**FIGURE 5.30** Current-mode (CM), left, and voltage-mode (VM), right, of the digital amplifier.

Figure 5.31 gives the spectra for both modes in the frequency domain in a mutual comparison. The spectrum of the CMCD at 1.3 GHz only yields the odd harmonics, the even harmonics are suppressed. The same is true for the 2.6-GHz square wave operation on the right. In the band pass delta sigma (BPDS) signal, all the harmonics appear and the spectrum is complex with significant noise contributions. The time-domain signal is given in Figure 5.32 for comparison. The good switching is proven by the open-eye diagram with unprecedented voltage swings of 6 V/division at this switching speed.

### 5.11.4  STATE-OF-THE-ART

Advanced and digital switch-mode RF-amplifiers leading to digital class-D and the class-S amplifier concepts gained increasing interest in recent years due to their characteristics to cope with flexible and multi-band and multistandard

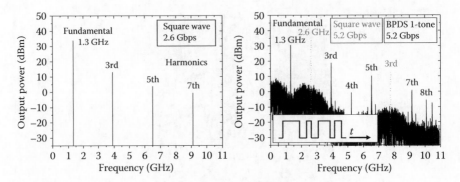

FIGURE 5.31   Spectra (left): current-mode class-D at 1.3GHz (square wave) and (right) bandpass D–S modulated (BPDS) at 1.3 GHz with the harmonic and 2.6 GHz square-wave operation for comparison. Courtesy Dr. Stephan Maroldt. (From Maroldt, S, Advanced Microwave Switch-Mode Amplifiers and Demonstrations Using GaN, Proceedings of WS European Microwave Conference, 26 September 2010, Paris, France [182]).

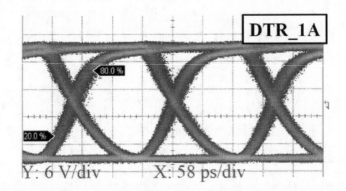

FIGURE 5.32   Time domain representation of the digital-PA mode. (From Maroldt, S. et al., High Efficiency Digital GaN MMIC Power Amplifiers for Future Switch-Mode Based Mobile Communication Systems, Compound Semiconductor Integrated Circuit Symposium (CISC), Greensboro, NC, pp. 1–4, 2009 [38].)

transceivers [186]. The realization of the class-S concept including the complex filtering aspect is given in the study of Leberer et al. [180]. The class-S amplifier requires a broadband digital signal to be an amplifier with an oversampling of at least 4:1 of the equivalent analogue center frequency. For example, a center frequency of 450 MHz relates to a digital bandwidth of 1.8 Gbit/s. Examples of GaN core chips for high frequencies for advanced class-S MMICs operation are presented in the study of Maroldt et al. [38]. They are based on gate lengths of 0.5 μm, and, especially, 0.25 μm to enable fast switching for typical microwave frequencies of 1 and 2 GHz, respectively.

## 5.11.5 DISCUSSION OF GaN DRIVER CAPABILITIES

The integration as GaN switch-mode IC is required as the broadband signals are transmitted with sampling rates of at least 4 Gbps for 2 GHz while maintaining high-power drive. However, GaN switch-mode driver amplifiers further require the input drive of this final switch-mode stage to achieve the right gain and output power to achieve power levels, for example, beyond 43 dBm. As silicon devices typically do not allow high-voltage swing of up to 3.5 V [181], additional consideration is needed for the input drive and has been addressed. For high-speed operation, an integrated 12-Gbps switch-mode driver MMIC with a peak-to-peak voltage of 5 V for digital transmitters is given in the study by Maroldt et al. [183]. A gate length of 100 nm is used and found suitable to overcome bandwidth issues and to allow operation up to 12 Gbps, which renders the BPDS-related frequency to 3 GHz.

To the same end, an active integrated digital switch-mode driver (DSMD) circuit is integrated with a GaN power transistor, instead of using Si-based-drivers. The GaN concept is also scalable in power for highly efficient switch-mode power amplifier (SMPA) applications as given in the study by Maroldt et al. [183]. An IC implementation of the various circuits avoids the use of bond wires with their filtering effect, which impairs the broadband operation. The GaN power transistor with integrated digital switch-mode driver layout yields a (digital) PAE of 76%, a digital DE beyond 80%, and output power levels of 6 W, which proofs the usefulness of the concept. A gain up to 35 dB and an operating frequency range from DC to 3 GHz were achieved with this driver concept.

## 5.11.6 DIGITAL FINAL AMPLIFIER STAGES

For the final-stage, output power levels of more than 20 W are required and have been demonstrated in the study by Maroldt et al. [36] for digital operation. The output power is in a strong trade-off with the switching speed, as both input as well as output capacitances. As they cannot be compensated in broadband operation, they have to be charged with each cycle. As shown below, GaN with very small gate lengths holds the promise to enable real highly-efficient switching at microwave frequencies.

The three-stage digital MMIC design using 0.25-µm-long gates can be flexibly used at any frequency operating in the ultra-high-frequency (UHF) band up to a maximum digital bit rate of 3 Gbps, equal to a square-wave frequency of 1.5 GHz. At 0.9 Gbps, a maximum broadband output power of 20.5 W was measured at a drain efficiency of 76%, and a PAE of 70%. The efficiencies are the switch-mode efficiencies, which require further consideration with respect to the classical analogue PAE values in the full transceiver chain. For a digital input signal with a peak-to-peak amplitude of 3.5 V, a large-signal gain of 25.3 dB was achieved in this broadband concept. Figure 5.33 gives the chip image of this final-stage MMIC.

Further, high-speed switching operation at 2 GHz was demonstrated in the work of Maroldt et al. [184]. Digital switch-mode MMIC PAs are reported for bit rates

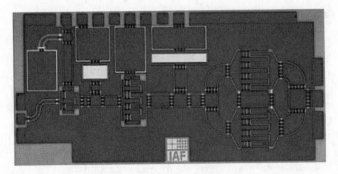

**FIGURE 5.33**   Chip image of a digital class-S final stage integrated circuit for UHF, chip size 1.5 × 3 mm². (From Maroldt, S. et al., Broadband GaN-Based Switch-Mode Core MMICs with 20 W Output Power Operating at UHF, Proceedings of Compound Semiconductor Integrated Circuit Symposium (CSIC), Waikoloa, HI, pp. 1–4, 2011 [36].)

related to mobile communication radio frequencies of up to 8 Gbps. This is performed using GaN FETs with a gate length of 0.15 µm gate length without field-plates. The two-stage MMIC PAs were compared for two different driver concepts, where the improved driver-based MMIC achieved a drain efficiency of 70% at an output power of 4.4 W for a 3.6 Gbit/s BPDS signal equivalent to a 900-MHz microwave frequency. This is used in the Class-S amplifier concepts and is thus suitable for the operation.

The efficiency of the voltage-mode concept is increased as given by Wentzel et al. [185] by additional envelope δ-Σ-modulation in combination with a voltage-mode class-S PA in back-off operation. The additional envelope enhances the linear efficiency over a 10-dB back-off range. A flexible module for future operation is presented by Wentzel et al. [186] with an increased digital content. A compact high-gain broadband voltage-mode PA reaches a large-signal gain of up to 40 dB. A single-chip H-bridge PA module for the 800-MHz band is reported, as well as a digital Doherty PA deduced from the voltage-mode MMIC. Again, this demonstrates the potential of GaN for high-power digital switching applications.

## ACKNOWLEDGMENTS

I would like to acknowledge the valuable contributions of Dr. Jutta Kühn, Dr. Stephan Maroldt, Dr. Markus Mußer, Dirk Schwantuschke, Rainer Weber, and Dr. Friedbert van Raay from the Microelectronics department at Fraunhofer IAF and thank them for the very good cooperation on the topic of MMIC design. Further, I would like to thank Dr. Michael Mikulla, Dr. Peter Brückner, Dr. Wolfgang Bronner, Dr. Patrick Waltereit, Dr. Michael Schlechtweg, Stefan Müller, and Professor Dr. Oliver Ambacher for their longstanding contributions to this work. The continuous support of colleagues the Fraunhofer IAF departments microelectronics, epitaxy, and process technology is greatfully acknowledged.

## REFERENCES

1. R. Quay, *Gallium Nitride Electronics*, Springer: Berlin, Heidelberg, 2008.
2. J.A. Archer, H.P. Weidlich, E. Pettenpaul, F.A. Petz, J. Huber, A GaAs monolithic low-noise broad-band amplifier, *IEEE Journal Solid-State Circuits*, 1981, vol. 16, no. 6, pp. 648–652.
3. P. Waltereit, S. Müller, L. Kirste, M. Prescher, S. Storm, A. Weber, B. Schauwecker, M. Hosch, and J. Splettstösser, Development of an Epitaxial Growth Process on European SiC Substrates for a Low Leakage GaN HEMT Technology with Power added Efficiencies around 65%, Proceedings of CS Mantech, 2013, New Orleans, LA, pp. 121–123.
4. R. Isaak , J. Diaz, M. Gerlach, J. Hulse, L. Schlesinger, P. Seekell, W. Zhu, W. Kopp, X. Yang, A. Stewart, K. Chu, P. C. Chao, X. Gao, M. Pan, D. Gorka, M. Oliver, and I. Eliashevich, The First 0.2 µm 6-Inch GaN-on-SiC MMIC Process, Proceedings of CS Mantech, 2014, Denver, CO, pp. 229–231.
5. M. Nishijima, T. Murata, Y. Hirose, M. Hikita, N. Negoro, H. Sakai, Y. Uemoto, K. Inoue, T. Tanaka, D. Ueda, A K-band AlGaN/GaN HFET MMIC Amplifier on Sapphire using Novel Superlattice Cap Layer, Proceedings of International Microwave Symposium Digest (IMS), 2005, Long Beach, CA, pp. 299–301.
6. B. Geller, A. Hanson, A. Chaudhari, A. Edwards, I.C. Kizilyalli, A Broadband Low Cost GaN-on-silicon MMIC Amplifier, Proceedings of IEEE Radio Frequency Integrated Circuits Symposium (RFIC), 2008, Atlanta, GA, pp. 527–530.
7. D. M. Fanning, L. C. Witkowski, C. Lee, D. C. Dumka, H. Q. Tserng, P. Saunier, W. Gaiewski, E. L. Piner, K. J. Linthicum, and J. W. Johnson, 25 W X-band GaN on Si MMIC, Proceedings of GaAs Mantech, 2005, Miami, FL, p. 8.3.
8. K.K. Chu, P.C. Chao, M.T. Pizzella, R. Actis, D.E. Meharry, K.B. Nichols, R.P. Vaudo, X. Xu, J.S. Flynn, J. Dion, G.R. Brandes, 9.4-W/mm power density AlGaN-GaN HEMTs on free-standing GaN substrates, *IEEE Microwave and Wireless Components Letters*, 2004, vol. 25, no. 9, pp. 596–598.
9. T. Paskova, D.A. Hanser, K.R. Evans, GaN substrates for III-nitride devices, *Proceedings of the IEEE*, 2010, vol. 98, no. 7, pp. 1324–1326.
10. B. Jacobs, B. van Straaten, P. de Hek, R. van Dijk, F. Karouta, and F. van Vliet, Coplanar Waveguides on AlN for AlGaN/GaN MMIC Applications, 2000. Proceedings of the 3rd Workshop Semiconductor Advances for Future Electronics (SAFE 2000), 2000, Veldhoven, the Netherlands, pp. 75–77.
11. E.J. Wilkinson, An N-way power divider, *IEEE Transactions on Microwave Theory and Techniques*, 1960, vol. 8, no. 1, pp. 116–118.
12. J. Lange, Interdigitated stripline quadrature hybrid, *IEEE Transactions on Microwave Theory and Techniques*, 1969, vol. 17, no. 12, pp. 1150–1151.
13. D.A. Gajewski, S. Sheppard, T. McNulty, J.B. Barner, J. Milligan, and J. Palmour, Reliability of GaN/AlGaN HEMT MMIC Technology on 100-mm 4H-SiC, 26th Annual JEDEC ROCS Workshop, Indian Wells, CA, 2011.
14. P. Waltereit, W. Bronner, R. Quay, M. Dammann, M. Caesar, S. Müller, R. Reiner, P Brückner, R. Kiefer, F. van Raay, J. Kühn, M. Musser, C. Haupt, M. Mikulla, and O. Ambacher, GaN HEMTs and MMICs for space applications, *Semiconductor Science and Technology*, 2013, vol. 28, no. 7, pp. 074010.
15. P. Brueckner, R. Kiefer, C. Haupt, A. Leuther, S. Müller, R. Quay, D. Schwantuschke, M. Mikulla, and O. Ambacher, Development of 100 nm gate AlGaN/GaN HEMT and MMIC technology suitable for mm-wave applications, *Physica Status Solidi*, 2012, vol. C 9, no. 3–4, pp. 903–906.
16. K. Bumjin, M. Greene, and M. Osmus, Broadband High Efficiency GaN Discrete and MMIC Power Amplifiers over 30—2700 MHz Range, IEEE MTT-S International Microwave Symposium (IMS), 2014, Tampa, FL, pp. 1–3.

17. C.F. Campbell, D.C. Dumka, and M.Y. Kao, Design Considerations for GaN Based MMICs, Comcas 2009, Tel Aviv, Israel, pp. 1–8.
18. J. Cheron, M. Campovecchio, R. Quéré, D. Schwantuschke, R. Quay, and O. Ambacher, High-efficiency Power Amplifier MMICs in 100 nm GaN Technology at Ka-band Frequencies, Proceedings of 8th European Microwave Integrated Circuits Conference (EuMIC), 2013, Nuremberg, Germany, pp. 492–495.
19. A.R. Djordjevic, A.G. Zajic, D.V. Tošic, and T. Hoang, A note on the modeling of Transmission-line losses, *IEEE Transactions on Microwave Theory Techniques*, 2003, vol. 51, no. 2, pp. 483–486.
20. S. Rezaei, L. Belostotski, F.M. Ghannouchi, P. Aflaki, Integrated design of a class-J power amplifier, *IEEE Transactions on Microwave Theory and Techniques*, 2013, vol. 61, no. 4, pp. 1639–1648.
21. R. Liu, D. Schreurs, W. De Raedt, F. Vanaverbeke, J. Das, M. Germain, R. Mertens, Integrated AlGaN/GaN HEMTs in MCM-D Technology, Proceedings of Electronic Components and Technology Conference, 2010, Las Vegas, pp. 1562–1567.
22. D. Marti, M. Vetter, A.R. Alt, H. Benedickter, and C.R. Bolognesi, 110 GHz Characterization of Coplanar Waveguides on GaN-on-Si Substrates, Proceedings of CS Mantech, 2011, Palm Springs, CA, pp 4.
23. Hsuan-ling Kao, Bo-Wen Wang, Chih-Sheng Yeh, Cheng-Lin Cho, Bai-Hong Wei, and Hsien-Chin Chiu, High power and low phase noise MMIC VCO using 0.35 μm GaN-on-Si HEMT, *International Journal of Computer Theory and Engineering*, 2013, vol. 5, no. 6, pp. 910–913.
24. U.K. Mishra, L. Shen, T.E. Kazior, and Y.F. Wu, GaN-based RF-power Devices and Amplifiers, Proceedings of the IEEE, 2008, vol. 96, no. 2, pp. 287–305.
25. S.C. Cripps, *Advanced Techniques in RF-Amplifiers Design*, Norwood, MA: Artech House, 2002.
26. P. Colantonio, G. Giannini, E. Limiti, *High Efficiency RF and Microwave Solid State Power Amplifiers*, John Wiley & Sons: Chichester, UK, 2009.
27. K. Inoue, S. Sano, Y. Tateno, F. Yamaki, K. Ebihara, N. Ui, A. Kawano, and H. Deguchi, Development of gallium nitride high electron mobility transistors for cellular base stations, *SEI Technical Review*, 2001, no. 71, pp. 88–93.
28. S. Wood, P. Smith, W. Pribble, R. Pengelly, and J. Crescenzi, High efficiency, high linearity GaN HEMT amplifiers for WiMAX applications, *High Frequency Electronics*, 2006, Vol. 5, No. 5, pp. 22–36.
29. E. Mitani, M. Aojima, A. Maekawa, and S. Sano, An 800-W AlGaN/GaN HEMT for S-band High-Power Application, Proceedings of CS Mantech, 2007, Austin, TX, pp. 213–215.
30. S. Xie, V. Paidi, R. Coffie, S. Keller, S. Heikman, B. Moran, A. Chini, S.P. DenBaars, U. Mishra, S. Long, and M.J.W. Rodwell, High-linearity class B power amplifiers in GaN HEMT Technology, *IEEE Microwave and Wireless Components Letters*, 2003, vol. 13, no. 7, pp. 284–286.
31. M. Litchfield, S. Schafer, T. Reveyrand, Z. Popovic, High-Efficiency X-Band MMIC GaN Power Amplifiers Operating as Rectifiers, Proceedings of International Microwave Symposium Digest (IMS), 2014, Tampa, FL, pp. 1–3.
32. W.H. Doherty, A New High Efficiency Amplifier for Modulated Waves, Proceeding IRE, vol. 24, pp. 1163–1182.
33. N. Ui, H. Sano, and S. Sano, A 80 W 2-stage GaN HEMT Doherty Amplifier with 50 dBc ACLR, 42% Efficiency 32 dB Gain with DPD for W-CDMA Base Station, Proceedings of International Microwave Symposium Digest (IMS), 2007, Honolulu, HI, pp. 1259–1262.
34. W.C.E. Neo, J. Qureshi, M.J. Pelk, J.R. Gajadharsing, L.C.N. de Vreede, A mixed-signal approach towards linear and efficient N-way Doherty amplifiers, *IEEE Transactions on Microwave Theory and Techniques*, 2007, vol. 55, no. 5, pp. 866–879.

35. S. Lin, A.E. Fathy, A 20 W GaN HEMT VHF/UHF Class-D Power Amplifier, IEEE 12th Annual Wireless and Microwave Technology Conference (WAMICON), 2011, Clearwater Beach, FL, pp. 1–4.

36. S. Maroldt, R. Quay, C. Haupt, and O. Ambacher, Broadband GaN-Based Switch-Mode Core MMICs with 20 W Output Power Operating at UHF, Proceedings of Compound Semiconductor Integrated Circuit Symposium (CSIC), 2011, Waikoloa, HI, pp. 1–4.

37. A. Wentzel, W. Heinrich, A GaN Voltage-mode Class-D MMIC with Improved OverallEfficiency for Future RRH Applications, Proceedings of European Microwave Conference (EUMW), 2013, Nuremberg, pp. 549–552.

38. S. Maroldt, C. Haupt, R. Kiefer, W. Bronner, S. Müller, W. Benz, R. Quay, O. Ambacher, High Efficiency Digital GaN MMIC Power Amplifiers for Future Switch-Mode Based Mobile Communication Systems, Compound Semiconductor Integrated Circuit Symposium (CISC), 2009, Greensboro, NC, pp. 1–4.

39. N.O. Sokal, A.D. Sokal, Class-E: A new class of high-efficiency tuned single-ended power amplifiers, *IEEE J. Solid-State Circuits*, 1975, vol. 10, no. 3, pp. 168–176.

40. N.O. Sokal, Class-E High-efficiency Power Amplifiers, from HF to Microwave, Proceedings of IEEE MTT-S International Microwave Symposium (IMS), 1998, Baltimore, MD, pp. 1109–1112.

41. N.O. Sokal, Class-E Switching-mode High-efficiency Tuned RF/Microwave Power Amplifier: Improved Design Equations Proceedings of IEEE MTT-S International Microwave Symposium (IMS), 2000, Boston, MA, pp. 779–782.

42. S. Jee, J. Moon, J. Kim, J. Son, and B. Kim, Switching behavior of class-E power amplifier and its operation above maximum frequency, *IEEE Transactions on Microwave Theory and Techniques*, vol. 60, no. 1, pp. 89–98, 2012.

43. V. Zomorrodian, U.K. Mishra, and R.A. York, A High-efficiency Class F MMIC Power Amplifier at 4.0 GHz using AlGaN/GaN HEMT Technology, Proceedings of Compound Semiconductor Integrated Circuit Symposium (CSIC), 2012, La Jolla, CA, pp. 1–4.

44. S. Gao, H. Xu, S. Heikman, U. Mishra, and R.A. York, Microwave Class-E GaN Power Amplifiers, Proceedings of Asia-Pacific Microwave Conference, 2005, Suzhou, Shanghai, pp. 4.

45. D.A. Calvillo-Cortes, Leo C.N. de Vreede, and M. de Langen, A Compact and Power-Scalable 70 W GaN Class-E Power Amplifier Operating from 1.7 to 2.6 GHz, Proceedings of Asia-Pacific Microwave Conference, 2011, Melbourne, Australia, pp. 1546–1549.

46. W.L. Pribble, J.M. Milligan, and R.S. Pengelly, High Efficiency Class-E Amplifier Utilizing GaN HEMT Technology, 2006 IEEE Radio and Wireless Symposium, 2006, San Diego, CA, pp. 3.

47. M.P. van der Heijden, M. Acar, J.S. Vromans, and D.A. Calvillo-Cortes, A 19 W High-Efficiency Wide-Band CMOS-GaN Class-E Chireix RF Outphasing Power Amplifier, Proceedings of IEEE MTT-S International Microwave Symposium (IMS), 2011, Baltimore, MD, pp. 1–4.

48. D.A. Calvillo-Cortes, M.P. van der Heijden, Mustafa Acar, M. de Langen, R. Wesson, F. van Rijs, and L.C.N. de Vreede, A package-integrated Chireix outphasing RF switch-mode high-power amplifier, *IEEE Transactions on Microwave Theory and Techniques*, 2013, vol. 61, no. 10, pp. 3721–3732.

49. H. Chireix, High Power Outphasing Modulation, Proceedings of IRE, 1935, vol. 23, no. 11, pp. 1370–1392.

50. J.H. Qureshi, M.J. Pelk, M. Marchetti, W.C.E. Neo, J.R. Gajadharsing, M.P. van der Heijden, L.C.N. de Vreede, A 90-W peak power GaN outphasing amplifier with optimum input signal conditioning, *IEEE Transactions on Microwave Theory and Techniques*, 2009, vol. 57, no. 8, pp. 1925–1935.

51. J.S. Moon, H. Moyer, P. Macdonald, D. Wong, M. Antcliffe, M. Hu, P. Willadsen, P. Hashimoto, C. McGuire, M. Micovic, M. Wetzel, and D. Chow, High Efficiency X-band Class-E GaN MMIC High-Power Amplifiers, IEEE Topical Conference on Power Amplifiers for Wireless and Radio Applications (PAWR), 2012, Santa Clara, CA, pp. 9–12.

52. F. H. Raab, Class-E, Class-C, and Class-F Power amplifiers based upon a finite number harmonics, *IEEE Transactions on Microwave Theory and Techniques*, 2001, vol. 49, no. 8, pp. 1462–1468.

53. D. Schmelzer and S.Long, A GaN HEMT class F Amplifier at 2 GHz with 80% PAE, Proceedings of Compound Semiconductor Integrated Circuit Symposium (CSIC), 2006, San Antonio, TX, pp. 96–99.

54. P. Saad, C. Fager, H. M. Nemati, H. Cao, H. Zirath and K. Andersson, A highly efficient 3.5 GHz inverse class-F GaN HEMT power amplifier, *International Journal of Microwave and Wireless Technologies*, 2010, vol. 2, no. 3–4, pp. 317–324.

55. A. Chin, C. Clark, Class-F GaN Power Amplifiers for CubeSat Communication Links, 2012 IEEE Aerospace Conference, Big Sky, MT, pp. 1–6.

56. P. Wright , J. Lees, J. Benedikt, P.J. Tasker, and S.C. Cripps, A methodology for realizing high efficiency class-J in a linear and broadband PA, *IEEE Transactions on Microwave Theory and Techniques*, 2009, vol. 57, no. 12, pp. 3196–3204.

57. S.C. Cripps, P. J. Tasker, A. L. Clarke, J. Lees, J. Benedikt, On the continuity of high efficiency modes in linear RF power amplifiers, *IEEE Microwave and Wireless Components Letters*, 2009, vol. 19, no. 10, pp. 665–667.

58. V. Carrubba, A.L. Clarke, M. Akmal, J. Lees, J. Benedikt, P.J. Tasker, and S.C. Cripps, The Continuous Class-F Mode Power Amplifier, Proceedings of European Microwave Conf. (EUMW), 2010, Paris, France, pp. 1674–1677.

59. T. Canning, P.J. Tasker, and S.C. Cripps, Continuous mode power amplifier design using harmonic clipping contours: Theory and practice, *IEEE Transactions on Microwave Theory and Techniques*, 2014, vol. 62, no. 1, pp. 100–110.

60. S. Rezaei, L. Belostotski, M. Helaoui, F.M. Ghannouchi, Harmonically tuned continuous class-C operation mode for power amplifier applications, *IEEE Transactions on Microwave Theory and Techniques*, 2014, vol. 62, no. 12, pp. 3017–3027.

61. J. R. Powell, M. J. Uren, T. Martin, A. McLachlan, P. Tasker, S. Woodington, J. Bell, R. Saini, J. Benedikt, and S. C. Cripps, GaAs X-Band High Efficiency (>65%) Broadband (>30%) Amplifier MMIC based on the Class B to Class J Continuum, Proceedings of IEEE MTT-S International Microwave Symposium Digest (IMS), 2011, Baltimore, MD, pp. 1–4.

62. W.L. Pribble, Large Signal GaN HEMT Models and their Application to Hybrid and Monolithic Circuit Designs, in IMS Workshop WMG: High Power Device Characterization and Modeling, 2007, Honolulu, HI.

63. S. Chen, E. Reese and T. Nguyen, A Compact 70 Watt Power Amplifier MMIC Utilizing S-band GaN on SiC HEMT Process, Proceedings of Compound Semiconductor Integrated Circuit Symposium (CSIC), 2012, La Jolla, CA, pp. 1–4.

64. C. Florian, R. Cignani, D. Niessen, and A. Santarelli, A C-Band AlGaN-GaN MMIC HPA for SAR, *IEEE Microwave and Wireless Components Letters*, 2012, vol. 22, no. 9, pp. 471–473.

65. T. Yamasaki, Y. Kittaka, H. Minamide, K. Yamauchi, S. Miwa, S. Goto, M. Nakayama, M. Kohno, and N. Yoshida, A 68% Efficiency, C-Band 100 W GaN Power Amplifier for Space Applications, IEEE MTT-S International Microwave Symposium Digest (IMS), 2010, Anaheim, CA, pp. 1384–1387.

66. S. Miwa, Y. Kamo, Y. Kittaka, T. Yamasaki, T. Tsukahara, T. Tanii, M. Kohno, S. Goto, and A. Shima, A 67% PAE, 100 W GaN Power Amplifier with On-Chip Harmonic Tuning Circuits for C-band Space Applications, IEEE MTT-S International Microwave Symposium Digest (IMS), 2011, Baltimore, MD, pp. 1–4.

67. J.C. Jeong, D.P. Jang, B.G. Han, and I.B. Yom, A compact C-band 50-W AlGaN/GaN high-power MMIC amplifier for radar applications, *ETRI Journal*, 2014, vol. 36, no. 3, pp. 498–501.

68. J. Kühn, M. Musser, F. van Raay, R. Kiefer, M. Seelmann-Eggebert, M. Mikulla, R. Quay, T. Rödle, and Oliver Ambacher, Design and realization of GaN RF-devices and circuits from 1 to 30 GHz, *International Journal of Microwave and Wireless Technologies*, 2010, vol. 2, no. 1, pp. 115–120.

69. R.S. Pengelly, S.M. Wood, J.W. Milligan, S.T. Sheppard, and W.L. Pribble, A review of GaN on SiC high electron-mobility power transistors and MMICs, *IEEE Transactions on Microwave Theory and Techniques*, 2012, vol. 60, no. 6, pp. 1764–1783.

70. P. Khan, L. Epp, and A. Silva, A Ka-Band Wideband-Gap Solid-State Power Amplifier: Architecture Identification, IPN Progress Report 42–162, August 15, 2005.

71. F. van Raay, R. Quay, R. Kiefer, W. Fehrenbach, W. Bronner, M. Kuri, F. Benkhelifa, H. Massler, S. Muller, M. Mikulla, M. Schlechtweg, G. Weimann, A Microstrip X-band AlGaN/GaN Power Amplifier MMIC on s.i. SiC Substrate, Proceedings of European Gallium Arsenide and other Compound Semiconductors Application Symposium (GAAS), 2005, Paris, France, pp. 233–236.

72. F. van Raay, F. van, R. Quay, R. Kiefer, F. Benkhelifa, B. Raynor, W. Pletschen, M. Kuri, H. Massler, S. Müller, M. Dammann, M. Mikulla, M. Schlechtweg, G. Weimann, A Coplanar X-band AlGaN/GaN power amplifier MMIC on s.i. SiC substrate, *IEEE Microwave and Wireless Components Letters*, 2005, vol. 15, no. 7, pp. 460–462.

73. F. van Raay, R. Quay, R. Kiefer, W. Bronner, M. Seelmann-Eggebert, M. Schlechtweg, M. Mikulla, G. Weimann, X-band High-power Microstrip AlGaN/GaN HEMT Amplifier MMICs, IEEE MTT-S International Microwave Symposium Digest (IMS), 2006, San Francisco, CA, pp. 1368–1371.

74. P. Schuh, R. Leberer, H. Sledzik, M. Oppermann, B. Adelseck, H. Brugger, R. Behtash, H. Leier, R. Quay, and R. Kiefer, 20 W GaN HPAs for Next Generation X-band T/R-modules, IEEE MTT-S International Microwave Symposium Digest (IMS), 2006, San Francisco, CA, pp. 726–729.

75. S. Piotrowicz, Z. Ouarch, E. Chartier, R. Aubry, G. Callet, D. Floriot, J.C. Jacquet, O. Jardel, E. Morvan, T. Reveyrand, N. Sarazin, and S.L. Delage, 43 W, 52% PAE X-Band AlGaN/GaN HEMTs MMIC Amplifiers, IEEE MTT-S International Microwave Symposium Digest (IMS), 2010, Anaheim, CA, pp. 505–508.

76. M. Ohtomo, Stability analysis and numerical simulation of multidevice amplifiers, *IEEE Transactions on Microwave Theory Techniques*, 1993, vol. 41, no. 6, pp. 983–991.

77. J. Kühn, F. van Raay, R. Quay, R. Kiefer, W. Bronner, M. Seelmann-Eggebert, M. Schlechtweg, M. Mikulla, O. Ambacher, and M. Thumm, Balanced Microstrip AlGaN/GaN HEMT Power Amplifier MMIC for X-Band Applications, Proceedings of European Microwave Integrated Circuit Conference (EUMIC), 2008, Amsterdam, the Netherlands, pp. 95–98.

78. J. Kühn, F. van Raay, R. Quay, R. Kiefer, T. Maier, R. Stibal, M. Mikulla, M. Seelmann-Eggebert, W. Bronner, M. Schlechtweg, O. Ambacher, and M. Thumm, Design of Highly-Efficient GaN X-Band-Power-Amplifier MMICs, IEEE MTT-S International Microwave Symposium Digest (IMS), 2009, Boston, MA, pp. 661–664.

79. J. Kühn, P. Waltereit, F. van Raay, R. Aidam, R. Quay, O. Ambacher, and M. Thumm, Harmonic Termination of AlGaN/GaN/(Al)GaN Single- and Double-heterojunction HEMTs, Proceedings of German Microwave Conf. (GEMIC), Berlin, Germany, 2010, pp. 122–125.

80. T. Canning, P. Tasker, and Steve Cripps, Waveform evidence of gate harmonic short circuit benefits for high efficiency X-band power amplifiers, *IEEE Microwave and Wireless Components Letters*, 2013, vol. 23, no. 8, pp. 439–441.

81. I.P. Smorchkova, M. Wojtowicz, R. Sandhu, R. Tsai, M. Barsky, C. Namba, P.S. Liu, R. Dia, M. Truong, D. Ko, J. Wang, H. Wang, and A. Khan, AlGaN/GaN HEMTs— Operation in the K-band and above, *IEEE Transactions on Microwave Theory and Techniques*, 2003, vol. 51, no. 2, pp. 665–668.

82. C. Friesicke, J. Kühn, P. Brückner, R. Quay, and A. F. Jacob, An Efficient AlGaN/GaN HEMT Power Amplifier MMIC at K-Band, Proceedings of the 7th European Microwave Integrated Circuits Conference (EUMIC), 2012, Amsterdam, the Netherlands, pp. 131–134.

83. C. Friesicke, R. Quay, B. Rohrdantz, and A. F. Jacob, A Linear 4 W Power Amplifier at K-Band Using 250 nm AlGaN/GaN HEMTs, Proceedings of the 8th European Microwave Integrated Circuits Conference (EUMIC), Nuremberg, Germany, pp. 157–160.

84. C.F. Campbell, K. Tran, K. Ming-Yih, and S. Nayak, A K-Band 5 W Doherty Amplifier MMIC Utilizing 0.15 µm GaN on SiC HEMT Technology, IEEE Compound Semiconductor Integrated Circuit Symposium (CSIC), La Jolla, CA, 2012, pp. 1–4.

85. T. Palacios, A. Chakraborty, S. Rajan, C. Poblenz, S. Keller, S.P. DenBaars, J.S. Speck, and U.K. Mishra, High-power AlGaN/GaN HEMTs for Ka-band applications, *IEEE Microwave and Wireless Components Letters*, 2005, vol. 26, no. 11, pp. 781–783.

86. A. M. Darwish, K. Boutros, B. Luo, B. Huebschman, E. Viveiros, and H. A. Hung, 4-Watt Ka-Band AlGaN/GaN Power Amplifier MMIC, Proceedings of IEEE MTT-S International Microwave Symposium (IMS), 2006, San Francisco, CA, pp. 730–733.

87. M. van Heijningen, F.E. van Vliet, R. Quay, F. van Raay, R. Kiefer, S. Müller, D. Krausse, M. Seelmann-Eggebert, M. Mikulla, and M. Schlechtweg, Ka-Band AlGaN/GaN HEMT High Power and Driver Amplifier MMICs, Proceedings of 15th European Gallium Arsenide and other Compound Semiconductors Application Symposium, 2006, Paris, France, pp. 237–240.

88. K. S. Boutros, W. B. Luo, Y. Ma, G. Nagy, and J. Hacker, 5 W GaN MMIC for Millimeter-Wave Applications, IEEE Compound Semiconductor Integrated Circuit Symposium (CSIC), 2006, San Antonio, TX, pp. 93–95.

89. C.F. Campbell, Y. Liu, M.Y. Kao, and S. Nayak, High Efficiency Ka-Band Gallium Nitride Power Amplifier MMICs, Proceedings of IEEE Comcas, 2013, Tel Aviv, Israel, pp 4.

90. S. Din, M. Wojtowicz, M. Siddiqui, High Power and High Efficiency Ka Band Power Amplifier, Proceedings of the IEEE MTT-S International Microwave Symposium (IMS), 2015, Phoenix, AZ, pp 4.

91. J. Cheron, M. Campovecchio, R. Quéré, D. Schwantuschke, R. Quay, and O. Ambacher, High-gain Over 30% PAE Power Amplifier MMICs in 100 nm GaN Technology at Ka-Band Frequencies, Proceedings of the 10th European Microwave Integrated Circuits Conference (EuMIC), 2015, Paris, France, pp 3.

92. J. Lucas, E. Geron, T. Ditchi, and S. Hole, Practical Use of the Kramers-Kronig relation at microwave frequencies. Application to photonic like lines and left handed materials, *Piers Online*, 2011, vol. 7, no. 4, pp. 387–393.

93. K.W. Kobayashi and K. Krishnamurthy, Broadband GaN MMICs: Multi-Octave Bandwidth PAs to Multi-Watt Linear LNAs, Proceedings of Power Amplifier Symposium, 2012, La Jolla, CA, pp. 4.

94. E. Reese, D. Allen, C. Lee, and T. Nguyen, Wideband Power Amplifier MMICs Utilizing GaN on SiC, IEEE MTT-S International Microwave Symposium Digest (IMS), 2010, Anaheim, CA, pp. 1230–1232.

95. P. Schuh, R. Leberer, H. Sledzik, D. Schmidt, M. Oppermann, B. Adelseck, H. Brugger, R. Quay, F. van Raay, M. Seelmann-Eggebert, R. Kiefer, and W. Bronner, Linear Broadband GaN MMICs for Ku-Band Applications, IEEE MTT-S International Microwave Symposium Digest (IMS), 2006, San Francisco, CA, pp. 1324–1326.

96. P. Dennler, R. Quay, and O. Ambacher, Novel Semi-reactively-matched Multistage Broadband Power Amplifier Architecture for Monolithic ICs in GaN Technology, IEEE MTT-S International Microwave Symposium (IMS), 2013, Seattle, Washington, DC, pp 4.

97. R.M. Fano, Theoretical limitations on the broadband matching of arbitrary impedances, *Journal of Franklin Institute*, 1950, vol. 249, no. 2, pp. 139–154.

98. K.W. Kobayashi, Y.C. Chen, I. Smorchkova, R. Tsai, M. Wojtowicz, and A. Oki, A 2 Watt, Sub-dB Noise Figure GaN MMIC LNA-PA Amplifier with Multi-octave Bandwidth from 0.2-8 GHz, IEEE MTT-S International Microwave Symposium (IMS), 2007, Honolulu, HI, pp. 619–622.

99. P. Dennler, F. van Raay, M. Seelmann-Eggebert, R. Quay, and O. Ambacher, Modeling and Realization of GaN-Based Dual-Gate HEMTs and HPA MMICs for Ku-Band Applications, IEEE MTT-S International Microwave Symposium (IMS), 2011, Baltimore, MD, pp 4.

100. S. Lin, M. Eron, and S. Turner, Development of Broadband Amplifier Based on GaN HEMTs, IEEE 12th Annual Wireless and Microwave Technology Conference (WAMICON), 2011, Clearwater, FL, 2011, pp 1–4.

101. S. Masuda, A. Akasegawa, T. Ohki, K. Makiyama, N. Okamoto, K. Imanishi, T. Kikkawa, and H. Shigematsu, Over 10 W C-Ku band GaN MMIC Nonuniform Distributed Power Amplifier with Broadband Couplers, IEEE MTT-S International Microwave Symposium Digest (IMS), 2010, Anaheim, CA, pp. 1388–1391.

102. J.J. Komiak, C. Kanin, P.C. Chao, Decade Bandwidth 2 to 20 GHz GaN HEMT Power amplifier MMICs in DFP and No FP Technology, IEEE MTT-S International Microwave Symposium Digest (IMS), 2011, Baltimore, MD, pp. 1–4.

103. P. Schuh, H. Sledzik, R. Reber, K. Widmer, M. Oppermann, M. Musser, M. Seelmann-Eggebert, and R. Kiefer, GaN-based amplifiers for wideband applications, *International Journal of Microwave and Wireless Technologies*, 2010, vol. 2 , no. 1, pp. 135–141.

104. G. Mouginot, Z. Ouarch, B. Lefebvre, S. Heckmann, J. Lhortolary, D. Baglieri, D. Floriot, M. Camiade, H. Blanck, M. Le Pipec, D. Mesnager, and P. Le Helleye, Three Stage 6–18 GHz High Gain and High Power Amplifier Based on GaN Technology, IEEE MTT-S International Microwave Symposium (IMS), 2010, Anaheim, CA, pp. 1392–1395.

105. U. Schmid, H. Sledzik, P. Schuh, J. Schroth, M. Oppermann, P. Brückner, F. van Raay, R. Quay, and M. Seelmann-Eggebert, Ultra-wideband GaN MMIC chip set and high power amplifier module for multi-function defense AESA applications, *IEEE Transactions on Microwave Theory and Techniques*, 2013, vol. 61, no. 8, pp. 3043–3051.

106. P. Schuh, H. Sledzik, M. Oppermann, R. Quay, J. Kühn, T. Lim, P. Waltereit, M. Mikulla, and O. Ambacher, InAlGaN/GaN MMICs in Microstrip Transmission Line Technology for Wideband Applications, 6th European Microwave Integrated Circuits Conference (EuMIC), 2011, Manchester, UK, pp. 69–72.

107. P. Saunier, M.L. Schuette, T.-M. Chou, H.Q. Tserng, A. Ketterson, E. Beam, M. Pilla, and X. Gao, InAlN barrier scaled devices for very high $f_T$ and for low-voltage RF applications, *IEEE Transactions on Electron Devices*, 2013, vol. 60, no. 10, pp. 3099–3104.

108. M. Musser, R. Quay, F. van Raay, M. Mikulla, and O. Ambacher, Analysis of GaN HEMTs for Broadband High-power Amplifier Design, Proceedings of European Microwave Integrated Circuits Conference (EuMIC), 2011, Manchester, UK, pp. 128–131.

109. S. Lin and A. E. Fathy, Development of a wideband highly efficient GaN VMCD VHF/UHF power amplifier, *Progress Electromagnetics Research C*, 2011, vol. 19, pp. 135–147.

110. U. Schmid, R. Reber, P. Schuh, M. Oppermann, Robust Wideband LNA Designs, Proceedings of the 9th European Microwave Integrated Circuits Conference (EUMIC), 2014, Rome, Italy, pp. 186–189.

111. K.W. Kobayashi, An 8-Watt 250-3000 MHz Low-noise GaN MMIC Feedback Amplifier with >+50 dBm OIP3, IEEE Compound Semiconductor Integrated Circuit Symposium (CSIC), 2011, Waikoloa, HI, pp. 1–4.

112. M. Chen, W. Sutton, I. Smorchkova, B. Heying, W.B. Luo, V. Gambin, F. Oshita, R. Tsai, M. Wojtowicz, R. Kagiwada, A. Oki, and J. Lin, A 1–25 GHz GaN HEMT MMIC low-noise amplifier, *IEEE Microwave and Wireless Components Letters*, 2010, vol. 20, no. 10, pp. 563–565.

113. S. Maroldt, B. Aja, F. van Raay, S. Krause, P. Brückner, R. Quay, QFN-packaged Highly-linear Cascode GaN LNA MMIC from 0.5 to 3 GHz, 8th European Microwave Integrated Circuits Conference (EUMIC), 2013, Nuremberg, Germany, pp. 428–431.

114. J. Gassmann, P. Watson, L. Kehias, and G. Henry, Wideband, High-Efficiency GaN Power Amplifiers Utilizing a Non- Uniform Distributed Topology, IEEE MTT-S International Microwave Symposium (IMS), 2008, Atlanta, GA, pp. 615–618.

115. D.E. Meharry, R.J. Lender, K. Chu, L.L. Gunter, and K.E. Beech, Multi-Watt Wideband MMICs in GaN and GaAs, IEEE MTT-S International Microwave Symposium (IMS), 2007, Fort Lauderdale, FL, pp 4.

116. A.G. Bert and D. Kaminsky, The traveling-wave power divider/combiner, *IEEE Transactions on Microwave Theory and Techniques*, 1980, vol. 28, no. 12, pp. 1468–1473.

117. Y. Ayasli, R.L. Mozzi, J. L. Vorhaus, L.D. Reynolds, and R.A. Pucel, A monolithic GaAs 1–13 GHz traveling-wave amplifier, *IEEE Transactions on Microwave Theory and Techniques*, 1982, vol. MTT-30, no. 7, pp. 976–981.

118. C.F. Campbell, C. Lee, V. Williams, M.Y. Kao, H.Q. Tserng, and P. Saunier, A Wideband Power Amplifier MMIC Utilizing GaN on SiC HEMT Technology, IEEE Compound Semiconductor Integrated Circuit Symposium (CSIC), 2008, Monterey, pp. 159–162.

119. P. Dennler, D. Schwantuschke, R. Quay, O. Ambacher, 8–42 GHz GaN Non-uniform Distributed Power Amplifier MMICs in Microstrip Technology, IEEE MTT-S International Microwave Symposium (IMS), 2012, Montreal, pp 3.

120. P. Dennler, R. Quay, P. Brückner, M. Schlechtweg, and O. Ambacher, Watt-Level Non-Uniform Distributed 6–37 GHz Power Amplifier MMIC with Dual-Gate Driver Stage in GaN Technology, Proceedings of Radio Wireless Week (RWW), 2014, Newport Beach, CA, pp. 37–39.

121. D. Schwantuschke, P. Brückner, R. Quay, M. Mikulla, O. Ambacher, and I. Kallfass, A U-band Broadband Power Amplifier MMIC in 100 nm AlGaN/GaN HEMT Technology, Proceedings of the 7th European Microwave Integrated Circuits Conference (EuMIC), 2012, Amsterdam, the Netherlands, pp. 703–706.

122. B.J. Millon, S.M. Wood, R.S. Pengelly, Design of GaN HEMT Transistor Based Amplifiers for 5 - 6 GHz WiMAX Applications, Proceedings of the 38th European Microwave Conference (EUMW), 2008, Amsterdam, the Netherlands, pp. 1090–1093.

123. E.M. Suijker, M. Rodenburg, J.A. Hoogland, M. van Heijningen, M. Seelmann-Eggebert, R. Quay, P. Brückner, and F.E. van Vliet, Robust AlGaN/GaN Low Noise Amplifier MMICs for C-, Ku- and Ka-band Space Applications, Proceedings of Compound Semiconductor Integrated Circuit Symposium (CSIC), 2009, Greensboro, NC, pp. 1–4.

124. D. Krausse, R. Quay, R. Kiefer, A. Tessmann, H. Massler, A. Leuther, T. Merkle, S. Müller, C. Schwörer, M. Mikulla, M. Schlechtweg, and G. Weimann, Robust GaN HEMT Low-noise Amplifier MMICs for X-band Applications Proceedings of the 12th European Gallium Arsenide and other Compound Semiconductors Application Symposium (GAAS), 2004, London, UK, pp. 71–74.

125. H. Xu, C. Sanabria, A. Chini, S. Keller, U.K. Mishra, and R.A. York, A C-band high-dynamic range GaN HEMT low-noise amplifier, *IEEE Microwave and Wireless Components Letters*, 2004, vol. 14, no. 6, pp. 262–264.

126. P. Schuh, H. Sledzik, R. Reber, A. Fleckenstein, R. Leberer, M. Oppermann, R. Quay, F. van Raay, M. Seelmann-Eggebert, R. Kiefer, and M. Mikulla, X-band T/R-module front-end based on GaN MMICs, *International Journal of Microwave and Wireless Technologies*, 2009, vol. 1, no. 4, pp. 387–394.

127. R. Pengelly, S. Sheppard, T. Smith, B. Pribble, S. Wood, and C. Platis, Commercial GaN Devices For Switching and Low Noise Applications, Proceedings of CS Mantech, 2011, Palm Springs, CA, pp 4.

128. P. Schuh and R. Reber, Robust X-band Low Noise Limiting Amplifiers, IEEE MTT-S International Microwave Symposium Digest (IMS), 2013, Seattle, Washington, DC, pp. 1–4.

129. M.V. Aust, A.K. Sharma, Y.C. Chen, M. Wojtowicz, Wideband Dual-Gate GaN HEMT Low Noise Amplifier for Front-End Receiver Electronics, Proceedings of Compound Semiconductor Integrated Circuit Symposium (CSIC), 2006, San Antonio, TX, pp. 89–92.

130. I. Kallfass, H. Massler, S. Wagner, D. Schwantuschke, P. Brückner, C. Haupt, R. Kiefer, R. Quay, and O. Ambacher, A Highly Linear 84 GHz Low Noise Amplifier MMIC in AlGaN/GaN HEMT Technology, IEEE MTT-S International Microwave Workshop on Millimeter Wave Integration Technologies, IMWS 2011, Sitges, Spain, pp. 707–710.

131. I. Kallfass, R. Quay, H. Massler, S. Wagner, D. Schwantuschke, C. Haupt, R. Kiefer, and O. Ambacher, A Single-Chip 77 GHz Heterodyne Receiver MMIC in 100 nm AlGaN/GaN HEMT Technology, IEEE MTT-S International Microwave Symposium Digest (IMS), 2011, Baltimore, MD, pp. 1–4.

132. H.T. Friis, Noise Figure of Radio Receivers, Proceedings of IRE, 1944, vol. 32, pp. 419–422.

133. H. Fukui, Optimal noise figure of microwave GaAs MESFETs, *IEEE Transactions on Electron Devices*, 1979, vol. 26, no. 7, pp. 1032–1037.

134. Y. Chen, R. Coffie, W.B. Luo, M. Wojtowicz, J. Smorchkova, B. Heying, Y.-M. Kim, M.V. Aust, and A. Oki, Survivability of AlGaN/GaN HEMT, Proceedings of IEEE MTT-S International Microwave Symposium (IMS), 2007, Honolulu, HI, pp. 307–310.

135. M. Rudolph, R. Behtash, R. Doerner, K. Hirche, J. Würfl, W. Heinrich, and G. Tränkle, Analysis of the survivability of GaN low-noise amplifiers, *IEEE Transactions on Microwave Theory Techniques*, vol. 55, no. 1, 2007, pp. 37–43.

136. M. Sudow, K. Andersson, M. Fagerlind, M. Thorsell, P.-A. Nilsson, and N. Rorsman, A single-ended resistive X-band AlGaN/GaN HEMT MMIC mixer, *IEEE Transactions on Microwave Theory and Techniques*, 2008, vol. 56, no. 10, pp. 2201–2206.

137. V.S. Kaper, IEEE, R.M. Thompson, T.R. Prunty, and J.R. Shealy, Signal Generation, Control, and Frequency Conversion AlGaN/GaN HEMT MMICs, *IEEE Trans. Microwave Theory and Techniques*, 2005, vol. 53, no. 1, pp. 55–65.

138. M.N. Do, M. Seelmann-Eggebert, R. Quay, D. Langrez, and J.L. Cazaux, AlGaN/GaN Mixer MMICs, and RF Front-end Receivers for C-, Ku-, and Ka-band Space Applications, Proceedings of the 5th European Microwave Integrated Circuits Conference (EuMIC), 2010, London, UK, pp. 57–60.

139. V. Di Giacomo, N. Thouvenin, C. Gaquière, A. Santarelli, and F. Filicori, Modelling and Design of a Wideband 6-18 GHz GaN Resistive Mixer, Proceedings of the 4th European Microwave Integrated Circuits Conference (EuMIC), Rome, Italy, 2009, pp. 459–462.

140. B. Mallet-Guy, L. Darcel, J.-P. Plaze, Y. Mancuso, First 0.25 μm GaN MMICs Dedicated to Compact, Wideband and High SFDR Receiver, Proceedings of the 7th European Microwave Integrated Circuits Conference (EuMIC), 2012, Amsterdam, the Netherlands, 2012, pp. 329–332.

141. M. van Heijningen, J.A Hoogland, A.P. de Hek, and F.E. van Vliet, 6-12 GHz Double-Balanced Image-Reject Mixer MMIC in 0.25 μm AlGaN/GaN Technology, Proceedings of the 9th European Microwave Integrated Circuits Conference (EuMIC), 2014, Rome, Italy, pp. 65–68.

142. I. Kallfass, G. Eren, R. Weber, S. Wagner, D. Schwantuschke, R. Quay, and O. Ambacher, High Linearity Active GaN-HEMT Down-Converter MMIC for E-Band Radar Applications, Proceedings of the European Microwave Conference (EUMW), 2014, Rome, Italy, pp. 4.

143. J. Janssen, K.P. Hilton, J.O. Maclean, D.J. Wallis, J. Powell, M. Uren, T. Martin, M. van Heijningen, and F. van Vliet, X-Band GaN SPDT MMIC with over 25 Watt Linear Power Handling, Proceedings of European Microwave Integrated Circuit Conference (EUMIC), 2008, Amsterdam, the Netherlands, pp. 190–193.

144. H. Ishida, Y. Hirose, T. Murata, Y. Ikeda, T. Matsuno, K. Inoue, Y. Uemoto, T. Tanaka, T. Egawa, and D. Ueda, A high-power RF Switch IC using AlGaN/GaN HFETs with Single-stage Configuration *IEEE Transactions on Electron Devices*, 2005, vol. 52, no. 8, pp. 1893–1899.

145. M.D. Yore, C. A. Nevers, and P. Cortese, High-isolation Low-loss SP7T pHEMT Switch Suitable for Antenna Switch Modules, Proceedings of European Microwave Integrated Circuit Conference (EUMIC), Paris, France, 2010, pp. 69–72.

146. P. Sun, P. Liu, P. Upadhyaya, D.H. Jeong, D. Heo, and E. Mina, Silicon-Based PIN SPST RF Switches for Improved Linearity, IEEE MTT-S International Microwave Symposium Digest (IMS), 2010, Anaheim, CA, pp. 948–951.

147. C.F. Campbell and D.C. Dumka, Wideband High Power GaN on SiC SPDT Switch MMICs, IEEE MTT-S International Microwave Symposium (IMS), 2010, Anaheim, CA, pp. 145–148.

148. B.Y. Ma, K.S. Boutros, J.B. Hacker, and G. Nagy, High Power AlGaN/GaN Ku-Band MMIC SPDT Switch and Design Consideration, IEEE MTT-S International Microwave Symposium (IMS), 2008, Atlanta, GA, pp. 1473–1476.

149. M. Yu, R. Ward, D.H. Hovda, G.M. , Hegazi, A.W. Hanson, and K. Linthicum, The Development of a High Power SP4T RF Switch in GaN HFET Technology, *IEEE Microwave and Wireless Components Letters*, vol. 17, no. 12, 2007, pp. 894–896.

150. S. Kaleem, J. Kühn, R. Quay, and M. Hein, A High-Power Ka-band Single-pole Single-throw Switch MMIC using 0.25 µm GaN on SiC, Proceedings of Radio Wireless Week (RWW), San Diego, CA, 2015, pp 4.

151. C.F. Campbell, Method to Reduce Control Voltage for High Power GaN RF Switches, IEEE MTT-S International Microwave Symposium (IMS), 2015, Phoenix, AZ, pp. 4.

152. M. Hangai, T. Nishino, Y. Kamo, M. Miyazaki, An S-band 100 W GaN Protection Switch, IEEE MTT-S International Microwave Symposium (IMS), 2007, Honolulu, HI, pp. 1389–1392.

153. A. Brown, K. Brown, J. Chen, K.C. Hwang, N. Kolias, and R. Scott, W-Band GaN Power Amplifier MMICs, IEEE MTT-S International Microwave Symposium Digest (IMS), 2011, Baltimore, MD, pp. 1–4.

154. M. Micoviv, GaN-HEMT Technologies for THz Applications Workshop Proceedings of Sub-millimeter-wave Monolithic Integrated Circuits, European Microwave Week (EUMW), 2012, Amsterdam, the Netherlands.

155. K. Shinohara, A. Corrion, D. Regan, I. Milosavljevic, D. Brown, S. Burnham, P.J. Willadsen, C. Butler, A. Schmitz, D. Wheeler, A. Fung, and M. Micovic, 220 GHz $f_T$ and 400 GHz $f_{MAX}$ in 40 nm GaN DH-HEMTs with Re-grown Ohmic, Proceedings of IEEE International Electron Device Meeting (IEDM), 2010, San Francisco, CA, pp. 672–675.

156. M. van Heijningen, G. van der Bent, M. Rodenburg, F.E. van Vliet, R. Quay, P. Brückner, D. Schwantuschke, P. Jukkala, and T. Narhi, 94 GHz Power Amplifier MMIC Development in State of the Art MHEMT and AlGaN/GaN Technology, Proceedoings of ESA Workshop Microwave Technology and Techniques Workshop, 2012, Noordwijk, the Netherlands.

157. D.L. Ingram, Y.C. Chen, J. Kraus, B. Brunner, B. Allen, H.C. Yen, K.F. Lau, A 427 mW, 20% Compact Wband InP HEMT MMIC Power Amplifier, IEEE Radio Frequency Integrated Circuits Symp. (RFIC), 1999, Anaheim, CA, pp. 95–98.
158. M. Micovic, A. Kurdoghlian, K. Shinohara, S. Burnham, I. Milosavljevic M. Hu, A. Corrion, A. Fung, R. Lin, L. Samoska, P. Kangaslahti, B. Lambrigtsen, P. Goldsmith, W.S. Wong, A. Schmitz, P. Hashimoto, P. J. Willadsen, and D. H. Chow, W-Band GaN MMIC with 842 mW Output Power at 88 GHz, IEEE MTT-S International Microwave Symposium Digest (IMS), 2010, Anaheim, CA, pp. 237–240.
159. M. Micovic, A. Kurdoghlian, A. Margomenos, D. F. Brown, K. Shinohara, S. Burnham, I. Milosavljevic, R. Bowen, A.J. Williams, P. Hashimoto, R. Grabar, C. Butler, A. Schmitz, P.J. Willadsen, and D.H. Chow, 92-96 GHz GaN Power Amplifiers, IEEE MTT-S International Microwave Symposium Digest (IMS), 2012, Montréal, Canada, pp 3.
160. S. Yoshida, M. Tanomura, Y. Murase, K. Yamanoguchi, K. Ota, K. Matsunaga, and H. Shimawaki, A 76 GHz GaN-on-Silicon Power Amplifier for Automotive Radar Systems, IEEE MTT-S International Microwave Symposium Digest (IMS), 2009, Boston, MA, pp. 665–668.
161. M. Micovic, A. Kurdoghlian, H.P. Moyer, P. Hashimoto, M. Hu, M. Antcliffe, P.J. Willadsen, W.S. Wong, R. Bowen, I. Milosavljevic, Y. Yoon, A. Schmitz, M. Wetzel, C. McGuire, B. Hughes, and D.H. Chow, GaN MMIC PAs for E-Band (71 GHz–95 GHz) Radio, Proceedings of Compound Semiconductor Integrated Circuits Symposium (CSIC), 2008, Monterey, CA, pp. 1–4.
162. A. Brown, K. Brown, J. Chen, D. Gritters, K. C. Hwang, E. Ko, N. Kolias, S. O'Connor, and M. Sotelo, High Power, High Efficiency E-Band GaN Amplifier MMICs, IEEE International Conference of Wireless Information Technology and Systems (ICWITS), 2012, Maui, HI, pp. 1–4.
163. Y. Nakasha, S. Masuda, K. Makiyama, T. Ohki, M. Kanamura, N. Okamoto, T. Tajima, T. Seino, H. Shigematsu, K. Imanishi, T. Kikkawa, K. Joshin, and N. Hara, E-band 85 mW Oscillator and 1.3 W Amplifier ICs using 0.12-µm GaN HEMTs for Millimeter-wave Transceivers, IEEE Compound Semiconductor Integrated Circuit Symposium (CSIC), 2010, Monterey, CA, pp. 1–4.
164. J. Schellenberg, B. Kim, and T. Phan, and W-Band, Broadband 2 W GaN MMIC, 2013 IEEE MTT-S International Microwave Symposium Digest (IMS), 2013, Seattle, Washington, DC, pp. 1–4.
165. J. Schellenberg, E. Watkins, M. Micovic, B. Kim, and K. Han, W-band, 5 W Solid-State Power Amplifier/Combiner, IEEE MTT-S International Microwave Symposium Digest (IMS), 2010, Anaheim, CA, pp. 240–242.
166. D. Schwantuschke, M. Seelmann-Eggebert, P. Brückner, R. Quay, M. Mikulla and O. Ambacher, A fully Scalable Compact Small-Signal Modeling Approach for 100 nm AlGaN/GaN HEMTs, Proceedings of the 8th European Microwave Integrated Circuits Conference (EUMIC), 2013, Nuremberg, Germany, pp. 284–287.
167. D. Schwantuschke, B. Aja, M. Seelmann-Eggebert, R. Quay, A. Leuther, P. Brückner, M. Schlechtweg, M. Mikulla, I. Kallfass, and O. Ambacher, Q- and E-band Amplifier MMICs for Satellite Communication, 2014 IEEE MTT-S International Microwave Symposium Digest (IMS), Tampa, FL, pp 3.
168. M. van Heijningen, M. Rodenburg, F.E. van Vliet, H. Massler, A. Tessmann, P. Brückner, S. Müller, D. Schwantuschke, R. Quay, and T. Narhi, W-band Power Amplifier MMIC with 400 mW Output Power in 0.1 µm AlGaN/GaN Technology, Proceedings of the 7th European Microwave Integrated Circuits Conference (EuMIC), 2012, Amsterdam, the Netherlands, pp. 135–138.
169. H. Xu, C. Sanabria, S. Heikman, S. Keller, U.K. Mishra, R.A. York, High power GaN Oscillators using Field-plated HEMT Structure, IEEE MTT-S International Microwave Symposium Digest (IMS), 2005, Long Beach, CA, pp. 4.

170. X. Lan, M. Wojtowicz, I. Smorchkova, R. Coffie, R. Tsai, B. Heying, M. Truong, F. Fong, M. Kintis, C. Namba, A. Oki, and T. Wong, A Q-band Low Phase Noise Monolithic AlGaN/GaN HEMT VCO, *IEEE Microwave and Wireless Components Letters*, 2006, vol. 16, no. 7, pp. 425–427.

171. X. Lan, M. Wojtowicz, M. Truong, F. Fong, M. Kintis, B. Heying, I. Smorchkova, and Y. C. Chen, A V-Band Monolithic AlGaN/GaN VCO, *IEEE Microwave and Wireless Components Letters*, 2008, vol. 18, no. 6, pp. 407–409.

172. R. Weber, D. Schwantuschke, P. Brückner, R. Quay, M. Mikulla, O. Ambacher, I. Kallfass, A 67 GHz GaN Voltage-controlled Oscillator MMIC with High Output Power, *IEEE Microwave and Wireless Components Letters*, 2013, vol. 23, no. 7, pp. 374–376.

173. R. Weber, D. Schwantuschke, P. Brückner, R. Quay, F. van Raay, and O. Ambacher, A 92 GHz GaN HEMT Voltage-Controlled Oscillator MMIC, IEEE MTT-S International Microwave Symposium (IMS), 2014, Tampa, FL, pp 4.

174. A. Tessmann, W. H. Haydl, T. Krems, M. Neumann, H. Massler, L. Verweyen, A. Hülsmann, M. Schlechtweg, A compact coplanar W-band variable gain amplifier MMIC with wide control range using dual-gate HEMTs, Proceedings of IEEE MTT-S International Microwave Symposium (IMS), 1998, Baltimore, MD, pp. 685–688.

175. S. Diebold, D. Müller, D. Schwantuschke, S. Wagner, R. Quay, T. Zwick, and I. Kallfass, AlGaN/GaN-based Variable Gain Amplifiers for W-band Operation, IEEE MTT-S International Microwave Symposium (IMS), 2013, Seattle, Washington, DC, pp 3.

176. T.N. Ross, K. Hettak, G. Cormier, and J.S. Wight, Design of X-Band GaN Phase Shifters, *IEEE Transactions on Microwave Theory and Techniques*, 2014, vol. 63, no. 1, pp. 244–255.

177. A.M.H. Kwan, L. Xiaosen Liu, K.J. Chen, Integrated Gate-protected HEMTs and Mixed-signal Functional Blocks for GaN Smart Power ICs, *IEEE Transactions on Electron Devices* Meeting (IEDM), 2012, San Francisco, CA, pp. 7.3.1–7.3.4.

178. J.D. Albrecht, C. Tsu-Hsi, A.S. Kane, and M.J. Rosker, DARPA's Nitride Electronic NeXt Generation Technology Program, Proceedings of Compound Semiconductor Integrated Circuit Symposium (CSIC), 2010, Monterey, CA, pp. 1–4.

179. A.L. Corrion, K. Shinohara, D. Regan, Y. Tang, D. Brown, J.F. Robinson, H.H. Fung, A. Schmitz, D. Le, S.J. Kim, T.C. Oh, and M. Micovic, High-speed 501-stage DCFL GaN ring oscillator circuits, *IEEE Microwave and Wireless Components Letters*, 2013, vol. 34, no. 7, pp. 846–848.

180. R. Leberer, R. Reber, M. and Oppermann, An AlGaN/GaN Class-S Amplifier for RF-Communication Signals, IEEE MTT-S International Microwave Symposium (IMS), 2008, Atlanta, GA, pp. 85–88.

181. S. Maroldt, P. Brückner, R. Quay, O. Ambacher, S. Maier, D. Wiegner, and A. Pascht, An Integrated 12 Gbps Switch-mode Driver MMIC with 5 V(PP) for Digital Transmitters in 100 nm GaN Technology, Proceedings of the 7th European Microwave Integrated Circuits Conference (EuMIC), 2012, Amsterdam, the Netherlands, pp. 115–118.

182. S. Maroldt, Advanced Microwave Switch-Mode Amplifiers and Demonstrations Using GaN, Proceedings of WS European Microwave Conference, 26 September 2010, Paris, France.

183. S. Maroldt, P. Brückner, R. Quay, and O. Ambacher, A Microwave High-power GaN Transistor with Highly-Integrated Active Digital Switch-mode Driver Circuit, IEEE MTT-S International Microwave Symposium (IMS), 2014, Tampa, FL, pp. 1–4.

184. S. Maroldt, D. Wiegner, S. Vitanov, V. Palankovski, R. Quay, and O. Ambacher, Efficient AlGaN/GaN linear and Digital-switch-mode Power Amplifiers for Operation at 2 GHz, *IEICE Transactions on Electronics* E93-C, 2010, vol. 8, pp. 1238–1244.

185. A. Wentzel, W. Heinrich, S. Hori, M. Hayakawa, and K. Kunihiro, Envelope Delta-Sigma-Modulated Voltage-Mode, Class-S PA, Proceedings of the 42nd European Microwave Conference, Amsterdam, the Netherlands, 2012, pp. 120–123.
186. A. Wentzel, S. Chevtchenko, P. Kurpas, and W. Heinrich, A Flexible GaN MMIC Enabling Digital Power Amplifiers for the Future Wireless Infrastructure, Proceedings of IEEE MTT-S International Microwave Symposium (IMS), 2015, Phoenix, AZ, pp 4.

# 6 GaN-Based Metal/ Insulator/Semiconductor- Type Schottky Hydrogen Sensors

*Ching-Ting Lee, Hsin-Ying Lee, and Li-Ren Lou*

## CONTENTS

## 6.1 INTRODUCTION

Hydrogen ($H_2$) gas has many interesting and useful properties and has been widely used as hydrogenating and reducing agents in numerous industrial branches, such as petroleum, chemical, metallurgical, and foodstuff industries. It is also used in power stations as a coolant in electric generators. Nowadays, environmental pollution has become a major issue. Hydrogen, as one of potentially clean energy sources, has attracted intensive attention. Hydrogen gas is expected to be one of the principal energy sources in the future and can be used in future power devices, solid oxide fuel cells, $H_2$ engine cars, and so on [1–6]. On the other hand, hydrogen gas is hazardous, is highly inflammable, has low ignition energy, and is explosive at room temperature when its concentration in air exceeds about 4%. However, hydrogen gas is odorless and colorless, so it is difficult to be detected by human sensing organs. Furthermore, because hydrogen's small molecular size promotes leaks and diffusion, leaking hydrogen gas warms, and may spontaneously ignite due to its negative Joule–Thomson coefficient at room temperature. Therefore, to ensure safety it is necessary to detect the leakage and monitor its existence and concentration in the surrounding atmosphere using specific detectors wherever hydrogen is produced, transported, stored, or used. Over time, much effort has been made to exploit various reliable, durable, accurate, sensitive, fast to respond, easy to operate, and inexpensive hydrogen sensors. Based on the variation of various physical–chemical properties of materials on exposure to hydrogen-containing ambience, different types of hydrogen sensors have been developed [7].

Among various types of hydrogen sensors, semiconductor-based Schottky sensors are of particular interest due to their prominent advantages, such as high response speed; high sensitivity; long lifetime; compact size; low cost; and simple fabrication and circuitry, especially, compatible with sophisticated integrated circuit technology. To fabricate Schottky sensors, various semiconductors have been used. Si-based Schottky hydrogen sensors were developed first about 40 years ago. A series of works on Si-based metal/insulator/semiconductor (MIS) sensors was carried out by Lundström et al. [8–12] and others [13], and the sensing mechanism was also analyzed for Si-based structures on hydrogen gas exposure [10–15]. However, the operating temperature of gas sensors based on silicon substrates is limited below 250°C due to the small energy bandgap of silicon. This restricts their use in specific environments such as automotive, aeronautical, and other harsh environmental areas. To solve the problem, recently, some wide energy bandgap compound semiconductors, such as GaN, SiC, and AlGaN, have been used as materials to fabricate hydrogen sensors allowing sensing operation at a high temperature [16–20]. In view of the high electron saturation velocity, high breakdown electric field, and superior thermal and chemical stability, GaN-based devices have attracted more attention for applications in microelectronic devices and optoelectronic devices [21–24]. High-temperature Schottky hydrogen sensor on n-type GaN was reported by Luther et al. [25] for the first time. After that, a series of GaN-based metal/semiconductor (MS) sensors were reported, including catalytic metal/GaN and catalytic metal/AlGaN/GaN MS sensors [26,27].

It was demonstrated that MIS structures have much better sensitivity in response to the hydrogen-containing atmosphere in comparison with the corresponding MS structures [28–31]. Several oxides, such as $SiO_2$, $SiN_x$, $Al_2O_3$, $Ga_2O_3$, $Sc_2O_3$, and $WO_3$, were used as the insulator materials for fabricating GaN-based catalytic MIS-structured (Schottky diode) hydrogen sensors [20,30–34]. Among the oxide materials, $SiO_2$, a very popular oxide used in various semiconductor devices, has been widely used to construct MIS hydrogen sensors. Irokawa et al. [32] fabricated MIS Pt-GaN diodes with a 10-nm-thick $SiO_2$ insulating layer deposited using radio frequency (RF) sputter system at room temperature. The resultant hydrogen sensors exhibited a remarkable improvement in hydrogen detection sensitivity in comparison with conventional Pt-GaN Schottky diodes. Besides, $Ga_2O_3$ is a promising material for sensor application and has been used to construct conductometric gas sensors for monitoring hydrogen [35,36]. In view of this fact, Trinchi et al. [17] chose $Ga_2O_3$ as the insulator layer material and fabricated $Pt/Ga_2O_3/SiC$ Schottky diodes as hydrogen sensors. In the sensors, the $Ga_2O_3$ layer was prepared through a sol-gel route starting from gallium isopropoxide and was deposited by spin-coating onto SiC substrates [37]. Thus, fabricated sensors exhibited excellent detecting sensibility.

It was demonstrated by a lot of reports that the MIS diodes with proper insulating layer inserted exhibited significantly better sensing performances, such as sensitivity and response time, than those of the catalytic MS sensors. Therefore, to exploit proper insulators, develop new deposition techniques of the insulator layers, and understand the role of insulator layers, GaN-based MIS Schottky hydrogen sensors have been made an important research topic and have attracted much effort.

In this chapter, the important problem of the insulator layer was focused. In Section 6.2, the general working principle of Schottky sensors is introduced, especially the general role of insulator layers in sensing processes. In Section 6.3, GaN-based MIS sensors fabricated by using novel deposition method are demonstrated, showing considerably improved sensing performances. Subsequently, in Section 6.4 the roles of the insulator layer of GaN-based MIS Schottky hydrogen sensors are further analyzed. Finally, a summary is given in Section 6.5.

## 6.2 SENSING MECHANISM OF SCHOTTKY HYDROGEN SENSORS

Typical MIS-structured Schottky hydrogen sensors generally have the configuration of catalytic MIS. When the thickness of the insulator layer is zero, MIS-structured sensors become MS-structured sensors. In the MIS structure, the metal layer serves not only as a Schottky metal contact but also as a catalyst. Among the used metals, the commonly used catalytic metals in this type of hydrogen sensors are Pd and Pt [38–42].

The working principle of Schottky sensors has been generally investigated previously [11,12,18,20,43]. According to the dipole model, originally proposed by Lundström [8], when the MS-type or MIS-type hydrogen sensors are exposed to dilute hydrogen ambiences the hydrogen molecules, $H_{2,gas}$, are dissociated into hydrogen atoms, H, at the catalytic metal surface. A part of the dissociated hydrogen atoms reacts with catalytic metal atoms and is adsorbed to the adsorption sites, $S_s$, on the metal surface. The corresponding reaction can be expressed as follows:

$$H_{2,gas} + 2S_s \leftrightarrows 2H\text{--}S_s \qquad (6.1)$$

In the case where oxygen coexists with hydrogen, such as in hydrogen-containing air ($H_2$/air), the oxygen ($O_{2,gas}$) is also dissociated and adsorbed on the catalytic metal surface. The adsorbed oxygen atoms react with the adsorbed hydrogen atoms to form $OH\text{--}S_s$, which in turn react with $H\text{--}S_s$ to form water ($H_2O$). The corresponding reactions are expressed as follows:

$$O_{2,gas} + 2S_s \leftrightarrows 2O\text{--}S_s \qquad (6.2)$$

$$O\text{--}S_s + H\text{--}S_s \leftrightarrows OH\text{--}S_s + S_s \qquad (6.3)$$

$$H–S_s + OH–S_s \leftrightarrows H_2O + 2S_s \tag{6.4}$$

$$2H–S_s + O–S_s \leftrightarrows H_2O + 3S_s \tag{6.5}$$

In addition to the reaction with surface oxygen and hydroxide, the adsorbed hydrogen atoms residing on the metal surface can also diffuse through the catalytic metal film and arrive at the MS interface or metal/insulator interface, where the hydrogen atoms are adsorbed at the corresponding adsorption sites, $S_i$. The reaction can be expressed as follows:

$$S_i + H–S_s \leftrightarrows S_s + H–S_i \tag{6.6}$$

For insulators, such as $\beta$-$Ga_2O_3$, which react with hydrogen, a part of the hydrogen atoms is absorbed at the corresponding adsorption sites, $S_i$, within the oxide layer. The absorption sites at the Pt/$\beta$-$Ga_2O_3$ interfaces are located at the oxygen atoms on the $\beta$-$Ga_2O_3$ surface [44]. Furthermore, the absorption sites in the $\beta$-$Ga_2O_3$ layer are located at the gallium or oxygen atoms [45]. The adsorption of hydrogen atoms at the insulator surface produces a dipole layer at the interface and correspondingly induces a potential jump across the interface [12]. Consequently, the dipole layer changes the effective work function of the metal. Therefore, the Schottky barrier height of the resulting diodes is changed (usually reduced, depending on the dipole orientation). The Schottky barrier height is changed from $\phi_{B,air}$ in the air ambience to $\phi_{B,H_2}$ in the $H_2$-containing ambience, as shown in Figure 6.1a and b for MS and MIS diodes, respectively. This causes variation in electric characteristics of the devices, such as the shift of the capacitance–voltage (C–V) curve or the current–voltage (J–V) curve of the diode, in which hydrogen detection is enabled.

From the aforementioned working mechanisms, it is known that Schottky hydrogen sensors are based on the hydrogen-induced change of the interface state and the properties initiated by metal catalysis reaction, in which the sensors are interface-controlled devices. The metal/insulator (or MS) interface is the key sensing component,

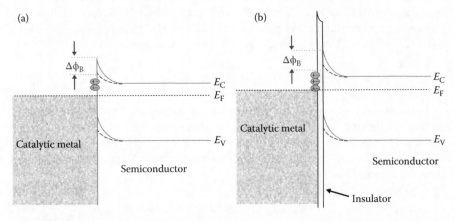

**FIGURE 6.1** Schematic energy band diagram for (a) metal/semiconductor-type and (b) metal/insulator/semiconductor-type diodes exposed to hydrogen-containing atmosphere.

whereas the semiconductor substrate is used to provide a proper Schottky barrier for responding to the hydrogen exposure. The variation in the Schottky barrier height of the diodes can be converted to electric signals as the responses to hydrogen.

For the MIS Schottky hydrogen sensors, the sensing performance is determined by the density of the hydrogen adsorption sites residing at the metal/insulator interface and the adsorption–desorption kinetics. As oxygen-terminated sites residing at the interface provide effective adsorption sites for hydrogen atoms, oxide materials are often used to construct the insulator layer in the MIS hydrogen sensors [46,47]. By properly choosing the insulator material and the insulator layer's parameter, MIS sensors can exhibit much better sensing performances compared with corresponding MS sensors. Therefore, the development of a high-quality insulator layer for fabricating catalytic MIS sensors becomes a key issue.

For GaN-based Schottky hydrogen sensors, apart from the catalytic metal layers, various aspects of the insulator layers, such as materials, layer deposition methods, layer geometrics, as well as layer interfaces, are of great importance. In this chapter, those various aspects of the Schottky hydrogen sensors are concentrated.

## 6.3 PROGRESS IN EXPLOITING NEW OXIDE MATERIALS AND DEPOSITION METHODS

Because of the key role of the insulator layers in GaN-based MIS hydrogen sensors, new oxide materials suitable for Schottky diode sensors, including the material itself and the interface between oxides and semiconductors, have been the focus point of many research groups. To improve the quality of the insulator layers and the related interfaces, the deposition methods and the processing techniques have been attracting intensive attention. Here, recent progress carried out on sensor fabrication, insulator layer deposition, and relevant kinetic process analysis is concentrated.

### 6.3.1 OXIDE LAYER GROWTH USING PHOTOELECTROCHEMICAL OXIDATION METHOD

In previous reports, the insulator layers used in MIS Schottky hydrogen sensors were deposited by various deposition methods. It is noticed that, conventionally, the insulator layer of MIS diodes was deposited externally on the semiconductor layer. Because of the contamination and defects inevitably existing on the semiconductor surface, the quality of the interface between the externally deposited insulator layer and the semiconductor deteriorates, which degrades the performances of the resulting sensors. Improving the interface quality is always one of intensive issues. A unique photoelectrochemical (PEC) oxidation method was developed to grow high-quality insulator layers of GaN-based MIS devices previously [48–51]. It was demonstrated that superior insulating layers with low interface-state density were obtained. The PEC oxidation method is certainly a potential method to fabricate GaN-based MIS hydrogen sensors. As reported previously [51], during the PEC oxidation processes etching and oxidation reactions occur at the interface between the semiconductor and the electrolyte simultaneously. By using an electrolyte with an appropriate pH value, the oxidation reaction dominates the process and, hence, the PEC reaction oxidizes the GaN-based semiconductor surface, forming the $Ga_2O_3$ oxide layer on the GaN surface or the mixed $Ga_2O_3$ and $Al_2O_3$ oxide layer on the AlGaN surface. Because the oxide layer was formed through chemical reaction with the GaN-based semiconductors, the interface between the PEC-grown oxide layer and the GaN-based semiconductor was nearly free of contamination, resulting in a low interface-state density. Moreover, the unique PEC oxidation method is different from the other conventional insulator deposition methods, it reveals obvious advantages including large growth rate, low cost, low deposition temperature, and damage-free plasma.

The schematic configuration of the PEC oxidation system is shown in Figure 6.2 [51]. The system mainly consists of an appropriate light source to generate electrons and holes in the semiconductor and a chemical solution serving as the electrolyte. The pH value of the electrolyte is properly chosen to ensure the domination of the oxidation reaction.

The PEC oxidation method was used for fabricating Pt/β-$Ga_2O_3$/GaN MIS-type hydrogen sensors. In this case, a He–Cd laser (325 nm), of which the photon energy is larger than the bandgap of GaN, was used as the illumination source to generate electron–hole pairs on the GaN surface. The pH value of the chemical solution ($H_3PO_4$) was selected to be 3.5. The work function $q\phi_E$ of the $H_3PO_4$ solution, depending on the pH value of the solution, could be determined by the following equation:

$$q\phi_E = 4.25 + 0.059 \times \text{pH value (eV)} \tag{6.7}$$

For a pH value of 3.5, the work function of the $H_3PO_4$ solution was 4.456 eV. When the n-type GaN semiconductor contacts with chemical solutions, a Schottky contact is formed at the interface owing to the difference in the work function between them. The energy band structure around the electrolytic solution/n-GaN interface is shown

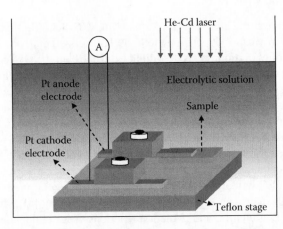

**FIGURE 6.2**  Schematic configuration of photoelectrochemical oxidation system.

in Figure 6.3. Owing to the built-in electrical field, the generated holes (h$^+$) are driven and move toward the interface. The PEC oxidation reactions can be described as follows [52]:

$$2GaN + 6h^+ + 3H_2O \rightarrow Ga_2O_3 + 6H^+ + N_2 \qquad (6.8)$$

$$Ga_2O_3 + 6H^+ \rightarrow 2Ga^{3+} + 3H_2O \qquad (6.9)$$

It can be seen that the hole density on the GaN surface is an important factor for controlling the oxidation reaction rate, which can be adjusted by varying the illuminating laser beam intensity.

The schematic configurations of GaN-based MS- and MIS-type hydrogen sensors are shown in Figure 6.4a and b, respectively [18]. Both the GaN-based MS- and MIS-type hydrogen sensors were fabricated using the same kind of epitaxial structure, which consisted of a 750-nm-thick GaN buffer layer and a 0.8-μm-thick undoped GaN layer grown on c-plane sapphire substrates using a metalorganic chemical vapor deposition (MOCVD) system. The insulator layer of the MIS-type hydrogen sensors shown in Figure 6.4b was grown on the undoped GaN surface using the PEC oxidation method with a H$_3$PO$_4$ chemical solution of a pH value of 3.5 under the illumination of a 10.0-mW/cm$^2$ He–Cd laser beam [48–52].

The fabricated Pt/β-Ga$_2$O$_3$/GaN MIS-type hydrogen sensors with the PEC oxidized β-Ga$_2$O$_3$ layer exhibited excellent hydrogen detection performances [18], indicating an effective reduction in Fermi pinning effect induced by interface states. The performance improvement in comparison with the corresponding MS-type hydrogen sensors was mainly attributed to more oxygen terminated sites at the β-Ga$_2$O$_3$ layer surface, which worked as hydrogen atom adsorption sites. As mentioned earlier, more hydrogen atoms adsorbed at the Pt/β-Ga$_2$O$_3$ interface induced a larger dipole moment density, resulting in a larger potential jump at the interface. Consequently, a larger Schottky barrier height change was induced in the Pt/β-Ga$_2$O$_3$/GaN MIS-type Schottky hydrogen sensors. As demonstrated from the experiments, the measured I-V characteristics of the MIS-type Schottky hydrogen sensors exhibited a voltage

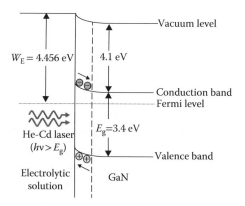

**FIGURE 6.3** The energy band diagram of electrolytic solution/n-GaN.

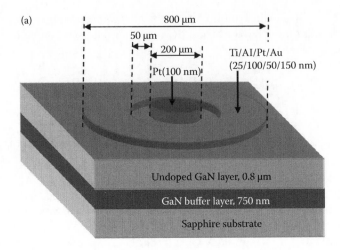

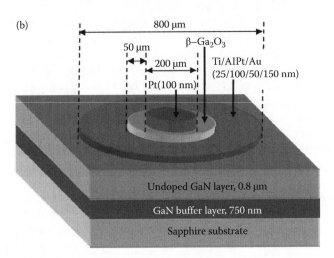

**FIGURE 6.4** Schematic configuration of (a) Pt/GaN (metal/semiconductor-type) and (b) Pt/β-Ga₂O₃/GaN (metal/insulator/semiconductor-type) hydrogen sensors. (Reproduced from Yan, J.T. and Lee, C.T., *Sens. Actuator B-Chem.*, 143, 192, 2009. With permission [18].)

shift (ΔV) on exposure to various hydrogen-containing ambiences. Figure 6.5 shows a typical example of the I-V curves measured at three temperatures for the fabricated MIS-type Schottky hydrogen sensors with a 4-nm-thick β-Ga₂O₃ layer. It was found that this particular sensor showed the best hydrogen detection sensitivity, much higher than that of the Pt/GaN MS-type Schottky hydrogen sensors [18] and also higher than that of the GaN-based MIS sensors with a thicker β-Ga₂O₃ layer.

The forward response, $S_F$, of the hydrogen sensors operated at a constant forward bias voltage mode is defined as follows:

$$S_F = \left(I_{H_2} - I_{air}\right)\big/I_{air} \tag{6.10}$$

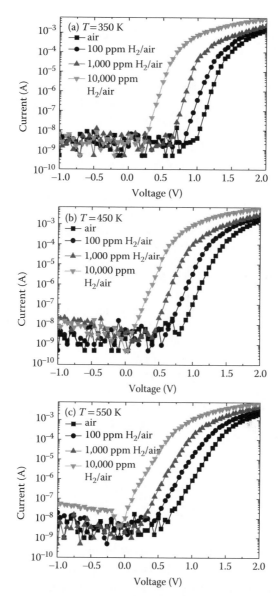

**FIGURE 6.5** Current–voltage characteristics of the Pt/β-Ga$_2$O$_3$ (4 nm)/GaN hydrogen sensors exposed to air and various dilute hydrogen ambiences at (a) 350 K, (b) 450 K, and (c) 550 K. (Reproduced from Lee, C.T. and Yan, J.T., *Sens. Actuator B-Chem.*, 147, 723, 2010. With permission [20].)

where $I_{H_2}$ and $I_{air}$ are the forward currents measured in a hydrogen-containing ambience and the air ambience, respectively. $I_{H_2}$ is generally a nonlinear function of the H$_2$ concentration in the ambience, as depicted in Figure 6.5. Therefore, it can be seen that the $S_F$ of the hydrogen sensors depends on the H$_2$ concentration of the ambience as well as the bias voltage. It was derived from the measured I-V characteristics that, at a forward bias voltage of 1.25 V, for the MIS-type hydrogen sensors with a 4-nm-thick $\beta$-$Ga_2O_3$ layer the $S_F$ on exposure to an ambience of 100 ppm $H_2$/air was 37.5, much larger than the value of 0.5 for Pt/GaN MS-type hydrogen sensors. Table 6.1 lists the forward response, $S_F$, values for various MIS-type hydrogen sensors with various thick $\beta$-$Ga_2O_3$ layers as well as MS-type hydrogen sensors on exposure to various hydrogen ambiences. The experimental results indicated that the MIS-type hydrogen sensors exhibited a capability of operating over a wide temperature range. The $S_F$ value increased with an increase in the hydrogen concentration and fell with an increase in the operating temperature. At a higher temperature, the forward response was reduced. The function of the oxide layer in hydrogen detection processes will be further discussed in Session 6.4.

Pt/mixed reactive insulator/AlGaN hydrogen sensors were also successfully fabricated with the insulator layer grown by directly oxidizing the AlGaN layer using the PEC oxidation method under the condition similar to that mentioned above [19]. The resultant oxide layer composed of $\beta$-$Ga_2O_3$ and $\alpha$-$Al_2O_3$ played a similar role as the $Ga_2O_3$ layer in the Pt/$\beta$-$Ga_2O_3$/GaN hydrogen sensors [49]. It was found that the AlGaN-based MIS-type hydrogen sensors operated at a forward bias exhibited more considerable sensitivity than that of the corresponding MS-type hydrogen sensors. The experimental results demonstrated an obvious lowering of the barrier height and the series resistance on exposure to hydrogen-containing atmosphere, which induced a shift in the measured I-V characteristics. This phenomenon will be discussed further in Section 6.4. In a higher hydrogen concentration, a faster response characteristic was observed. The obtained good performances of the hydrogen sensors indicated that the mixed $\beta$-$Ga_2O_3$ and $\alpha$-$Al_2O_3$ reactive insulator grown by PEC oxidation method was expected to be a promising candidate for using in the AlGaN-based hydrogen sensors. Besides, the associated series resistance and the ideality factor of the MIS-type hydrogen sensors were decreased as exposed to a hydrogen-containing ambience, which was attributed to the change in electrical properties of the $\beta$-$Ga_2O_3$ oxide layer as exposed to a hydrogen-containing ambience. The aforementioned examples demonstrated that because of the high performances and

---

**TABLE 6.1**

**Forward Response, $S_F$, of Pt/GaN MS-type Hydrogen Sensors and Pt/$\beta$-$Ga_2O_3$/GaN MIS-Type Hydrogen Sensors on Exposure to Air Ambience and Various Dilute Hydrogen Ambiences at Room Temperature**

|  |  | MS-Type Schottky Sensor Diode | MIS-Type Sensor Diode ($\beta$-$Ga_2O_3$ = 4 nm) | MIS-Type Sensor Diode ($\beta$-$Ga_2O_3$ = 15 nm) | MIS-Type Sensor Diode ($\beta$-$Ga_2O_3$ = 60 nm) |
|---|---|---|---|---|---|
| $S_F$ | 100 ppm $H_2$/air | 0.5 | 37.5 | 20.5 | 0.7 |
|  | 1,000 ppm $H_2$/air | 1.7 | 426.9 | 228.8 | 3.5 |
|  | 10,000 ppm $H_2$/air | 6 | 22,985.6 | 12,999 | 2,361.7 |

*Source:* Yan, J.T. and Lee, C.T., *Sens. Actuator B-Chem.*, 143, 192, 2009. With permission [18].

low interface-state density of the β-Ga$_2$O$_3$ films and the mixed oxide films directly grown using the PEC oxidation method, thus derived reactive insulator layers were expected to improve the hydrogen forward response of the resulting MIS-type hydrogen sensors.

### 6.3.2 Pt/i-ZnO/GaN MIS Hydrogen Sensors with Intrinsic ZnO Film Deposited Using Vapor Cooling Condensation Method

ZnO-based materials are important oxide materials for toxic and combustible gas sensing applications. Furthermore, the ZnO- and GaN-based semiconductors possess the same wurtzite crystalline structure and the similar lattice constant. It implies that intrinsic ZnO (i-ZnO) is an appropriate oxide material to be used for the insulator layer of GaN-based MIS sensors. Several methods have been used to deposit ZnO films on various substrates [53–55]. However, a high-quality i-ZnO film, which is required for MIS structures, is difficult to obtain due to the compensation effect of shallow donors induced by oxygen vacancies and zinc interstitials [56,57]. This problem was also suffered in the fabrication of various ZnO-related devices. To overcome the difficulty, a unique vapor cooling condensation method was developed, especially for i-ZnO deposition [58]. The mechanisms of the high-quality i-ZnO thin films deposited using the vapor cooling condensation system at low temperature were also proposed, previously [59]. Furthermore, this method has been used to deposit different kinds of structures of various devices [60–62]. By using this method, the GaN-based Pt/i-ZnO/GaN MIS hydrogen sensors were successfully fabricated [63].

The schematic configuration of Pt/i-ZnO/GaN MIS hydrogen sensors is shown in Figure 6.6. The epitaxial structure of the MIS hydrogen sensors consists of a 750-nm-thick GaN buffer layer and a 0.8-μm-thick undoped GaN layer; all of the layers were grown on c-plane sapphire substrate using a MOCVD system. To deposit various thick i-ZnO films on the undoped GaN layer, the sample was attached underneath a liquid nitrogen–cooled stainless steel plate and the ZnO powder was put on the tungsten boat and then heated in the vapor cooling condensation system. The ZnO vapor material that sublimated from the heated tungsten boat was coolly condensed and deposited on the surface of the undoped GaN layer, which was cooled by liquid nitrogen [58,59]. Using Hall measurement at room temperature, the electron concentration and the electron mobility of the deposited i-ZnO films were measured

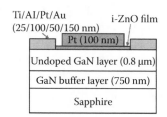

**FIGURE 6.6** Schematic configuration of Pt/i-ZnO/GaN hydrogen gas sensors. (Reproduced from Lee, H.Y. et al., *Sens. Actuator B-Chem.*, 157, 460, 2011. With permission [63].)

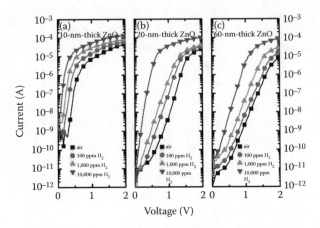

**FIGURE 6.7** Current–voltage characteristics of Pt/i-ZnO/GaN hydrogen sensors with (a) 10-nm-thick, (b) 20-nm-thick, and (c) 60-nm-thick i-ZnO layers for exposing to air ambience, 100 ppm $H_2$, 1,000 ppm $H_2$, and 10,000 ppm $H_2$ at room temperature. (Reproduced from Lee, H.Y. et al., *Sens. Actuator B-Chem.*, 157, 460, 2011. With permission [63].)

to be $7.6 \times 10^{15}$ cm$^{-3}$ and 2.1 cm$^2$/V·s, respectively. After depositing a 100-nm-thick catalytic metal Pt on the surface of the i-ZnO layer, Pt/i-ZnO/GaN hydrogen sensors with various thick i-ZnO layers were fabricated.

Figure 6.7a through c shows the current–voltage (I-V) characteristics of the Pt/i-ZnO/GaN MIS hydrogen sensors with i-ZnO thicknesses of 10, 20, and 60 nm measured in various hydrogen ambiences at room temperature, respectively. When the MIS hydrogen sensors are exposed to dilute hydrogen ambience, the current is increased in response to an increase in hydrogen concentration. The I–V characteristics can be decomposed into an exponential region and a resistance-limited region corresponding to low and high current limit, respectively. As shown in Figure 6.7, the current of the hydrogen sensors decreases with an increase in i-ZnO thickness.

Because a larger area of i-ZnO film can be deposited using the simple system at a low temperature and without post-annealing process, the cost and fabrication process of the hydrogen sensors can be reduced and simplified. Compared with the Ga$_2$O$_3$ insulator layer directly grown on the GaN layer using the PEC methods [18,20], the merit of the i-ZnO as the insulator layer of hydrogen sensors lies in the practical applications.

## 6.4 DISCUSSION ON THE SENSING MECHANISM OF GaN-BASED MIS HYDROGEN SENSORS

### 6.4.1 ROLE OF INSULATOR LAYER OF MIS HYDROGEN SENSORS IN SENSING PROCESS

As aforementioned, it is generally accepted that the oxide layer of the MIS hydrogen sensors provides more oxygen terminated sites at the metal/oxide interface for hydrogen adsorption. Consequently, a more intense interface dipole layer in MIS hydrogen sensors can be induced on exposure to hydrogen-containing ambience, resulting in a reduction of the Schottky barrier height and the change in electric properties of the hydrogen sensors. To understand further insight into the role of the oxide layer in the sensing process, the I-V characteristics of the MIS hydrogen sensors were analyzed in detail. The following discussion is focused on Pt/β-Ga$_2$O$_3$ GaN hydrogen sensors, as an example.

It was noticed that all the MIS hydrogen sensors discussed earlier exhibited a good rectifying behavior both in an air ambience and in diluted hydrogen containing ambiences. The I-V characteristics consisted of an exponential part in the low-current region and a resistance-limited part in the high-current region. This phenomenon suggested that the MIS hydrogen sensors could be approximately thought to be composed of an ideal MS diode in series with an electric resistor. Based on the thermionic emission model, the I-V characteristics of this kind of MIS-type hydrogen sensors operated at $V > 3kT/q$ and could be described by the following equation [64–66]:

$$I_F = I_0 \exp\left[q(V - I_F R_s)/nkT\right]$$
(6.11)

where $I_F$ is the forward current, $I_0$ is the saturation current, $T$ is the absolute temperature, $k$ is the Boltzmann constant, $q$ is the electron charge, $n$ is the ideality factor, $V$ is the applied voltage, and $R_s$ is the series resistance. The saturation current $I_0$ is defined as follows:

$$I_0 = SA^{**}T^2 \exp\left[(-q\phi_B)/(kT)\right]$$
(6.12)

where $S$ is the area of the Pt gate, $A^{**}$ is the Richardson constant (~24 A·cm$^{-2}$·K$^{-2}$ for GaN), and $\phi_B$ is the barrier height between Pt and GaN. With the hydrogen sensors exposed to air and dilute hydrogen ambiences, the associated barrier height $\phi_B$ is denoted as $\phi_{B,air}$ and $\phi_{B,H_2}$, respectively. Correspondingly, the saturation current is $I_{0,air}$ and $I_{0,H_2}$ and the series resistance is $R_{s,air}$ and $R_{s,H_2}$, respectively. For hydrogen detection, the barrier height change $\Delta\phi_B = \phi_{B,air} - \phi_{B,H_2}$ is of more interest. According to Equation 6.12, the barrier height change $\Delta\phi_B$ can be expressed using the saturation currents $I_{0,air}$ and $I_{0,H_2}$ as follows:

$$\Delta\phi_B = \frac{kT}{q} \ln\left(\frac{I_{0,H_2}}{I_{0,air}}\right)$$
(6.13)

By fitting I–V characteristics with Equations 6.11 through 6.13, the value of the barrier height change $\Delta\phi_B$, series resistance $R_s$, and ideality factor $n$ can be extracted. For Pt/β-Ga$_2$O$_3$(4 nm)/GaN hydrogen sensors operated at various ambiences and temperatures, the corresponding results [20] derived from the experimental I-V curves shown in Figure 6.5 are listed in Table 6.2.

The barrier height change rose with an increase in hydrogen concentration and fell with an increase in operating temperature. The deduced barrier height change interpreted the variation of forward response [20], which increased with an increase in hydrogen concentration and decreased with an increase in operating temperature, as shown in Figure 6.8. Moreover, it was noticed that when MIS-type hydrogen sensors were exposed to various dilute hydrogen ambiences the associated series resistance (approximately equal to $dV/dI$ in the resistance-limited region) decreased with the hydrogen concentration and operating temperature. It indicated that the series resistance variation also contributed to the forward response of the MIS hydrogen sensors. The decreased series resistance of the MIS-type Pt/β-Ga$_2$O$_3$/GaN

### TABLE 6.2
### Barrier Height Change and Series Resistance of Pt/β-Ga$_2$O$_3$/GaN Hydrogen Sensors Exposed to Air Ambience and Various Dilute Hydrogen Ambiences at Different Temperatures

|  | Temperature Ambience | 350 K | 450 K | 550 K |
|---|---|---|---|---|
| Barrier height change (V) | 100 ppm H$_2$/air | 0.204 | 0.138 | 0.089 |
|  | 1,000 ppm H$_2$/air | 0.272 | 0.216 | 0.165 |
|  | 10,000 ppm H$_2$/air | 0.581 | 0.515 | 0.465 |
| Series resistance (Ω) | Air | 327 | 284 | 257 |
|  | 100 ppm H$_2$/air | 312 | 265 | 233 |
|  | 1,000 ppm H$_2$/air | 293 | 242 | 211 |
|  | 10,000 ppm H$_2$/air | 176 | 166 | 155 |

*Source:* Lee, C.T. and Yan, J.T., *Sens. Actuator B-Chem.*, 147, 723, 2010. With permission [20].

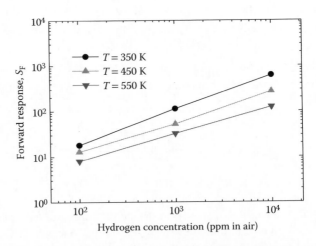

**FIGURE 6.8** Forward response $S_F$ as a function of hydrogen concentration of Pt/β-Ga$_2$O$_3$/ GaN hydrogen sensors exposed to dilute hydrogen ambiences at various temperatures. (Reproduced from Lee, C.T. and Yan, J.T., *Sens. Actuator B-Chem.*, 147, 723, 2010. With permission [20].)

hydrogen sensors on exposure to hydrogen ambiences was attributed to the reduction of β-Ga$_2$O$_3$ resistance induced by the absorption of hydrogen atoms inside the β-Ga$_2$O$_3$ oxide layer. The absorbed hydrogen atoms reacted with β-Ga$_2$O$_3$ and produced localized defect levels in the oxide and consequently reduced the resistivity of the β-Ga$_2$O$_3$ oxide layer [18,20,28].

### 6.4.2 PARAMETER DETERMINATION OF HYDROGEN ADSORPTION PROCESS

#### 6.4.2.1 Analysis of Hydrogen Atom Adsorption at the Interface

As mentioned earlier, the static response of GaN-based MIS hydrogen sensors depended not only on the hydrogen concentration of the ambience but also on the temperature [63,67]. The two important parameters barrier height at the interface and series resistance related to the oxide layer of the MIS hydrogen sensors determined the hydrogen detection characteristics. Deeply understanding the mechanisms and kinetic processes is certainly of great importance.

The kinetic processes of hydrogen atoms adsorbed at the Pt/β-Ga$_2$O$_3$ interface could be understood from the Langmuir adsorption. According to the Langmuir isothermal equation, the fractional coverage θ of the adsorbed hydrogen atoms at the interface under the steady-state condition is described as follows [68]:

$$\frac{\theta}{1-\theta} = K_e \left( \frac{P_{H_2}}{P_{O_2}^{\beta}} \right)^{1/2} \tag{6.14}$$

where $K_e$ is a temperature-dependent equilibrium constant, which depends on the adsorption difference between the surface and the interface; β is the reaction order; $P_{H_2}$ is the hydrogen partial pressure; and $P_{O_2}$ is the oxygen partial pressure. The θ value varies between 0 and 1, where θ = 1 implies that all the available hydrogen adsorption sites are occupied. In the case of hydrogen-containing atmosphere, the oxygen partial pressure $P_{O_2}$ is 20,265 Pa, whereas the hydrogen partial pressures of hydrogen $P_{H_2}$ are 10.1 Pa, 101.3 Pa, and 1013.3 Pa corresponding to 100 ppm H$_2$/air, 1,000 ppm H$_2$/air, and 10,000 ppm H$_2$/air, respectively. The barrier height change depends on the dipole moment density residing at the Pt/oxide interface produced by the interface-adsorbed hydrogen atoms. It is a function of the adsorbed amount of hydrogen atoms at the interface. The barrier height change induced by the hydrogen adsorption can be assumed to be proportional to the hydrogen fractional coverage θ at the interface. Therefore, the barrier height change $\Delta\phi_B$ can be expressed as follows:

$$\Delta\phi_B = \Delta\phi_{B,max} \times \theta \tag{6.15}$$

where $\Delta\phi_{B,max}$ is the maximum barrier height change. By substituting it into Equation 6.13, the maximum barrier height change $\Delta\phi_{B,max}$ is described as follows:

$$\Delta\phi_{B,max} = \frac{kT}{q} \ln \left( \frac{I_{0,H_2,max}}{I_{0,air}} \right) \tag{6.16}$$

where $I_{0,H_2,max}$ is the maximum saturation current for the MIS hydrogen sensors exposed to the dilute hydrogen ambience at a fixed temperature. By combining with Equations 6.13 through 6.16, the Langmuir isothermal equation above the temperature of 350 K can be rewritten as follows [20,67]:

$$\frac{1}{\ln(I_{0,H_2}/I_{0,air})} = \frac{\sqrt{P_{O_2}^\beta}}{K_e \ln(I_{0,H_2,max}/I_{0,air})} \times \frac{1}{\sqrt{P_{H_2}}} + \frac{1}{\ln(I_{0,H_2,max}/I_{0,air})}$$

or                                                                                          (6.17)

$$\frac{1}{\Delta\phi_B} = \frac{\sqrt{P_{O_2}^\beta}}{K_e \Delta\phi_{B,max}} \times \frac{1}{\sqrt{P_{H_2}}} + \frac{1}{\Delta\phi_{B,max}}$$

The reaction order $\beta$ is one for operating temperatures above 348 K [11,12]. For Pt/$\beta$-Ga$_2$O$_3$(4 nm)/GaN MIS-type hydrogen sensors, the plots of $1/\ln(I_{0,H_2}/I_{0,air})$ versus $(P_{H_2})^{-1/2}$ in ambiences at various operating temperatures are shown in Figure 6.9. From the slopes and the intercepts, the temperature-dependent equilibrium constant $K_e$ values of the Pt/$\beta$-Ga$_2$O$_3$/GaN MIS hydrogen sensors were obtained as 28.1, 17.0, and 9.5 at the temperatures 350 K, 450 K, and 550 K, respectively. Similarly, for Pt/i-ZnO/GaN MIS hydrogen sensors the equilibrium constant $K_e$ values were derived to be 12.8, 7.8, 6.6, and 3.5 at the temperatures 350 K, 400 K, 450 K, and 500 K, respectively. The $K_e$ value, in fact, decreased with an increase in the operating temperature. By substituting $K_e$ value into Equation 6.14, the fractional coverage $\theta$ of the adsorbed hydrogen atoms on the interface could be obtained. As an example, for the Pt/$\beta$-Ga$_2$O$_3$/GaN MIS-type hydrogen sensors, on exposure to 10,000 ppm H$_2$/air, the fractional coverage $\theta$ values of the adsorbed hydrogen atoms at the temperatures 350 K, 450 K, and 550 K were deduced to be 0.86, 0.79, and 0.68, respectively. Correspondingly, the $\theta$ values on exposure to 1000 ppm H$_2$/air were 0.67, 0.55, and 0.40, respectively, and the corresponding values induced by 100 ppm H$_2$/air were 0.39, 0.27, and 0.17, respectively.

The equilibrium constant $K_e$ is related to a series of fundamental processes, such as dissociation of molecules, adsorption/desorption for both hydrogen and oxygen atoms, reaction between them, atom diffusion, and so on. It is an important parameter for the adsorption process.

According to the adsorption thermodynamics, the thermal equilibrium constant is related to the adsorption heat and can be described by the van't Hoff relation [69]:

$$\frac{d \ln K_e}{d(1/T)} = -\frac{\Delta H^0}{R}$$                                     (6.18)

where $\Delta H^0$ and $R$ are the hydrogen adsorption enthalpy and gas constant (i.e., 8.315 J/mol·K), respectively. The natural logarithmic value of the equilibrium constant $K_e [\ln(K_e)]$ as a function of the reciprocal temperature (1000/$T$) is shown in the inset of Figure 6.9, from which the slope of about 1.02 is extracted and the hydrogen adsorption enthalpy at the interface is derived to be -8.5 kJ/mol. The negative

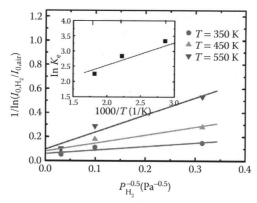

**FIGURE 6.9** Reciprocal natural logarithmic current variation $1/\ln\left(I_{0,\mathrm{H_2}}/I_{0,\mathrm{air}}\right)$ as a function of the square root of the reciprocal hydrogen pressure $(P_{\mathrm{H2}})^{-1/2}$ at 350 K, 450 K, and 550 K. The inset shows natural logarithmic equilibrium constant $\left[\ln\left(K_e\right)\right]$ as a function of reciprocal temperature $(1000/T)$. (Reproduced from Lee, C.T. and Yan, J.T., *Sens. Actuator B-Chem.*, 147, 723, 2010. With permission [20].)

enthalpy obtained is an indication that the kinetic reaction is an exothermic reaction. The results verify why the hydrogen fractional coverage θ of the adsorbed hydrogen atoms decreases at a higher operating temperature. Similar analysis is also carried out for Pt/i-ZnO/GaN MIS hydrogen gas sensors [63] and Pt/reactive insulator/ AlGaN hydrogen sensors with mixed $Ga_2O_3$ and $Al_2O_3$ insulator oxide [67].

### 6.4.2.2 Analysis of Hydrogen Atom Absorption in Oxide Layer

Now, we discuss the series resistance of MIS hydrogen sensors and its change on exposure to hydrogen-containing ambience. As mentioned earlier, the discussion is mainly focused on Pt/β-$Ga_2O_3$/GaN MIS-type hydrogen sensors. When the hydrogen sensors are exposed to dilute hydrogen ambiences, a part of the hydrogen atoms is catalytically dissociated at the surface of the Pt layer and diffuses through the Pt layer to the Pt/β-$Ga_2O_3$ interface. A part of them was then subsequently diffused into the β-$Ga_2O_3$ oxide layer, absorbed or dissolved in the oxide layer. The absorbed hydrogen atom concentration [H] in the β-$Ga_2O_3$ oxide layer, according to Henry's law, is dependent on the hydrogen partial pressure, $P_{\mathrm{H_2}}$, in the ambience as follows [70]:

$$P_{\mathrm{H_2}}^{\alpha} = K_{\mathrm{H}}[\mathrm{H}] \tag{6.19}$$

where $K_{\mathrm{H}}$ is the Henry constant and α is a constant depending on the temperature in which it proceeds to the balancing process. The series resistance changes with the hydrogen concentration in the oxide layer. It is usually assumed that the series resistance change $\Delta R_s$ is proportional to the concentration of the adsorbed hydrogen atoms and can be expressed as follows:

$$\Delta R_s = \Delta R_{s,\max} \times \frac{[\mathrm{H}]}{[\mathrm{H_{\max}}]} \tag{6.20}$$

where $\Delta R_{s,max}$ is the maximum series resistance change, and $[H_{max}]$ is the maximum absorbed hydrogen concentration. By combining Equations 6.19 and 6.20 together, the series resistance change $\Delta R_s$ can be rewritten as follows:

$$\Delta R_s = P_{H_2}^{\alpha} \frac{\Delta R_{s,max}}{K_H[H_{max}]} \tag{6.21}$$

By taking the natural logarithm of both sides, Equation 6.21 can be expressed as follows:

$$\ln \Delta R_s = \alpha \ln P_{H_2} + \ln \frac{A}{K_H} \tag{6.22}$$

where $A = \Delta R_{s,max}/[H]$. The plots of $\ln(\Delta R_s)$ versus $\ln(P_{H_2})$ in ambiences at temperatures of 350 K, 450 K, and 550 K for the Pt/β-Ga$_2$O$_3$/GaN MIS-type hydrogen sensors are shown in Figure 6.10. From the slopes of the plots shown in Figure 6.10, the α values are obtained as 0.50, 0.39, and 0.32 corresponding to the temperatures 350 K, 450 K, and 550 K, respectively. The α value decreases with an increase in operating temperature, indicating that the absorbed hydrogen concentration decreases with an increase in operating temperature as shown in Equation 6.19.

From the intercepts of the plots shown in Figure 6.10, the $\left(K_H/A\right)$ values of the Pt/β-Ga$_2$O$_3$/GaN MIS-type hydrogen sensors induced by balancing hydrogen with air were obtained as 0.24, 0.13, and 0.09 corresponding to the temperatures 350 K,

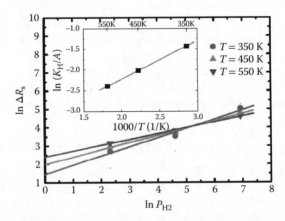

**FIGURE 6.10** Natural logarithmic series resistance change $\ln(\Delta R_s)$ as a function of the natural logarithmic hydrogen partial pressure $\ln(P_{H_2})$ at the temperatures 350 K, 450 K, and 550 K. The inset shows natural logarithmic $\left(K_H/A\right)$ as a function of reciprocal temperature (1000/$T$). (Reproduced from Lee, C.T. and Yan, J.T., *Sens. Actuator B-Chem.*, 147, 723, 2010. With permission [20].)

450 K, and 550 K, respectively. According to the absorption thermodynamics, $K_H$ is related to the absorption heat and, therefore, can be described by the van't Hoff relation. The natural logarithmic value of $K_H/A$ $\left(\ln K_H/A\right)$ as a function of the reciprocal temperature (1000/$T$) is shown in the inset of Figure 6.10. Because the value of $\left[H_{max}\right]/\Delta R_{s,max}$ is kept a constant, the hydrogen absorption enthalpy can be extracted from the slope of the curve. From Figure 6.10, a slope of about 0.92 is obtained. Correspondingly, the obtained enthalpy value is about –7.65 kJ/mol. The negative enthalpy value indicates that the kinetic reaction is an exothermic reaction. Therefore, the hydrogen absorption concentration at the β-Ga$_2$O$_3$ oxide layer is reduced at a higher operating temperature.

### 6.4.2.3 Transient Behavior of Hydrogen Sensors

The sensing process of hydrogen sensors involves a series of elementary processes, which take place simultaneously and compete with each other. It is usually assumed that the response rate is predominantly controlled by the hydrogen adsorption step. To characterize the transient behavior of the hydrogen sensors, the hydrogen sensor response on a stepwise change in hydrogen concentration of the ambience at various temperatures is usually detected. Under this condition (say, from air to H$_2$/air), the rate equation of the hydrogen coverage variation at the metal/oxide interface can be approximately expressed as follows [71]

$$\frac{d\theta}{dt} = k_d(\theta_{eq} - \theta)$$

(6.23)

where $t$ is the measuring time, $k_d$ is the rate constant depending on the hydrogen partial pressure as well as the operating temperature, and $\theta_{eq}$ is the equilibrium fractional coverage of adsorbed hydrogen atoms under the H$_2$/air ambience. Integrating Equation 6.23 and taking into account Equations 6.13 and 6.15, the following relationship is obtained:

$$\ln\left[1 - \frac{\ln\left(I_g/I\right)}{I_{g,eq}/I}\right] = -k_d t$$

(6.24)

where $I$ is the current at the initial state, $I_g$ is the hydrogen response current at time $t$, and $I_{g,eq}$ is the response current at the steady-state condition. That is, the response current (as well as the fractional coverage) changes exponentially toward a new equilibrium value. Using the aforementioned equation, the pressure-dependent rate constant $k_d$ can be obtained from the measured $I_g$–$t$ characteristics. For example, the $k_d$ values of the Pt/β-Ga$_2$O$_3$/GaN hydrogen sensors, on exposure to 100 ppm H$_2$/air, are obtained as 0.105, 0.107, and 0.137 s$^{-1}$ corresponding to the temperatures 350 K, 450 K, and 550 K, respectively. The reciprocal of $k_d$ gives the characteristic response time. Table 6.3 lists the obtained response times for Pt/β-Ga$_2$O$_3$/GaN MIS-type hydrogen sensors on exposure to ambiences of various hydrogen concentrations. It shows that the response time decreases with an

**TABLE 6.3**

**Response Time of Pt/β-Ga₂O₃/GaN MIS-Type Hydrogen Sensors Exposed to Air Ambience and Various Hydrogen-Containing Ambiences at Various Operation Temperatures**

|  | Temperature | 100 ppm H₂/Air | 1,000 ppm H₂/Air | 10,000 ppm H₂/Air |
|---|---|---|---|---|
| Response time (s) | 350 K | 8.1 | 5.0 | 1.0 |
|  | 450 K | 3.7 | 3.1 | 0.8 |
|  | 550 K | 3.0 | 2.0 | 0.7 |

*Source:* Lee, C.T. and Yan, J.T., *Sens. Actuator B-Chem.*, 147, 723, 2010. With permission [20].

increase in operating temperature and hydrogen partial pressure, as expected. In other words, a higher detection speed is achievable at a higher operating temperature or a higher hydrogen concentration.

## 6.5 SUMMARY

GaN-based hydrogen sensors have attracted extensive attention for operation at high temperature. Various types of devices, such as MS diodes, MIS diodes, high-electron-mobility transistors (HEMTs), have been developed and explored. Among them, MIS structures are of particular interest, for which the insulator layers, especially the reactive oxide layers, play an important role in the hydrogen-detection process. In this chapter, we introduced two unique methods, PEC oxidation and vapor cooling condensation methods, for the deposition of high-quality oxide layers. The deposited oxide layers were successfully used to construct MIS hydrogen sensors. The fabricated sensors exhibited excellent performances, which indicated that two unique methods were very promising for future application in the fabrication of hydrogen sensors. As rapidly developing fields, various aspects of hydrogen sensors have been studied, including nanostructured sensors, which are expected to be of higher sensitivity. Even in this case, the introduced methods have also exhibited promising applications.

## REFERENCES

1. I. Lundström, H. Sundgren, F. Winquist, M. Eriksson, C.K. Rulcker, and A.L. Spetz, Twenty-five years of field effect gas sensor response in Linkoping, *Sens. Actuator B-Chem.* **121**, 247 (2007).
2. K. Luongo, A. Sine, and S. Bhansali, Development of a highly sensitive porous Si based hydrogen sensor using Pd nano-structures, *Sens. Actuator B-Chem.* **111**, 125 (2005).
3. A. Trinchi, S. Kandasamy, and W. Wlodarski, High temperature field effect hydrogen and hydrocarbon gas sensors based on SiC MOS devices, *Sens. Actuator B-Chem.* **133**, 705 (2008).
4. S.Y. Chiu, H.W. Huang, T.H. Huang, K.C. Liang, K.P. Liu, J.H. Tsai, and W.S. Lour, Comprehensive study of Pd/GaN metal–semiconductor-metal hydrogen sensors with

symmetrically bi-directional sensing performance, *Sens. Actuator B-Chem.* **138**, 422 (2009).

5. S. Petrescu, C. Petre, M. Costea, O. Malancioiu, N. Boriaru, A. Dobrovicescu, M. Feidt, and C. Harman, A methodology of computation, design and optimization of solar stirling power plant using hydrogen/oxygen fuel cells, *Energy* **35**, 729 (2010).

6. I. Cumalioglu, A. Ertas, Y. Ma, and T. Maxwell, State of the art: Hydrogen storage, *J. Fuel Cell Sci. Technol.* **5**, 034001 (2008).

7. T. Huberta, L. Boon-Brettb, G. Blackb, and U. Banacha, Hydrogen sensors—A review, *Sens. Actuator B-Chem.* **157**, 329 (2011).

8. I. Lundström, S. Shivaraman, C. Svensson, and L. Lundkvist, A hydrogen-sensitive MOS field-effect transistor, *Appl. Phys. Lett.* **26**, 55 (1975).

9. I. Lundström, S. Shivaraman, and C. Svensson, A hydrogen-sensitive Pd-gate MOS transistor, *J. Appl. Phys.* **46**, 3876 (1975).

10. I. Lundström and T. Distefano, Hydrogen induced interfacial polarization at Pd-SiO$_2$ interfaces, *Surf. Sci.* **59**, 23 (1976).

11. I. Lundström, Hydrogen sensitive MOS-structures part 1: Principles and applications, *Sens. Actuator B-Chem.* **1**, 403 (1981).

12. I. Lundström, Hydrogen sensitive MOS-structures part 2: Characterization, *Sens. Actuator B-Chem.* **2**, 105 (1981).

13. L. Stiblert and C. Svensson, Hydrogen leak detector using a Pd-gate MOS transistor, *Rev. Sci. Instrum.* **46**, 1206 (1975).

14. L.G. Petersson, H.M. Dannetun, J. Fogelberg, and I. Lundström, Hydrogen adsorption states at the external and internal palladium surfaces of a palladium-silicon dioxide-silicon structure, *J. Appl. Phys.* **58**, 404 (1985).

15. I. Lundström and T. Distefano, Influence of hydrogen on Pt-SiO$_2$-Si structures, *Solid State Commun.* **19**, 871 (1976).

16. C.K. Kim, J.H. Lee, S.M. Choi, I.H. Noh, H.R. Kim, N.I. Cho, C. Hong, and G.E. Jang, Pd and Pt-SiC Schottky diodes for detection of H$_2$ and CH$_4$ at high temperature, *Sens. Actuator B-Chem.* **77**, 455 (2001).

17. A. Trinchi, W. Wlodarski, and Y.X. Li, Hydrogen sensitive Ga$_2$O$_3$ Schottky diode sensor based on SiC, *Sens. Actuator B-Chem.* **100**, 94 (2004).

18. J.T. Yan and C.T. Lee, Improved detection sensitivity of Pt/β-Ga$_2$O$_3$/GaN hydrogen sensor diode, *Sens. Actuator B-Chem.* **143**, 192 (2009).

19. C.T. Lee and J.T. Yan, Investigation of a metal–insulator–semiconductor Pt/mixed Al$_2$O$_3$ and Ga$_2$O$_3$ insulator/AlGaN hydrogen sensor, *J. Electrocehm. Soc.* **157**, J281 (2010).

20. C.T. Lee and J.T. Yan, Sensing mechanisms of Pt/β-Ga$_2$O$_3$/GaN hydrogen sensor diodes, *Sens. Actuator B-Chem.* **147**, 723 (2010).

21. C.T. Lee, H.W. Chen, and H. Y. Lee, Metal-oxide-semiconductor devices using Ga$_2$O$_3$ dielectrics on n-type GaN, *Appl. Phys. Lett.* **82**, 4304 (2003).

22. C.T. Chang, S.K. Hsiao, E.Y. Chang, C.Y. Lu, J.C. Huang, and C.T. Lee, Changes of electrical characteristics for AlGaN/GaN HEMTs under uniaxial tensile strain, *IEEE Electron Devices Lett.* **30**, 213 (2009).

23. Y.L. Chiou, L.H. Huang, and C.T. Lee, GaN-based p-type metal-oxide-semiconductor devices with a gate oxide layer grown by a bias-assisted photoelectrochemical oxidation method, *Semicond. Sci. Technol.* **25**, 045020 (2010).

24. H.Y. Lee, X.Y. Huang, and C.T. Lee, Light output enhancement of GaN-based roughened LEDs using bias-assisted photoelectrochemical etching method, *J. Electrochem. Soc.* **155**, H707 (2008).

25. B.P. Luther, S.D. Wolter, and S.E. Mohney, High temperature Pt Schottky diode gas sensors on n-type GaN, *Sens. Actuator B-Chem.* **56**, 164 (1999).

26. M. Ali, V. Cimalla, V. Lebedev, H. Romanus, V. Tilak, D. Merfeld, P. Sandvik, and O. Ambacher, Pt/GaN Schottky diodes for hydrogen gas sensors, *Sens. Actuator B-Chem.* **113**, 797 (2006).

27. J. Schalwig, G. Muller, U. Karrer, M. Eickhoff, O. Ambacher, M. Stutzmann, L. Gorgens, and G. Dollinger, Hydrogen response mechanism of Pt–GaN Schottky diodes, *App. Phys. Lett.* **80**, 1222 (2002).

28. A. Trinchi, S. Kaciulis, L. Pandolfi, M.K. Ghantasala, Y.X. Li, W. Wlodarski, S. Viticoli, E. Comini, and G. Sberveglieri, Characterization of $Ga_2O_3$ based MRISiC hydrogen gas sensors, *Sens. Actuator B-Chem.* **103**, 129 (2004).

29. B.S. Kang, F. Ren, B.P. Gila, C.R. Abernathy, and S.J. Pearton, AlGaN/GaN-based metal–oxide–semiconductor diode-based hydrogen gas sensor, *App. Phys. Lett.* **84**, 1123 (2004).

30. B.S. Kang, S. Kim, F. Ren, B.P. Gila, C.R. Abernathy, and S.J. Pearton, Comparison of MOS and Schottky W/Pt–GaN diodes for hydrogen detection, *Sens. Actuator B-Chem.* **104**, 232 (2005).

31. T.H. Tsai, J.R. Huang, K.W. Lin, W.C. Hsu, H.I. Chen, and W.C. Liu, Improved hydrogen sensing characteristics of a Pt/SiO_2/GaN Schottky diode, *Sens. Actuator B-Chem.* **129**, 292 (2008).

32. Y. Irokawa, Y. Sakuma, and T. Sekiguchi, Effect of dielectrics on hydrogen detection sensitivity of metal–insulator–semiconductor Pt–GaN diodes, *Jpn. J. Appl. Phys.* **46**, 7714 (2007).

33. O. Weidemann, M. Hermann, G. Steinhoff, H. Wingbrant, A.L. Spetz, M. Stutzmann, and M. Eickhoff, Influence of surface oxides on hydrogen-sensitive Pd:GaN Schottky diodes, *Appl. Phys. Lett.* **83**, 773 (2003).

34. Y. Irokawa, Interface states in metal-insulator-semiconductor Pt-GaN diode hydrogen sensors, *J. Appl. Phys.* **113**, 026104 (2013).

35. M. Fleischer and H. Meixner, Sensitive, selective and stable $CH_4$ detection using semiconducting $Ga_2O_3$ thin films, *Sens. Actuator B-Chem.* **26**, 81 (1995).

36. T. Weh, J. Frank, M. Fleischer, and H. Meixner, On the mechanism of hydrogen sensing with $SiO_2$ modified high temperature $Ga_2O_3$ sensors, *Sens. Actuator B-Chem.* **78**, 202 (2001).

37. Y.X. Li, A. Trinchi, W. Wlodarski, K. Galatsis, and K. Kalantar-zadeh, Investigation of the oxygen gas sensing performance of $Ga_2O_3$ thin films with different dopants, *Sens. Actuator B-Chem.* **93**, 431 (2003).

38. L.M. Lechuga, A. Calle, D. Golmayo, P. Tejedor, and F. Briones, A new hydrogen sensor based on a Pt/GaAs Schottky diode, *J. Electrochem. Soc.* **138**, 159 (1991).

39. L.M. Lechuga, A. Calle, D. Golmayo, and F. Briones, Different catalytic metals (Pt, Pd and Ir) for GaAs Schottky barrier sensors, *Sens. Actuator B-Chem.* **7**, 614 (1992).

40. W.C. Liu, H.J. Pan, H.I. Chen, K.W. Lin, S.Y. Cheng, and K.H. Yu, Hydrogen-sensitive characteristics of a novel Pd/InP MOS Schottky diode hydrogen sensor, *IEEE Trans. Electron Devices* **48**, 1938 (2001).

41. M. Yousuf, B. Kuliyev, B. Lalevic, and T.L. Poteat, Pd-InP Schottky diode hydrogen sensors, *Solid-State Electron.* **25**, 753 (1985).

42. C.K. Kim, J.H. Lee, H.Y. Lee, N.I. Cho, and D.J. Kim, A study on a platinum–silicon carbide Schottky diode as a hydrogen gas sensor, *Sens. Actuator B-Chem.* **66**, 116 (2000).

43. P.K. Karin, Semiconductor junction gas sensors, *Chem. Rev.* **108**, 367 (2008).

44. M. Eriksson, A. Salomonsson, I. Lundström and A. E. Åbom, The influence of the insulator surface properties on the hydrogen response of field-effect gas sensors, *J. Appl. Phys.* **98**, 034903 (2005).

45. W. Jochum, S. Penner, K. Föttinger, R. Kramer, G. Rupprechter, and B. Klötzer, Hydrogen on polycrystalline β-$Ga_2O_3$: Surface chemisorption, defect formation, and reactivity, *J. Catal.* **256**, 268 (2008).

46. G. Steinhoff, M. Hermann, W.J. Schaff, L. F. Eastman, M. Stutzmann, and M. Eickhoff, pH response of GaN surfaces and its application for pH-sensitive field-effect transistors, *Appl. Phys. Lett.* **83,** 177 (2003).
47. S.J. Pearton, B.S. Kang, S. Kim, F. Ren, B.P. Gila, C.R. Abernathy, J. Lin, and S.N.G. Chu, GaN-based diodes and transistors for chemical, gas, biological and pressure sensing, *J. Phys.: Condens. Matter* **16**, R961 (2004).
48. L.H. Huang, S.H. Yeh, C.T. Lee, H.P. Tang, J. Bardwell, and J.B. Webb, AlGaN/GaN metal-oxide-semiconductor high-electron mobility transistors using oxide insulator grown by photoelectrochemical oxidation method, *IEEE Electron Device Lett.* **29**, 284 (2008).
49. L.H. Huang and C.T. Lee, Investigation and analysis of AlGaN MOS devices with an oxidized layer grown using the photoelectrochemical oxidation method, *J. Electrochem. Soc.* **154**, H862 (2007).
50. Y.L. Chiou, L.H. Huang, and C.T. Lee, Photoelectrochemical function in gate-recessed AlGaN/GaN metal-oxide-semiconductor high-electron-mobility transistors, *IEEE Electron Device Lett.* **31**, 183 (2010).
51. C.T. Lee, H.Y. Lee, and H.W. Chen, GaN MOS device using $SiO_2$-$Ga_2O_3$ insulator grown by photoelectrochemical oxidation method, *IEEE Electron Device Lett.* **24**, 54 (2003).
52. C.T. Lee, H.W. Chen, F.T. Hwang, and H.Y. Lee, Investigation of Ga oxide films directly grown on n-type GaN by photoelectrochemical oxidation using He-Cd laser, *J. Electron. Mater.* **34**, 282 (2005).
53. C.T. Lee, Y.K. Su, and H.M. Wang, Effects of r.f. sputtering parameters on ZnO films deposited on to Ga as substrates, *Thin Solid Films* **150**, 283 (1987).
54. F.X. Xiu, Z. Yang, L.J. Mandalapu, J.L. Liu, and W.P. Beyermann, P-type ZnO films with solid-source phosphors doping by molecular-beam epitaxy, *Appl. Phys. Lett.* **88**, 052106 (2006).
55. L.W. Lai and C.T. Lee, Investigation of optical and electrical properties of ZnO thin films, *Mater. Chem. Phys.* **110**, 393 (2008).
56. S.J. Pearton, C.R. Abernathy, M.E. Overberg, G.T. Thaler, D.P. Norton, N. Theodoropoulou, A.F. Hebard, Y.D. Park, F. Ren, J. Kim, and L.A. Boatner, Wide band gap ferromagnetic semiconductors and oxides, *J. Appl. Phys.* **93**, 1 (2003).
57. L.W. Lai, J.T. Yan, C.H. Chen, L.R. Lou, and C.T. Lee, Nitrogen function of aluminum nitride codoped ZnO films deposited using cosputter system, *J. Mater. Res.* **24**, 2252 (2009).
58. R.W. Chuang, R.X. Wu, L.W. Lai, and C.T. Lee, ZnO-on-GaN heterojunction light emitting diodes grown by vapor cooling condensation technique, *Appl. Phys. Lett.* **91**, 231113 (2007).
59. H.Y. Lee, S.D. Xia, W.P. Zhang, L.R. Lou, J.T. Yan, and C.T. Lee, Mechanisms of high quality i-ZnO thin films deposition at low temperature by vapor cooling condensation technique, *J. Appl. Phys.* **108**, 073119 (2010).
60. C.T. Lee, Y.L Chiou, and C.S. Lee, AlGaN/GaN MOS-HEMTs with gate ZnO dielectric layer, *IEEE Electron Device Lett.* **31**, 1220 (2010).
61. C.T. Lee, Fabrication methods and luminescent properties of ZnO materials for light-emitting diodes, *Mater.* **3**, 2218 (2010).
62. C.T. Lee, Y.S. Chiu, L.R. Lou, S.C. Ho, and C.T. Chuang, Integrated pH sensors and performance improvement mechanism of ZnO-based ion-sensitive field-effect transistors, *IEEE Sens. J.* **14**, 490 (2014).
63. H.Y. Lee, H.L. Huang, and C.T. Lee, Hydrogen sensing performances of Pt/i-ZnO/GaN metal–insulator–semiconductor diodes, *Sens. Actuator B-Chem.* **157**, 460 (2011).
64. E.H. Rhoderick and R.H. Williams, *Metal-Semiconductor Contacts*, 2nd ed., Clarendon Press, Oxford, United Kingdom (1988).

65. S. Nakagomi, K. Okuda, and Y. Kokubum, Electrical properties dependent on $H_2$ gas for new structure diode of Pt–thin $WO_3$–SiC, *Sens. Actuator B-Chem.* **96**, 264 (2003).

66. Y.Y. Tsai, K.W. Lin, H.I. Chen, I.P. Liu, C.W. Hung, T.P. Chen, T.H. Tsai, L.Y. Chen, K.Y. Chu, and W.C. Liu, Hydrogen sensing properties of a Pt-oxide–GaN Schottky diode, *J. Appl. Phys.* **104**, 024515 (2008).

67. H.Y. Lee and C.T. Lee, Thermodynamic sensing mechanisms of AlGaN-based metal/reactive insulator/semiconductor-type hydrogen sensors, *IEEE Sens. J.* **12**, 1450 (2012).

68. W.P. Kang and Y. Gurbuz, Comparison and analysis of Pd- and Pt-GaAs Schottky diodes for hydrogen detection, *J. Appl. Phys.* **75**, 8175 (1994).

69. R.J. Silbey and R.A. Alberty, *Physical Chemistry*, 3rd ed., Wiley, New York (2001).

70. M. Pasturel, R.J. Wijngaarden, W. Lohstroh, H. Schreuders, M, Slaman, B. Dam, and R. Griessen, Influence of the chemical potential on the hydrogen sorption kinetics of $Mg_2Ni/TM/Pd$ (TM = transition metal) trilayers, *Chem. Mater.* **19**, 624 (2007).

71. H.I. Chen, Y.I. Chou, and C.K. Hsiung, Comprehensive study of adsorption kinetics for hydrogen sensing with an electronless-plated Pd-InP, *Sens. Actuator B-Chem.* **92**, 6 (2003).

# 7 InGaN-Based Solar Cells

*Ezgi Dogmus and Farid Medjdoub*

## CONTENTS

Climate change is one of the most significant phenomena of the twenty-first century. Most of the climate scientists agree that global warming is caused by the extensive exploitation of fossil fuels, such as coal, crude oil, and natural gas, which produce increased levels of greenhouse gases [1]. Over the last decades, our world has already confronted the melting of icecaps and glaciers, accelerated rise of sea level, and abrupt changes in weather patterns as a result of the inexorable global warming. In addition to the environmental concerns, it should be seriously considered that the supply of these fossil fuels is finite despite our progressively growing dependence on them due to the increasing energy demand of the world population. Given the environmental, social, and economic problems associated with the continued use of natural resources, it becomes crucial to switch to sustainable and renewable energy sources, which can adequately meet potential energy needs. Renewable electricity generated by renewable sources of energy, which include solar, wind, hydro, tidal, geothermal heat, and biomass, accounted for 20.1% share [2] of European (E-27) gross electricity consumption in 2010 and 9.3% share [3] of total energy consumption in the United States in 2011.

## 7.1   SOURCE OF ALL LIFE: THE SUN

Among the various options of renewable energy sources listed earlier, sunlight is the most promising one. It is a remarkable fact that the solar energy received on the top of the atmosphere in an hour ($1.7 \times 10^5$ TW) is much more than the energy annually consumed by the whole world [4]. The sun is not only the origin of almost all of the renewable and conventional energy sources but also an inexhaustible, absolutely safe, omnipresent, and free energy source that is required by the nature to sustain life. In addition to its functionality in heating, photosynthesis, and formation of fossil fuels, sunlight can also be directly converted into electricity by a process called photovoltaic (PV) process. This way of harnessing solar energy is highly reliable, environmentally safe, extremely clean, and easy to install, and it operates at low cost once the solar technologies are installed without producing any greenhouse gas emission. Solar panel electricity systems incorporate PV cells that consist of semiconductor materials that capture the sun's energy and convert it into electricity.

## 7.2   PRINCIPLE OF SOLAR CELL OPERATION

Direct PV energy conversion in solar cells (SCs) results from the PV effect, and it has fundamentally two essential steps, charge generation and charge separation. A large part of the solar spectrum is targeted to be absorbed in a light-absorbing semiconductor called emitter where the charge carriers are excited to higher energy states and electron–hole pairs are generated. Different absorption properties can be obtained by making use of different types of emitter materials such as inorganic or organic semiconductors, dye molecules, quantum dots (QDs), or quantum wells (QWs). Once the photoexcited charge carriers are created, they have to be separated by means of a driving force provided by the pn junction. Finally, owing to the presence of conductive contact layers, the electrons are collected at the n-contact and the holes at the p-contact, thus generating electrical power.

Basically, SCs are based on the formation of a built-in electrical field created at the pn junction to separate the charges. The junction between the emitter and base materials can be a pn homojunction when it is formed between the same semiconducting material with different doping or a pn heterojunction when it is between two different semiconductors with different doping. Today, the diversity of SCs with different PV performances results simply from the choices of the emitter material and the charge separation mechanism used to create the built-in electric field.

## 7.3   SHORT HISTORY OF PHOTOVOLTAICS

Since the discovery of PV effect by Becquerel [5] and the development of the first silicon SC with 6% efficiency by Chapin et al. [6], the progress in higher efficiency and lower cost PV technologies has expanded very rapidly with a significant potential for long-term growth in the world. In the last few decades, technological leaps and continual innovations both in material science and in solar panel production have dramatically lowered solar manufacturing costs and PV prices. Figure 7.1 shows that over the past decade, the regional PV cell and module shipments from the United States,

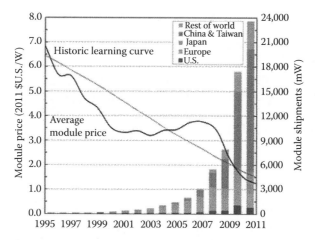

**FIGURE 7.1** Photovoltaic price and performance trends over the past decade. (From Goodrich, A.C. et al., *Energy Environ. Sci.*, 6, 2811–2821, 2013. With permission [7].)

Japan, Europe, China/Taiwan, and the rest of the world (ROW) achieved a compound annual growth rate of 52% and attained a reduction in the average global crystalline silicon modules from 2008 to 2011 [7]. In 2012, the world's cumulative PV electricity capacity passed 100 gigawatts (GW) [8], which corresponds to an annual capacity of production of electricity from 16 coal power plants or nuclear reactors of 1 GW each.

In solar technologies, conversion efficiency and cost are two features that set the road map of SCs. The main goal is always to acquire volume production of high-efficiency and high-performance solar modules at low cost. The search for the ideal material or material systems began with crystalline silicon-based SCs, the first-generation SCs, which represent 90% of the global market today. Currently, the most efficient first-generation SCs are based on monocrystalline silicon solar (c-Si) cells, which benefited greatly from a high-quality production technology called Czochralski process. With innovative technologies such as surface texturing [9], using buried contacts [10], and reducing wafer thickness, high efficiencies up to 27.6% have been achieved by c-Si SCs [11]. Although c-Si SCs offer high carrier mobilities and a large spectral absorption range, single junction silicon devices have almost approached their detailed balance limit of 30% and their manufacturing technology remains laborious and expensive [12].

In accordance with the purpose of lowering the cost of production and using less material, second-generation SCs have been developed in the form of thin-film SCs of amorphous silicon (a-Si), polycrystalline silicon (poly-Si), cadmium telluride (CdTe), or copper indium gallium selenide (CIGS). However, Si-based second-generation SCs suffer mostly from the structural defects and stability (for a-Si) that make them less efficient compared to first-generation cells. On the other hand, CIGS SCs, in which 99% of the sunlight is absorbed in the first micron of the cell owing to high optical absorption coefficient of $CuInSe_2$ compound [13], have been able to reach efficiencies over 20% in laboratory studies on flexible substrates [14,15]. For CdTe SCs that have high quality and less expensive large-scale production, best research efficiencies of 20.4% have been achieved [16], although toxicity is one of the big obstacles for the production and the recycling of the solar modules [17].

Even though the costs were progressively reduced as the second-generation SCs became mature, so far none of them have been able to compete with the bulk crystalline silicon SCs in terms of high performance, stability, abundance, and nontoxicity. On the other hand, the main driving force for the development of third-generation SCs results from the fact that the need for low-cost solar modules with efficiencies higher than the theoretical limit of the traditional single-junction Si PV devices [12] is still not fulfilled.

### 7.3.1 High-Efficiency Solar Cell Designs

With the evolution of new materials technology and thin-film growth techniques, third-generation SCs offer several promising approaches to exceed the detailed balance limit of single junction SCs. Today, high-efficiency SC designs include the use of concentrated sunlight with the help of mirrors and lenses and/or stacking of multiple cells with semiconductors of different bandgaps in tandem cell arrangements. In addition, intermediate bandgap approach of incorporating multiple absorption bands by low dimensional structures such as multiple QDs or QWs is also promising for high-efficiency SCs. Further approaches of the third-generation SCs include the creation of multiple electron pairs per incident photon, or the use of hot carriers, which allow excess energy in the bandgap to be stored in the photoexcited carrier populations and to be converted into electricity- or thermoelectric-based SCs to transform the extracted energy of a heated absorber to electrical energy [18].

#### 7.3.1.1 Tandem Cells

Single junction SCs suffer from two major losses: First, the photons having less energy than the bandgap of the semiconductor are not absorbed (transmission losses); second, the excess energy of high-energy photons are lost as heat (thermalization losses). Tandem SCs can revolutionarily eliminate these losses by splitting the spectrum and absorbing efficiently different sections of the spectrum by stacking several cells with different bandgaps. Semiconductors having higher bandgaps are designed to be stacked on top to absorb the high-energy photons by filtering and transmitting the lower energy photons to the subcells beneath with lower bandgaps. Tandem cells are classified into two groups: "stacked cells," which consist of isolated and individually probed cells, and "monolithic cells," in which the subcells are series connected internally by tunnel junctions (TJs) on one substrate. For an infinite stack of independently operated cells, tandem cell conversion efficiency is calculated to be 86.8% [19]. In addition, the tandem cell performance increases by adding new subcells in the stack by keeping the ohmic contact resistances low between the cells.

Tandem SC designs based on different III-V semiconductors are the most successful method to achieve high efficiencies. Currently, the monolithic triple junction of GaInP-GaInAs-Ge tandem holds the world's efficiency record of 44.7% under concentrated sunlight. Using multijunction SCs based on III-V semiconductors in concentrating photovoltaics (CPVs) exhibits the benefit of minimizing the expensive material usage while increasing the conversion efficiency compared to traditional Si SCs [20].

### 7.3.1.2  Intermediate Band Cells

Intermediate SCs have recently emerged as third-generation SCs enabling absorption and utilization of photons having energies lower than the bandgap of the semiconductor material. To allow sub-bandgap absorption within the system, multiple quasi-Fermi levels can be created by the incorporation of defects, impurities [21], QDs, or QWs [22]. The quantum confinement approach that uses QDs [23] and QWs having lower bandgap material in a higher bandgap host material is quite promising for efficiency improvements in SCs. QD SCs, which are based on carrier multiplication and production of hot carriers, have the main challenge of defect-free high growth quality of multiple layers of QDs for efficient connection of individual dots [24]. Alternatively, QW SC designs employ a p-i-n junction with QWs located in the intrinsic region to confine the carriers into two dimensions [25]. The QWs generate quantized energy levels in which lower energy photons are captured as they shift the spectral response of the cell to lower energy ranges. Although, lower bandgap wells decrease the open-circuit voltage, higher overall power output is achieved by the increased photocurrent leading to higher conversion efficiencies. The absorption edge and efficiency of the SC can be tailored by bandgap, size, and number of incorporated QWs. Recent studies with the use of concentrated sunlight for GaAsP/InGaAs single junction QW SCs and QWs in InGaP/GaAs tandem SCs [26] have reached high efficiencies (higher than 28%), making the QW approach attractive for high-performance SCs.

### 7.3.1.3  Very-High-Efficiency Solar Cells

Very-high-efficiency solar cell (VHESC) design is based on the codesign of an optical system, interconnects, and PV cells and targets a practical efficiency greater than 50% [27]. This approach focuses on the integration of high-performance multiple junction optical/solar modules while retaining low area costs. The design includes the use of current state-of-the-art PV technologies and materials that have demonstrated either higher performance at similar cost or lower cost at the same performance in novel cell architectures.

A VHESC six-junction SC is proposed to have subcells of different material systems that show the highest performance for their bandgap range, that is, Si subcell for energy, $E_g$, near 1.0 eV and GaInAsP subcells for the energy range between 1.4 and 1.9 eV. This design can achieve a theoretical efficiency of 54% at 20 suns, as shown in Table 7.1.

However, to achieve the target efficiencies higher than 50%, this VHESC design needs a sub-module including a bandgap of 2.4 eV. In this frame, indium gallium nitride (InGaN) material system is one of the most promising material to provide such a wide bandgaps of 2.4 eV or greater.

In addition, Figure 7.2 exhibits the U.S. Department of Energy's National Renewable Energy Laboratory's (NREL) latest chart of record efficiency SCs [28]. The current efficiency record, which is held by an InGaP/GaAs/InGaAs-based triple-junction tandem SC at 44.4% under a light-concentrating magnification of

**TABLE 7.1**

**Predicted Contributions of Each Solar Cell in the Proposed Very High-Efficiency Solar Cell Six-Junction Design**

| 6J SC Bandgaps | Thermodynamic Efficiency | Derating Factor | Ideal Target Efficiency |
|---|---|---|---|
| High $E_g$ 2.4 eV | 14.9% | 0.89 | 13.3% |
| GaInP ($E_g$ = 1.84 eV) | 16.6% | 0.86 | 14.3% |
| GaAs ($E_g$ = 1.43 eV) | 13.9% | 0.84 | 11.7% |
| Si ($E_g$ = 1.12 eV) | 9.7% | 0.80 | 7.8% |
| 0.95 eV | 5.0% | 0.74 | 3.7% |
| 0.7 eV | 4.1% | 0.70 | 2.9% |
| | Total η = 64.2% | | Total η = 53.5% |

*Source:* Barnett, A. et al., *Prog. Photovoltaics Res. Appl.*, 17(1), 75–83, 2009. With permission [27].

302 times [29], shows explicitly the domination of multijunction concentrator SCs over other technologies. However, modeling results and material limitations indicate that such structures are reaching their theoretical efficiency limits. Consequently, exploration of new material systems becomes indispensable to overcome the 50% efficiency barrier. The InGaN material system, which demonstrates the versatility, promises a successful high-efficiency PV material. As a result, in the following parts of this chapter, III-nitride semiconductors will be described in the frame of the development of high-efficiency PV. The current status and main challenges of InGaN-based SCs will be discussed.

## 7.4 III-NITRIDE SEMICONDUCTORS FOR PHOTOVOLTAIC APPLICATIONS

In the last 50 years, PV technologies have revealed numerous breakthrough efficiencies with the single junction crystalline Si SCs and the triple GaInP-GaInAs-Ge tandem cells. Even though, evolutionary process technologies and light-trapping methods lead to higher conversion efficiencies with current material systems, the present state-of-the-art efficiency of the single and triple junction SCs have almost reached their maximum achievable limit [30] and hence, the search for new approaches or materials that boost efficiencies beyond 50% remains to be the main challenge. Table 7.2 shows detailed achievable balance limit of multijunction SCs under black body radiation at 6000 K under 500 suns. To achieve practical PV efficiencies greater than 50%, the tandem cells designs of four or more stacks having a material with bandgap greater than 2.4 eV are required.

Over the last decade, III-V nitride material system that include GaN, InN, AlN, and its alloys, has attracted many researchers for the full-solar spectrum PV applications owing to its direct bandgap range between 0.65 eV and 3.4 eV [32]. The Ga-rich and high bandgap InGaN compound have already been extensively used in high-efficiency light-emitting diodes (LEDs) and lasers with mature growth and processing techniques [33,34]. The recently emerging green LEDs made of lower bandgap

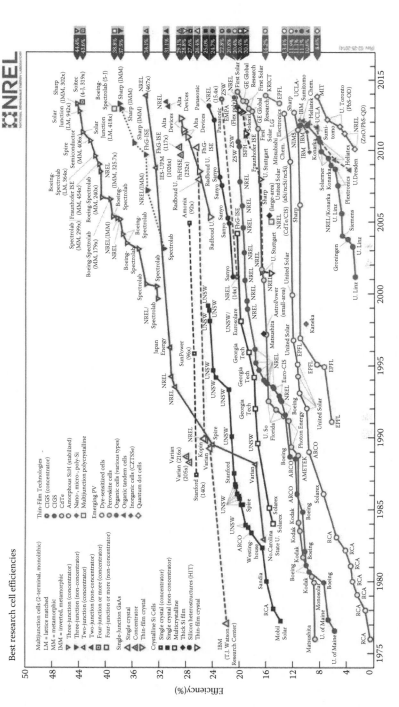

**FIGURE 7.2** Progressive development of best research efficiency solar cells from 1977 to 2014 illustrated by U.S. Department of Energy's National Renewable Energy Laboratory. (From http://www.nrel.gov/ncpv/images/efficiency_chart.jpg.)

**TABLE 7.2**

**Detailed Balance Calculations of Bandgaps and Achievable Efficiencies of Three to Eight Stack Tandem Cells**

| Number of Stacks | Values of Bandgap (eV) | | | | | | | Maximum Limit Efficiency, $\eta$ (%) | Achievable Efficiency, $0.8 \times \eta$ (%) |
|---|---|---|---|---|---|---|---|---|---|
| 3 | 0.7 | 1.37 | 2 | | | | | 56 | 44.8 |
| 4 | 0.6 | 1.11 | 1.69 | 2.48 | | | | 62 | 49.6 |
| 5 | 0.53 | 0.95 | 1.4 | 1.93 | 2.68 | | | 65 | 52 |
| 6 | 0.47 | 0.84 | 1.24 | 1.66 | 2.18 | 2.93 | | 67.3 | 53.84 |
| 7 | 0.47 | 0.82 | 1.19 | 1.56 | 2 | 2.5 | 3.21 | 68.9 | 55.12 |
| 8 | 0.44 | 0.78 | 1.09 | 1.4 | 1.74 | 2.14 | 2.65 3.35 | 70.2 | 56.16 |

*Source:* Jani, O.K., Development of wide-bandgap InGaN solar cells for high efficiency photovoltaics. PhD Thesis, Georgia Institute of Technology, 2008. With permission [31].

$In_xGa_{1-x}N$ alloy with indium (In) content ($x$) higher than 0.2 have been served as useful guidance for the development of InGaN-based SCs that can be employed in ultrahigh efficiency tandems with five or greater junctions.

InGaN material system has numerous advantages such as high carrier mobility, high drift velocity, low effective mass of electrons and holes, high thermal conductivity, and high temperature stability, which makes it promising for potential use under concentrated sunlight [35]. Besides, the III-nitrides are also promising for radiation-hard high-efficiency SC for space applications and operation under harsh environments [36]. In addition, the III-nitride semiconductors possess large spontaneous and piezoelectric polarization [37]. This introduces electric fields and surface dipoles, which provide various useful designs of polarization, induced TJs [38,39].

### 7.4.1 CURRENT STATUS OF INGaN-BASED SOLAR CELLS

Since the InGaN was proposed as a potential material for SCs by Wu et al. [36] in 2003, significant progress has been established while many challenges remain to be overcome to meet the requirements of high-efficiency III-nitride PV. In the initial stages, InGaN alloy was examined extensively for single junction SC applications with Ga-rich materials owing to mature growth and doping technologies of GaN. The first clear demonstrations of PV effect were achieved by designs of Jani et al. [40] and Yang et al. [41] having Ga-rich p-n $In_xGa_{1-x}N$ homojunction and p-GaN/InGaN/n-GaN heterojunctions [42] grown by metal-organic vapor phase epitaxy (MOCVD). These preliminary studies resulted in the evolution of fabrication of InGaN-based SCs in two different axes: homojunction InGaN SC sand heterojunction SCs including $In_xGa_{1-x}N$ sandwiched between p-GaN and n-GaN. Further advances and efforts, which will be detailed throughout this chapter, include the growth of In-rich InGaN layers by molecular beam epitaxy (MBE) [43,44], the introduction of InGaN/GaN multiple quantum wells (MQWs) in the absorption region [45], the demonstration [46] and fabrication [47] of two-junction InGaN/Si tandem SCs, the optimization of the top contact grid [48], formation of low-resistance p-type ohmic contacts by the introduction of different current spreading layers (CSL) [49,50], and the utilization of different substrates with various antireflection methods to enhance the external quantum efficiencies (EQE) and internal quantum efficiencies (IQE) of the devices [51,52].

### 7.4.1.1 Homojunction InGaN Solar Cells

High-efficiency SCs based on InGaN can be achieved only by exploiting $In_xGa_{1-x}N$ layers incorporating In contents higher than 0.2 [53]. The primary studies based on InGaN SCs reported the development of $In_xGa_{1-x}N$ homojunction SCs due to the large mismatch between In-rich InGaN and GaN layers in sandwich-type devices. Jani et al. [40] first demonstrated single-junction 100-nm-thick MOCVD grown $In_{0.28}Ga_{0.72}N$ SC with an open-circuit voltage of 2.1 V, which successfully corresponds to its bandgap of 2.5 eV. They reported the degradation of SC performance by the phase separation in InGaN layers thicker than 200 nm and poor p-type InGaN doping. Yamamoto et al. [54,55] reported a MOCVD grown p-n $In_xGa_{1-x}N$ homojunction SC with high In content (25%) that showed a clear photo-response with open-circuit voltage $V_{oc}$ of 1.5 V and a short-circuit current density $J_{sc}$ of 0.5 mA/cm². The origin of such low values of $V_{oc}$ and $J_{sc}$ were explained by radiative and non-radiative recombination of carriers due to highly defective thick InGaN layers (700 nm). The SCs suffered also from the low-doping capability of p-InGaN causing high series resistance, $R_{series}$, in the device. On the other hand, MBE grown p-n $In_{0.31}Ga_{0.69}N$ SC by Misra et al. [44] showed very poor performances of $V_{oc}$ of 0.55 V and $J_{sc}$ of 0.24 mA/cm² under one sun of zero atmosphere (AM0) illumination. Chen et al. [43] observed very poor photo-response with high leakage current in MBE grown p-i-n InGaN devices with 0.2 and 0.3 In compositions. Boney et al. [56], who reported the highest In content, 39%, in InGaN homojunction SCs so far, mentioned an increase in $J_{sc}$ while an overall degradation in other SC characteristics with increasing In contents.

The major challenges of epitaxial growth and p-type doping of In-rich InGaN layers caused the limitation of the In contents to lower values in further studies with homojunction InGaN SCs. Jampana et al. [57] reported p-n junction InGaN SC with approximately 17% In that resulted in $V_{oc}$ of 1.73 V and $J_{sc}$ of 0.91 mA/cm² under concentrated AM0 illumination. Cai et al. [58] indicated that the crystalline quality and hence, the performance of the homojunction InGaN SCs deteriorate with the increase in the In content. Therefore, good crystalline quality and efficient p-type InGaN doping are key factors to achieve high-performance homojunction InGaN SCs.

### 7.4.1.2 Heterojunction InGaN Solar Cells

III-Nitride-based SCs with heterojunction structures, that is, p-GaN/ InGaN/n-GaN have been extensively studied and most of the notable progress and promising results have been accomplished in this type of SCs. InGaN absorption layer in the form of a single bulk $In_xGa_{1-x}N$ or MQW/superlattice (SL) structure is sandwiched by p-GaN and n-GaN layers by benefiting from the mature growth and fabrication LED technology for GaN compared to InGaN. First of all, Jani et al. [59] demonstrated a clear photo-response in a p-GaN/i-In$_x$Ga$_{1-x}$N/n-GaN single-layer heterojunction cell with an In composition of 0.04–0.05. The devices resulted in high $V_{oc}$ of 2.4 V and consistent IQEs of 60% at their band edge but showed relatively low $J_{sc}$. Zheng et al. [60] then reported an improvement in the same structure with increased In content of 0.1 that demonstrated a high $V_{oc}$ of 2.1 V and high fill factor (FF) of up to 81% under an illumination of AM 1.5 solar light.

One of the first innovations in InGaN-based SCs was reported by Neufeld et al. [48] by optimizing the p-contact grid spacing on MOCVD grown In$_{0.12}$Ga$_{0.88}$N/GaN p-i-n double heterojunction SCs. As a result, the absorption of a larger fraction of incident light allowed the demonstration of a FF of 75%, high EQE of 63%, $V_{oc}$ of 1.81 V, and $J_{sc}$ of 4.2 mA/cm$^2$ under concentrated AM0 illumination. Besides, Shim et al. [49] compared performances of the p-i-n In$_x$Ga$_{1-x}$N /GaN ($x$~0.108) SCs with the CSL of Ni/Au and indium tin oxide (ITO) under AM 1.5 conditions. The devices with ITO showed superior $V_{oc}$ characteristics of 2.0 V, $J_{sc}$ of 0.64 mA/cm$^2$, EQE of 74.1%, and 1% conversion efficiency while devices without any CSL and Ni/Au CSL resulted in conversion efficiency ($\eta$) of 0.93 and 0.75, respectively. In a further study, Shim et al. [50] reported an improved $\eta$ (1.2%) and $J_{sc}$ (0.83 mA/cm$^2$) by utilizing a CSL of graphene on the same structure.

However, the studies revealing poor crystalline quality and high dislocation density in thick single layers of InGaN (with In content ≥0.15) grown on GaN pointed out the fact that the growth thickness of InGaN should be below a critical thickness to suppress the defects. As an alternative, MQWs or SL structure consisting of few nanometers InGaN layers has been proposed for the active region of InGaN-based SCs. Dahal et al. [45] first fabricated In$_x$Ga$_{1-x}$N/GaN MQW ($x$~0.30) SCs that delivered unprecedented EQE of 40% at long operating wavelength of 450 nm. Then, they reported In$_x$Ga$_{1-x}$N/GaN MQW ($x$~0.35) concentrator SCs having 12 periods of 3-nm-thick InGaN QWs and 17-nm-thick GaN barriers that demonstrated $V_{oc}$, $J_{sc}$, FF, and maximum power of 1.80 V, 2.56 mA/cm$^2$, 64%, and 2.95 mW/cm$^2$, respectively under one sun [61]. The novel approach of InGaN/GaN MQWs certainly led to an improvement of overall performance of the InGaN-based SCs by enabling growth of high In content In$_x$Ga$_{1-x}$N ($x$~0.35) material. The device was reported to achieve overall $\eta$ of 2.95 and 3.03% under one sun and 30 suns, respectively. However, the total thickness of the InGaN wells (36 nm) was mentioned to be insufficient for efficient light absorption limiting the conversion efficiencies lower than expected theoretical efficiency of a single junction SC at the same optical energy bandgap (8%). This was evidenced by the increase in photocurrent density by more than 15% when an aluminum back reflector was deposited on the sapphire side of the device. However, it was stated to be rather challenging to obtain In-rich InGaN/GaN MQW structures with total InGaN absorption thicker than 150 nm.

Furthermore, Sheu et al. [62] reported InGaN SCs having 28 pairs of 3- to 4-nm In$_{0.25}$Ga$_{0.75}$N/GaN SL structures sandwiched between p-In$_{0.19}$Ga$_{0.81}$N top layer and n-GaN. The SC resulted in low $V_{oc}$ of 1.4 V and $J_{sc}$ of 0.8 mA/cm$^2$ due to the high defect density in the SL structure. Yang et al. [63] then reported a study comparing SCs with GaN/In$_{0.21}$Ga$_{0.79}$N SL structures to Al$_{0.14}$Ga$_{0.86}$N/ In$_{0.21}$Ga$_{0.79}$N (4nm/3nm) SL structures in which the polarization-induced electric fields enhanced by the incorporation of Al in barrier layers. The reduction in $R_{series}$ and the enhanced extraction efficiency of photo-generated carriers from the active layer in devices with AlGaN SLs were attributed to the increased polarization fields.

Besides, some of the studies have focused on the choice of the substrate for the growth and the texturing of the surface to enhance the light absorption of the SC. The most commonly used substrate is sapphire where InGaN growth is performed on a GaN buffer layer on top of the substrate. Horng et al. [64] and Tsai et al. [65] both reported exploiting the laser liftoff technique to remove the sapphire substrate of the thin-film p-i-n $In_xGa_{1-x}N$ /GaN SCs with In content of 10% and 8.5%, respectively. Horng et al. [64] then transferred the device on Ti/Ag mirror-coated Si substrate via wafer bonding and observed an increment in η from 0.55% to 0.8%, whereas Tsai et al. [64] mounted the device on AlN sub-mount by coating reflector Ag in between and achieved an 13.6% increase in short-circuit current density. Lee et al. [66] reported the fabrication of $In_{0.23}Ga_{0.77}N$/GaN MQW SCs on chemically etched patterned sapphire substrates that achieved an enhancement in $J_{sc}$ by 60%. On the other hand, Kuwahara et al. [67] reported thin-film $In_{0.11}Ga_{0.89}N$/GaN p-i-n double heterojunction SCs grown on freestanding c-plane GaN substrate and a c-plane sapphire substrate covered with a low-temperature-deposited buffer layer. The device on GaN substrate showed a superior performance with a conversion efficiency of 1.41% under 1.5 suns owing to lower pit density and higher shunt resistance ($R_{shunt}$) compared to sapphire substrates. They also fabricated SCs with active region including 50 pairs of $In_{0.17}Ga_{0.83}N$ (3 nm)/$In_{0.07}Ga_{0.93}N$ (0.6 nm) and 10 pairs of doped $In_{0.17}Ga_{0.83}N$: Si (3 nm)/GaN: Si (3 nm) after optimizing the InGaN barrier thickness for the pit density [68]. As a result, the device exhibited $V_{oc}$, $J_{sc}$, FF, and η of 1.78V, 3.08 mA/cm$^2$, 70%, and 2.5%, respectively, under 1.5 suns.

Recently, Liou [69] has reported an impressive study on 20-period $In_xGa_{1-x}N$/ GaN MQWs SCs with various In compositions grown on SiCN/Si (111) substrate. The devices were the first to achieve such high conversion efficiencies, that is, η = 5.95%, 5.87%, 5.52%, and 4.97% for the In content of 0.19, 0.29, 0.34, and 0.36, respectively. It was shown that the η strongly depends on the number and width of the QWs. The proposed device and fabrication technology was reported to be applicable to the realization of SCs with high $V_{oc}$ of 2.72–2.92 V, high $J_{sc}$ of 2.72–2.97 mA/cm$^2$, and high FF of 61.51%–74.89%.

Finally, InGaN/Si tandem SCs were modeled by Hsu and Walukiewicz [70] and Ager III et al. [71] bearing in mind that InGaN-based SCs are highly promising. $In_{0.45}Ga_{0.55}N$/Si tandem cells were proposed for SC applications that may have η higher than 31% owing to the low resistance ohmic junctions between n-$In_{0.45}Ga_{0.55}N$ and p-Si as shown in Figure 7.3 [70]. Hence, no TJ is necessary to connect the $In_{0.45}Ga_{0.55}N$ and Si subcells. Demonstration of n-$In_{0.4}Ga_{0.6}N$/p-Si tandem cells was reported by Tran et al. [47] in which Al and ITO (or Ti/Al/Ni/Au) materials for p- and n-type contacts were investigated. The SCs with ITO as n-type contact was reported to demonstrate an enhanced $J_{sc}$ due to the increased amount of light that was absorbed by the SC as compared to the device using Ti/Al/Ni/Au as n-type contact. The device with ITO contact showed good PV performances and clear spectral response with an FF of about 54%, an external quantum efficiency of 20.8 and 14% at 375 and 390 nm, respectively and the overall conversion efficiency of 7.12%. It is

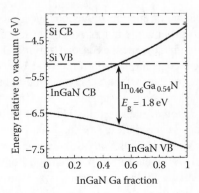

**FIGURE 7.3**  Calculated band alignments for the n-In$_{0.45}$Ga$_{0.55}$N/p-Si junction where electron concentration in InGaN and hole concentration in Si was set at $1 \times 10^{19}$ and $3 \times 10^{15}$ cm$^{-3}$, respectively. (From Hsu, L. and Walukiewicz, W., *J. Appl. Phys.*, 10(2), 024507-1–024507-7, 2008 [70].)

important to note that this study, which achieved the highest conversion efficiency in InGaN-based SCs, demonstrates the potential of InGaN material system for high energy PV.

To conclude, it can be noted that several promising InGaN-based SC results have been established so far, but to compete with mature existing PV technologies, higher conversion efficiencies need to be achieved.

### 7.4.2  MAJOR CHALLENGES IN InGaN-BASED SOLAR CELLS

Although many InGaN homojunction and InGaN/GaN heterojunction SCs have been fabricated and characterized, the efficiencies and performances are still low mainly due to use of Ga-rich InGaN material and the SC structures with poor crystalline quality. The main limitation comes from the fact that the device performance tends to degrade at higher In compositions and longer wavelengths. In addition, establishing decent p-type conduction in GaN and InGaN has been one of the main challenges in III-nitride PV. Even though various efforts have been accomplished toward the goal of high-efficiency InGaN SCs, many of the challenges, which will be described in the following section, still remain to be overcome.

#### 7.4.2.1  Challenges Based on the Material Quality

One of the major challenges in III-nitride-based PV is clearly the material quality. First of all, due to the large lattice mismatch (−16% for GaN on sapphire and +29% for InN on sapphire) and thermal expansion coefficient mismatch (−34% for GaN on sapphire and −100% for InN on sapphire), III-nitride epitaxial films on sapphire substrates have generally high misfit and threading dislocation (TD) densities [53]. Second, although GaN buffer layers are included between the sapphire and the In$_x$Ga$_{1-x}$N material to reduce the misfits, the large lattice mismatch between InN and GaN hinders the growth of high In content In$_x$Ga$_{1-x}$N epilayers with good crystalline quality. It was shown that an epilayer can be grown pseudomorphically on a substrate below a critical thickness ($h_C$), while beyond the $h_C$, plastic relaxation occurs resulting in misfit dislocations [72]. On the other hand, the thermal strain, due to the difference in the thermal expansion coefficients of the substrate and the epilayers, causes the formation of TDs and the nucleation as well as propagation of V-shaped defects on the TDs. Both misfit dislocations and TDs act as non-radiative recombination centers (NRCs), which are detrimental to the device performance. Also, phase separation is a key challenge that has to be overcome to achieve efficient III-nitride PV devices with thick $In_xGa_{1-x}N$ ($x > 0.2$) active layers. Phase separation in high In content InGaN,

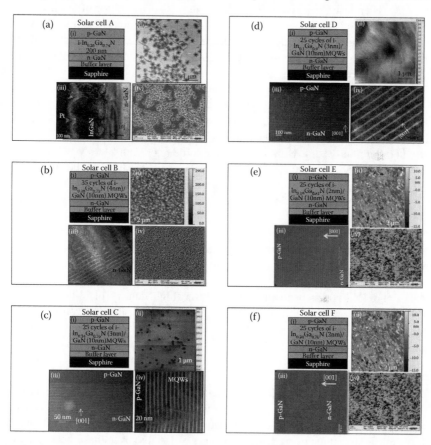

**FIGURE 7.4** Summary of structural characterization by atomic force microscopy (AFM), high angle annular dark field scanning transmission electron microscopy (HAADF-STEM), and scanning electron microscopy (SEM) methods of all solar cell designs. a(i) The schematic design of layers, a(ii) 5 x 5 µm2 AFM image, a(iii) TEM cross section image along [110] zone axis, and a(iv) SEM micrograph of Solar Cell design A; b(i) The schematic design of layers, b(ii) 5 x 5 µm2 AFM image, b(iii) STEM cross section image along [001] zone axis, and b(iv) SEM micrograph of Solar Cell design B; c(i) The schematic design of layers, c(ii) 5 x 5 µm2 AFM image, STEM cross section image along [001] zone axis c(iii) at lower and c(iv) higher magnification of Solar Cell design C; d(i) The schematic design of layers d(ii) 5 x 5 µm2 AFM image , STEM cross section image along [001] zone axis d(iii) at lower and d(iv) higher magnification of Solar Cell design D; e(i) the schematic design of layers e(ii) 5 x 5 µm2 AFM image , e(iii) 1 x 1 µm2 AFM image and e(iv) SEM micrograph of Solar Cell design E.

which is generally seen in form of InGaN clusters having different size, distribution, and In composition, degrades the $V_{oc}$, $J_{sc}$, and FF by generating recombination centers.

In our study, four generations of InGaN-based SC designs having different i-InGaN active regions (bulk or MQWs) sandwiched between 100-nm-thick p-GaN ([Mg] ~ $5 \times 10^{18}$ $cm^{-3}$) and 2-μm-thick n-GaN ([Si] ~ $10^{18}$ $cm^{-3}$) were grown by MOCVD on c-plane (0001) sapphire substrate by the company Novagan. Figure 7.4a(i) shows the schematic structure first-generation SC design (SC-A) which includes 200 nm of $In_{0.26}Ga_{0.74}N$ epilayer. Figure 7.4b(i) presents the second generation of SC design (SC-B) that consists of 35 cycles of $In_{0.28}Ga_{0.72}N$ (4-nm thick)/GaN (10-nm thick) MQWs as the active region that were grown with the aim of reaching an In content of 30%. The third generation of SC designs, SC-C (Figure 7.4c[i]) and SC-D (Figure 7.4d[i]), include 25 cycles of $In_xGa_{1-x}N$ (3-nm thick)/GaN (10-nm thick) MQWs, with $x = 0.24$ and $x = 0.1$, respectively. The last generation of SC design, SC-E, presented in Figure 7.4e(i) and Figure 7.4f(i) were grown by targeting 25 cycles of $In_xGa_{1-x}N$ (2-nm thick)/GaN (10-nm thick) MQWs with $x = 0.19$ and 0.24, respectively. The surface quality of InGaN epilayers in bulk and MQWs forms was assessed by means of scanning electron microscopy (SEM) and atomic force microscopy (AFM). The analysis of the morphological evolution and micro-compositional study of the InGaN/GaN MQWs were carried out through high angle annular dark field (HAADF)–scanning transmission electron microscopy (STEM) and energy-dispersive X-ray spectroscopy (EDX) analyses.

Figure 7.4a, 7.4b, and 7.4c present a comprehensive comparison of InGaN-based SCs with remarkably high In content ($x\sim0.3$) in different growth forms, that is bulk, 35 cycles of 4-nm-thick QWs and 25 cycles of 3-nm-thick QWs, respectively using AFM, STEM, and SEM methods. The AFM and SEM images, shown in Figure 7.4a(ii) and 7.4a(iv), respectively, illustrate the high density of V-defects on the surface of fully relaxed bulk $In_{0.26}Ga_{0.74}N$ layer. Transmission electron microscopy (TEM) cross-sectional image of SC-A along (110) zone axis exhibits clearly the local phase separation within the InGaN bulk layer and the generation of numerous TDs from the $In_{0.26}Ga_{0.74}N$/GaN interface due to high lattice mismatch leading to V-shaped defects.

Similarly, many studies with single thick layers of InGaN with In contents above 20% have reported high dislocation density and phase separation that resulted in poor device performances [53–56]. Consequently, to improve the structural quality, MQW or SL structures using few nanometers of InGaN layers have been exploited as an alternative for the active regions of the SCs. Such structures aim to reduce the defect density in devices by incorporating very thin InGaN layers below the $h_C$ depending on the composition of In.

As shown in Figure 7.5, for instance, $h_C$ for the formation of misfit dislocations in $In_{0.2}Ga_{0.8}N$/GaN systems is only around 1–2 nm, whereas hundred nanometers of InGaN are required for full absorption in the active region. As a result, in InGaN/GaN MQW SCs, higher number of InGaN QWs is needed for efficient absorption of the photons. However, the growth of high number of InGaN MQWs is also challenging as it leads to a transition of the growth mode of QWs over growth time as investigated in detail in our study.

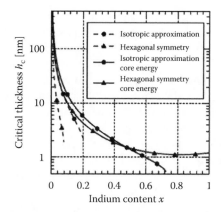

**FIGURE 7.5** The variation of critical thickness against composition for single-layer In$_x$Ga$_{1-x}$N /GaN systems. (From Holec, D., *J. Appl. Phys.*, 104(12), 123514-1–123514-7, 2008. With permission [72].)

As a matter of fact, with the aim of growing thinner and defect-free InGaN epilayers by targeting same In content (x~0.3) in second-generation SCs, 35 cycles of In$_{0.28}$Ga$_{0.72}$N (4 nm thick)/GaN (10-nm thick) MQWs were grown and characterized (Figure 7.4b). The cross-sectional HAADF-STEM characterization in Figure 7.4b(iii) points out that as the growth proceeds, the pseudomorphically grown two-dimensional (2D) InGaN epilayers tend to get thicker and richer in In and eventually the three-dimensional (3D) growth is favored on the inclined planes. The EDX analysis revealed that the primary 2D pseudomorphic InGaN QWs demonstrate strong lateral fluctuations in In composition over growth time with an average of 20% causing an increase in both the growth rate and the elastic strain at In-rich regions.

On the other hand, the average In content of the 3D-grown InGaN regions, which represent the majority of the structure, was extracted to be 28% by EDX and high resolution X-ray diffraction (HRXRD) analysis on the same sample. Similarly, Pantzas et al. [73] reported the lateral fluctuations and increase in elastic strain in 2D-grown InGaN MQWs which trigger the faceting of the growth surface and the transition to a 3D growth mode. Furthermore, the high surface roughness of 41.5 nm in Figure 7.4b(iii) displays very high density of V-defects due to propagation of TDs on certain QWs. Wu et al. [74] reported that the formation of V-defects is promoted by the increase in the inelastic relaxation in strain as a result of increase in the number of QWs in a MQW structure or increase in the In content in the QWs.

Therefore, in the third-generation SC-C design, both the number of cycle of MQWs and the growth thickness of InGaN QWs were decreased by keeping the In content same as in designs SC-A and SC-B. Figure 7.4c(ii) and (iii) show high-quality cross-sectional HAADF-STEM and AFM images with a root-mean-square (rms) of 6.4 nm, respectively, of fully strained 25 pairs of In$_{0.24}$Ga$_{0.76}$N (3 nm)/GaN (10 nm) MQWs with relatively low density of V-defects. Figure 7.4c(iv) displays one defect initiated by a misfit dislocation within the MQW stack and some local thickness fluctuations within the epilayers.

In addition, Figure 7.4d(iii) and (iv) show strain-balanced 25 cycles of In$_{0.1}$Ga$_{0.9}$N (3 nm)/GaN (10 nm) MQWs with very low defect density and abrupt interfaces. In Figure 7.4d(ii), remarkably flat surface with rms of 0.45 nm obtained by AFM measurements also confirms the perfect growth quality of the MQWs. Furthermore, photoluminescence (PL) measurements that were conducted for the third-generation SCs, revealed PL emission lines corresponding to bandgaps of 2.28 eV and 3.21 eV for SC-C and SC-D designs, respectively. It is important to note that InGaN-based SCs having In content more than 20% are promising to be used in PV devices to absorb efficiently at bandgaps unreachable ($E_g > 2.4$ eV) with standard Si or GaAs-based SCs.

As a matter of fact, the fourth generation of SCs E and F were designed to include 25 cycles of In$_x$Ga$_{1-x}$N (2-nm thick)/GaN (10-nm thick) MQWs of $x = 0.19$ and 0.24, respectively, as an active region. The surface roughness of SC-E from AFM analysis (Figure 7.4e[ii]) was 3.3 nm indicating low defect density. The microstructural analysis by STEM and EDX revealed perfectly strained InGaN/GaN MQWs with homogeneous In composition and well thickness. On the other hand, SC-F that has higher In content compared to SC-E, showed poorer material quality as shown in Figure 7.4f(iii). The STEM and EDX analyses indicate a transition from 2D to 3D growth of InGaN epilayers as well as the fluctuations in In content and InGaN well thickness.

In addition to the structural characterization, the design SC-D was fabricated into a SC and its PV response was characterized under one sun. The SC-D design was processed into SCs with $1 \times 1$ mm$^2$ mesa size by Cl$_2$ based inductively coupled plasma etching. Ohmic contacts to n-GaN and p-GaN were formed by electron-beam evaporation of Ti/Al/Ni/Au (12/200/40/100 nm) and Ni/Au (30/80 nm) grid with 200-µm-wide fingers spaced 90 µm. Before the formation of the p-type ohmic contacts, a CSL of Ni/Au (5nm/5nm) was deposited by electron-beam evaporation. The following Figure 7.6 displays the preliminary current density versus voltage

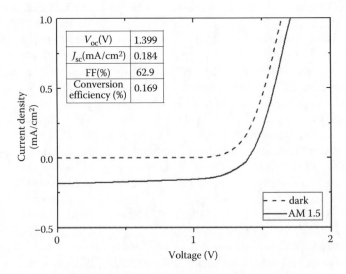

| $V_{oc}$(V) | 1.399 |
|---|---|
| $J_{sc}$(mA/cm$^2$) | 0.184 |
| FF(%) | 62.9 |
| Conversion efficiency (%) | 0.169 |

**FIGURE 7.6**  Current density versus voltage characteristics of the SC with In$_{0.1}$Ga$_{0.9}$N/GaN MQWs as the active region (design SC-D).

characteristics of the SC in dark and under illumination under AM 1.5 G illumination with intensity of 1 kW/m$^2$ (one sun).

Owing to its low defective structure, the SC having 25 cycles of In$_{0.1}$Ga$_{0.9}$N (3 nm)/GaN (10 nm) MQWs, which was fabricated into a SC, showed a poor PV response. To enhance the EQE of the device, the p-contact grid spacing shall be optimized in addition to making use of antireflection coating or backside reflector layer.

On the other hand, it is important to note that to achieve better device performances with high IQEs, the material quality has to be improved by eliminating the V-defects. As a solution, low growth temperature of GaN cap layer on InGaN QWs was proposed following the high growth temperature of GaN quantum barriers

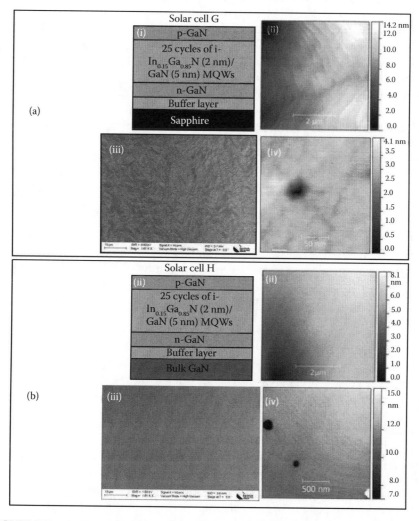

**FIGURE 7.7** (i) The schematic design of layers (ii) 5 × 5 μm$^2$ atomic force microscopy image, (iii) scanning electron microscope micrograph, and (iv) micrograph of (a) SC-G and (b) SC-H, respectively.

(QBs) [75]. The low temperature GaN cap layer protects the integrity of In in QWs, although the subsequent GaN QB layer, which was grown at higher temperature improves surface morphology and effectively fills in the V-defects. In a further study, Young et al. [76] reported the optimized growth of InGaN/GaN MQWs on sapphire and bulk (1000) GaN substrates. Owing to much lower TD density of GaN grown on bulk GaN substrates that is, at least two orders of magnitude, an improvement in performance was observed compared to sapphire substrates, especially with an increasing number of QWs. The best SC performance exhibited EQEs up to 60%, $V_{oc}$ as high as 2.28 V, FF up to 80%, and a maximum conversion efficiency of 2.4% under one sun AM0 equivalent illumination. However, as MQW thickness increased, the device performance degraded with the onset being delayed on bulk (0001) GaN substrates. This study indicated that V-defects form on preexisting TDs during MQW growth on sapphire, whereas new defects nucleate during MQW growth on bulk (0001) GaN.

As a matter of fact, our study also includes a comparison of structural quality and PV performance of the SC structures grown on sapphire and bulk (0001) GaN substrates with $In_{0.15}Ga_{0.85}N$ (2 nm)/GaN (5 nm) MQWs (Figure 7.7a[i] and b[i]). The preliminary SEM and AFM analysis reveal enhanced crystalline quality with less dislocations density compared to the SC-E and SC-F with higher In contents ($x >$ 0.19). In addition, the surface roughness of SC-G and SC-H from AFM analysis (Figure 7.7a and b[ii]) was 3.0 nm and 0.5 nm indicating lower TD density on bulk GaN substrate compared to sapphire substrate. Hence, the devices grown on bulk GaN substrates are expected to outperform those grown on sapphire. The microstructural analysis by STEM and EDX as well as the compositional analysis by HRXRD are ongoing in this study to assess the structural quality of the InGaN/GaN MQWs on sapphire and bulk GaN substrates.

In conclusion, the realization of thick InGaN SCs with high In contents remains challenging due to the large lattice mismatch between InN and GaN, resulting generally in phase separation, In fluctuation and high density of TDs, misfit dislocations, and V-defects. Besides, the ultimate success of the InGaN PVs highly depends on the achievement of lattice matched InGaN-based multijunction/tandem SCs, which still needs to be further developed.

### 7.4.2.2 Challenges Due to Polarization of III-Nitrides

One of the unique features of the III-nitride material system is its polarization effect as a consequence of the non-centrosymmetry of its wurtzite structure and the high degree of ionicity of the covalent metal–nitrogen bonds. Experimental and theoretical investigations have shown that macroscopic polarization effects have a great influence on the characteristics, performance, and response of the low-dimensional nitride heterostructure devices grown on c-plane [77,78]. The total polarization, which is a sum of spontaneous and strain-induced piezoelectric polarization, and the sheet charges existing at the heterointerfaces have been reported to be detrimental to the device performances [79,80]. Wurtzite-structured nitrides grown along the (0001) axis are known to possess large spontaneous polarization ($P_{SP}$) (~MV/cm), which is intrinsic to the material and points along (000(000$\bar{1}$)). On the other hand, the lattice mismatch strain between GaN and $In_xGa_{1-x}N$ ($x > 0.2$) layers leads to the introduction of strain-induced piezoelectric polarization ($P_{PZ}$) in the (0001) orientation as

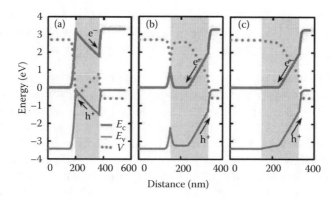

**FIGURE 7.8** Modeled band diagrams and potentials for (a) p-GaN/i-InGaN/n-GaN SC, (b, c) p-InGaN/i-InGaN/n-InGaN. The shaded region delineates the higher In composition $In_xGa_{1-x}N$ layers ($x$ = 8% and 9%). (From Raring, J.W., *Appl. Phys. Express*, 3, 112101-1–112101-3, 2010. With permission [83].)

well as the formation of high dislocation density, which were indicated previously. Consequently, the discontinuities in $P_{SP}$ and $P_{PZ}$ cause generation of polarization sheet charges at GaN/InGaN heterointerfaces that creates electric fields opposing the built-in pn-junction electric field in Ga polar devices. Wierer et al. [81] reported the effect of both strain-induced $P_{PZ}$ and NRCs on the performance of the p-i-n GaN/InGaN SCs with single thick $In_{0.09}Ga_{0.91}N$ layers. The SC with i-InGaN (170 nm) sandwiched between p-GaN and n-GaN (Figure 7.8a) resulted in lower $J_{sc}$ than the two other devices with n-type and p-type InGaN layers above and below the i-InGaN (Figure 7.8b and c). Figure 7.8a shows the creation of strain-induced $P_{PZ}$ charges at GaN/InGaN heterointerfaces and a potential within the i-InGaN opposite to the direction of the depletion field tilting the energy bands of InGaN layer in a detrimental manner. Photo-generated carriers drift in the wrong direction and eventually recombine without contributing to the current, resulting in poor carrier collection and poor $J_{sc}$. Meanwhile, Figure 7.8b and c show the band diagrams at zero bias of the samples with the designs of n-$In_{0.08}Ga_{0.92}N$ (80 nm)/i-$In_{0.08}Ga_{0.92}N$ (80 nm) /p-$In_{0.015}Ga_{0.985}N$ (50 nm) and graded n-$In_{0-0.09}Ga_{1-0.91}N$ (80 nm)/i-$In_{0.09}Ga_{0.91}N$ (80 nm)/p-$In_{0.015}Ga_{0.985}N$ (50 nm), respectively. They reported that sandwiching i-InGaN reduces and/or redistributes the $P_{PZ}$ leading to increased $J_{sc}$. In addition, the growth of InGaN p-i-n homojunction SCs eliminates the formation of high density of defects, which would originate from the lattice mismatch between GaN and InGaN.

Furthermore, the polarization effects in InGaN/GaN MQWs were shown to be dependent on the composition of the incorporated In in the MQWs and the inclination angle of the growth plane from the polar c-plane on GaN. Recently, nonpolar and semipolar III-nitrides have been investigated for possible substrates for the growth of In-rich InGaN/GaN MQWs [82,83]. The growths in semipolar $11\bar{2}2$ and $20\bar{2}1$ plane and nonpolar m-orientation on free standing GaN substrates were reported to reduce or eliminate the polarization related to electric-field effects. Besides, Zhao et al. [84] reported that the semipolar planes $11\bar{2}2$ and $20\bar{2}1$ exhibit higher In incorporation than that of the polar c-plane or nonpolar m-plane. However, the high growth quality of MQWs with high In composition on semipolar and nonpolar GaN substrates remain currently challenging due to formation of high density of defects.

### 7.4.2.3 Challenges Based on the p-Type Doping and Ohmic Contact

Achieving low resistance and transparent ohmic contacts to p-type GaN or p-type InGaN with high thermal stability and reliability plays a key role in the fabrication of high-performance GaN-based optoelectronic devices such as LEDs, laser diodes, and InGaN-based SCs [85]. However, low resistivity ohmic contacts to p-GaN still remain challenging due to relatively deep Mg acceptor levels and low incorporation/activation efficiency. It was also reported that the degradation in the crystalline quality by increasing the In content in p-InGaN material, leads to lower acceptor concentrations and hence, lower p-type InGaN doping capability [86].

In addition to challenges in doping, there is readily no available metal that could form low resistivity ohmic contacts on p-GaN and p-InGaN, which both possess high work functions and high electron affinity. Consequently, transparent p-type ohmic contacts are generally realized by metal- (e.g., Ni/Au) or transparent conductive oxide (e.g., ITO)–based CSL schemes, which result in specific contact resistance in the range of approximately $10^{-4}$ to $10^{-6}$ $\Omega \cdot cm^2$ [47–50]. This method requires deposition of metal grid electrodes to increase carrier collection whereas it strongly reduces the efficient photo absorption in InGaN active layer by decreasing the EQE of the cell. Furthermore, ITO is reported to have specific resistivity of $1.19 \times 10^{-4}$ $\Omega \cdot cm^2$ and transparency of 94.17% [87] whereas the best transmittance of Ni/Au CSL represent only 83% in blue wavelength range [85]. Thus, these contact layers eventually exhibit limited transparency and resistivity in $In_xGa_{1-x}N$-based SCs with bandgaps lower than 2.30 eV (for $x > 0.2$).

As an alternative, Krishnamoorthy et al. [88] proposed the replacement of highly resistive p-contact with low resistance n-type top contact layer by making use of regrown polarization-engineered GaN/InGaN/GaN TJ diodes with high tunneling probability and low tunneling resistance. The polarization-engineered TJ approach with a specific resistivity of $1.2 \times 10^{-4}$ $\Omega \cdot cm^2$ [39] is very attractive for GaN-based LEDs and SCs as it would require very little metal footprint on the surface, resulting in enhanced EQE by eliminating any current spreading layer on p-type GaN.

## 7.5 CONCLUSION

In conclusion, over the last decade III-V nitride material system, in particular InGaN alloys, are reported by many studies to be an ideal material system to fabricate high efficiency SCs owing to their tunable direct wide bandgap, outstanding physical and chemical stability in harsh environment, superior resilience to high-energy particle radiation, and high thermal stability. Even though noteworthy work has been established so far in InGaN homojunction and InGaN/GaN heterojunction SCs, the device performances with high In content InGaN are still rather low mainly due to challenges in material growth and crystalline quality in addition to challenges in p-type conduction in GaN and InGaN. The innovative methods such as 2D growth of InGaN/GaN epilayers, incorporation of GaN cap layers into GaN QBs, growth of InGaN/GaN MQW layers on free-standing GaN substrates would lead to better material quality hence, better SC performance.

Furthermore, to attain high efficiency PV based on InGaN, multijunction SCs should be fabricated by connecting several junctions of InGaN subcells with different In contents. Besides, possible integration of silicon and InGaN/GaN SCs would combine high efficiency with low-cost production PV technology. The InGaN-on-Si tandem SCs can also be used under concentration to enhance the device performances.

## REFERENCES

1. I. Allison, N.L. Bindoff, R.A. Bindschadler, P.M. Cox, N. de Noblet, M.H. England, J.E. Francis et al. *The Copenhagen Diagnosis: Updating the World on the Latest Climate Science*, UNSW Climate Change Research Center, Sydney, Australia, 2009, p. 7.
2. Eurostat-European Commission. *Renewable Energy Statistics: Share of Renewable Energy in Gross Final Energy Consumption.* 18 February 2014, (http://epp.eurostat .ec.europa.eu/statistics_explained/index.php/Renewable_energy_statistics).
3. R. Gelman. *2011-Renewable Energy Data Book,* Energy Efficiency & Renewable Energy (EERE), Golden, Colorado, USA, 2012, p.7.
4. K. Tanabe. A review of ultrahigh efficiency III-V semiconductor compound solar cells: Multijunction tandem, lower dimensional, photonic up/down conversion and plasmonic nanometallic structures. *Energies*, 2: 504–530, 2009.
5. A.E. Becquerel. Mémoire sur les effets électriques produits sous l'influence des rayons solaires [Memory on the electrical effects generated by solar rays]. *Comt. Rend. Acad. Sci.*, 9: 561–597, 1839.
6. D.M. Chapin, C.S. Fuller, and G.L. Pearson. A new silicon p-n junction photocell for converting solar radiation into electrical power. *J. Appl. Phys.*, 25: 676–677, 1954.
7. A.C. Goodrich, D.M. Powell, T.L. Jamesa, M. Woodhousea, and Tonio Buonassisi. Assessing the drivers of regional trends in solar photovoltaic manufacturing. *Energy Environ. Sci.*, 6: 2811–2821, 2013.
8. European Photovoltaic Industry Association. 2013. *World's Solar Photovoltaic Capacity Passes 100-Gigawatt Landmark after Strong Year.* [Press release] 11 February 2013.
9. J. Zhao, A. Wang, M.A. Green, and F. Ferrazza. 19.8% efficient "honeycomb" textured multicrystalline and 24.4% monocrystalline silicon solar cells. *Appl. Phys. Lett.*, 73(14): 1991–1993, 1998.
10. C.B. Honsberg, J.E. Cotter, K.R. McIntosh, S.C. Pritchard, B.S. Richards, and S.R. Wenham. Design strategies for commercial solar cells using the buried contact technology. *IEEE Trans. Electron. Dev.*, 46(10): 1984–1992, 1999.
11. A. Slade and V. Garboushian. *27.6% Efficient Silicon Concentrator Solar Cells for Mass Production.* Technical Digest of 15th International Photovoltaic Science and Engineering Conference, Shangai, October 2005, p.701–702.
12. W. Shockley and H.J. Queisser. Detailed balance limit of efficiency of p-n junction solar cells. *J. Appl. Phys.*, 32: 510–519, 1961.
13. J.J. Loferski. The role of multinary alloys of I-III-VI2 chalcopyrite semiconductors in solar cells. *Cryst. Res. Technol.*, 31(1): 419–429, 1996.
14. EMPA Materials Science & Technology. 2013. *A New World Record for Solar Cell Efficiency.* [Press release] 17 January 2013.
15. P. Jackson, D. Hariskos, E. Lotter, S. Paetel, R. Wuerz, R. Menner, W. Wischmann, and M. Powalla. New world record efficiency for Cu (In,Ga)Se$_2$ thin-film solar cells beyond 20%. *Prog. Photovoltaics: Res. Appl.*, 19:894–897, 2013.

16. First Solar. 2014. *First Solar Sets World Record for CdTe Solar Cell Efficiency.* [Press release] 25 February 2014.

17. Nomination of cadmium telluride to the national toxicology program, memorandum of the US National Renewables Energy Laboratory (NREL) and the Brookhaven National Laboratory (BNL). 11 April 2003, (http://ntp-server.niehs.nih.gov/ntp/htdocs/Chem _Background/ExSumPdf/CdTe_508.pdf).

18. M.A. Green. Third generation photovoltaics: Solar cells for 2020 and beyond. *Physica E*, 14: 65–70, 2002.

19. A. Marti and G.L. Araujo. Limiting efficiencies for photovoltaic energy conversion in multigap systems. *Sol. Energy Mater. Sol. Cells*, 43: 203–222, 1996.

20. H. Cotal, C. Fetzer, J. Boisvert, G. Kinsey, R. King, P. Hebert, H. Yoon, and N. Karam. III–V multijunction solar cells for concentrating photovoltaics. *Energy Environ. Sci.*, 2(2): 174–192, 2009.

21. M.J. Keevers and M.A. Green. Extended infrared response of silicon solar cells and the impurity photovoltaic effect. *Sol. Energy Mater. Sol. Cells*, 41(1): 195–204, 1996.

22. G.F. Brown and J. Wu. Third generation photovoltaics. *Laser Photonics Rev.*, 3(4): 394–405, 2009.

23. A.J. Nozik. Quantum dot solar cells. *Physica E*, 14: 115–120, 2002.

24. N. Lopez, A. Marti, A. Luque, C. Stanley, C. Farmer, and P. Diaz. Experimental analysis of the operation of quantum dot intermediate band solar cells. *J. Sol, Energy Eng.*, 129: 319–322, 2007.

25. K.W.J. Barnham and G. Duggan. A new approach to high-efficiency multi-band-gap solar cells. *J. Appl. Phys.*, 67: 3490–3493, 1990.

26. J.G.J. Adams, B.C. Browne, I.M. Ballard, J.P. Connolly, N.L.A. Chan, A. Ioannides, W. Elder, P.N. Stavrinou, K.W.J. Barnham, and N.J. Ekins-Daukes. Recent results for single-junction and tandem quantum well solar cells. *Prog. Photovoltaics Res. Appl.*, 19(7): 865–877, 2011.

27. A. Barnett, D. Kirkpatrick, C. Honsberg, D. Moore, M. Wanlass, K. Emery, R. Schwartz et al. Very high efficiency solar cell modules. *Prog. Photovoltaics Res. Appl.*, 17(1): 75–83, 2009.

28. NREL: Best Research Cell Efficiencies 2013. 15 April 2014, (http://www.nrel.gov /ncpv/images/efficiency_chart.jpg).

29. Sharp. 2014. *Sharp Develops Concentrator Solar Cell with World's Highest Conversion Efficiency of 44.4%.* [Press release] 14 June 2014.

30. A.S. Brown and M.A. Green. Detailed balance limit for the series constrained two terminal tandem solar cell. *Physica E*, 14: 96–100, 2002.

31. O.K. Jani. Development of wide-band gap InGaN solar cells for high efficiency photovoltaics. PhD Thesis, Georgia Institute of Technology, Aug, 2008.

32. J. Wu, W. Walukiewicz, K.M. Yu, J.W. Ager III, E.E. Haller, H. Lu, W.J. Schaff, Y. Saito, and Y. Nanishi. Unusual properties of the fundamental band gap of InN. *Appl. Phys. Lett.*, 80: 3967–3969, 2002.

33. S. Nakamura, S. Peartona, and G. Fasol. *The Blue Laser Diode.* Spring Verlag, Berlin, Heidelberg, GmbH, 2000.

34. M.R. Krames, O.B. Shchekin, R. Mueller-Mach, O.G. Mueller, L. Zhou, G. Harbers, and M.G. Craford. Status and future of high-power light-emitting diodes for solid-state lighting. *J. Display Technol.*, 3(2): 160–175, 2007.

35. S. Strite and H. Morkoç. GaN, AlN, and InN: A review. *J. Vac. Sci. Technol.*, B10: 1237–1266, 1992.

36. J. Wu, W. Walukiewicz, K.M. Yu, W. Shan, J.W. Ager III, E.E. Haller, H. Lu, W.J. Schaff, W.K. Metzger, and S. Kurtz. Superior radiation resistance of alloys: Full-solar-spectrum photovoltaic material system. *J. Appl. Phys.*, 94(10): 6477–6782, 2003.

37. V. Fiorentini and F. Bernardini. Spontaneous versus piezoelectric polarization in III-V nitrides: Conceptual aspects and practical consequences. *Phys. Stat. Sol.*, B 216(1): 391–398, 1999.

38. J. Simon, Z. Zhang, K. Goodman, H. Xing, T. Kosel, P. Fay, and D. Jena. Polarization induced Zener tunnel junctions in wide bandgap heterostructures. *Phys. Rev. Lett.*, 103: 026801-1–026801-4, 2009.

39. S. Krishnamoorthy, F. Akyol, P. Sung Park, and S. Rajan. Low resistance GaN/InGaN/GaN tunnel junctions. *Appl. Phys. Lett.*, 102: 113503-1–113503-5, 2013.

40. O. Jani, H. Yu, E. Trybus, B. Jampana, I. Ferguson, A. Doolittle, and C. Honsberg. Effect of phase separation on performance of III-V nitride solar cells. 22nd European Photovoltaic Solar Energy Conference, 2007.

41. C. Yang, X. Wang, H. Xiao, J. Ran, C. Wang, G. Hu, X. Wang, X. Zhang, J. Li, and J. Li. Photovoltaic effects in InGaN structures with p–n junctions. *Phys. Stat. Sol. (a)*, 204(12): 4288–4291, 2007.

42. O. Jani, C. Honsberg, Y. Huang, J.-O. Song, I. Ferguson, G. Namkoong, E. Trybus, A. Doolittle, and S. Kurtz. Design, growth, fabrication and characterization of high-band GaP InGaN/GaN solar cells. *Conf. Rec. Proc. 4th World Conf. Photovoltaic. Energy Convers*, 2006.

43. X. Chen, K.D. Matthews, D. Hao, W.J. Schaff, and L.F. Eastman. Growth, fabrication, and characterization of InGaN solar cells. *Phys. Stat. Sol. (a).*, 205(5): 1103–1105, 2008.

44. P. Misra, C. Boney, N. Medelci, D. Starikov, A. Freundlich, and A. Bensaoula, Fabrication and characterization of 2.3eV InGaN photovoltaic devices. *33rd IEEE Photovoltaic Specialists Conf.*, 2008.

45. R. Dahal, B. Pantha, J. Li, J.Y. Lin, and H.X. Jiang. InGaN/GaN multiple quantum well solar cells with long operating wavelengths. *Appl. Phys. Lett.*, 94(6): 063 505-1–063 505-3, 2009.

46. L.A. Reichertz, I. Gherasoiu, K.M. Yu, V.M. Kao, W. Walukiewicz, and J.W. Ager III. Demonstration of a III–nitride/silicon tandem solar cell. *Appl. Phys. Express.*, 2(12): 122 202-1–122 202-3, 2009.

47. B.-T. Tran, E.-Y. Chang, H.-D. Trinh, C.-T. Lee, K.C. Sahoo, K.-L. Lin, M.-C. Huang et al. Fabrication and characterization of n-In0.4Ga0.6N/p-Si solar cell. *Energy Mater. Sol. Cells*, 102: 208–211, 2012.

48. C.J. Neufeld, N.G. Toledo, S.C. Cruz, M. Iza, S.P. DenBaars, and U.K. Mishra. High quantum efficiency InGaN/GaN solar cells with 2.95 eV band gap. *Appl. Phys. Lett.*, 93(14): 143 502-1–143 502-3, 2008.

49. J.P. Shim, S.R. Jeon, Y.K. Jeong, and D.-S. Lee. Improved efficiency by using transparent contact layers in InGaN-based p-i-n solar cells. *IEEE Electron Device Lett.*, 31(10): 1140–1142, 2010.

50. J.-P. Shim, M. Choe, S.-R. Jeon, D. Seo, T. Lee, and D.-S. Lee. InGaN based p–i–n solar cells with graphene electrodes. *Appl. Phys. Express.*, 4(5): 052302-1–052302-3, 2011.

51. Y. Kuwahara, T. Fujii, T. Sugiyama, D. Iida, Y. Isobe, Y. Fujiyama, Y. Morita et al. GaInN-based solar cells using strained-layer GaInN/GaInN superlattice active layer on a freestanding GaN substrate. *Appl. Phys. Express*, 4(2): 021001-1–021001-3, 2011.

52. B.W. Liou. Design and fabrication of $In_xGa_{1-x}N$/GaN solar cells with a multiple-quantum well structure on SiCN/Si (111) substrate. *Thin Solid Films*, 520(3): 1084–1090, 2011.

53. A.G. Bhuiyan, K. Sugita, A. Hashimoto, and A. Yamamoto. InGaN solar cells: Present state of the art and important challenges. *IEEE J. Photovoltaics*, 2(3), 276-293, 2012.

54. A. Yamamoto, K. Sugita, M. Horie, Y. Ohmura, Md. R. Islam, and A. Hashimoto, Mg-doping and N+ -P junction formation in MOVPE grown $In_xGa_{1-x}N$ ($x\sim0.4$). *33rd IEEE Photovoltaic Specialists Conf.*, 2008.

55. A. Yamamoto, Md. R. Islam, T.T. Kang, and A. Hashimoto. Recent advances in InN-based solar cells: Status and challenges in InGaN and InAlN solar cells. *Phys. Stat. Sol. (c)*, 7(5): 1309–1316, 2010.

56. C. Boney, I. Hernandez, R. Pillai, D. Starikov, A. Bensaoula, M. Henini, M. Syperek, J. Misiewicz, and R. Kudrawiec. Growth and characterization of InGaN for photovoltaic devices. *Phys. Stat. Sol. (c)*, 8(7–8): 2466–2668, 2011.

57. B.R. Jampana, A.G. Melton, M. Jamil, N.N. Faleev, R.L. Opila, I.T. Ferguson, and C.B. Honsberg. Design and realization of wideband-gap (~2.67 eV) InGaN p-n junction solar cell. *IEEE Electron Device Lett.*, 31(1): 32–34, 2010.

58. X.-M. Cai, S.-W. Zeng, and B.-P. Zhang. Fabrication and characterization of InGaN p-i-n homojunction solar cell. *Appl. Phys. Lett.*, 95(17): 173 504-1–173 504-3, 2009.

59. O. Jani, I. Ferguson, C. Honsberg, and S. Kurtz. Design and characterization of GaN/InGaN solar cells. *Appl. Phys. Lett.*, 91(13): 132117-1–132117-3, 2007.

60. X. Zheng, R.-H. Horng, D.-S. Wuu, M.-T. Chu, W.-Y. Liao, M.-H. Wu, R.-M. Lin, and Y.-C. Lu. High-quality InGaN/GaN heterojunctions and their photovoltaic effects. *Appl. Phys. Lett.*, 93(26): 261108-3, 2008.

61. R. Dahal, J. Li, K. Aryal, J.Y. Lin, and H.X. Jiang. InGaN/GaN multiple quantum well concentrator solar cells. *Appl. Phys. Lett.*, 97(7): 073115-1–073115-3, 2010.

62. J.-K. Sheu, C.-C. Yang, S.-J. Tu, K.-H. Chang, M.-L. Lee, W.-C. Lai, and L.-C. Peng. Demonstration of GaN-based solar cells with GaN/InGaN superlattice absorption layers. *IEEE Electron Device Lett.*, 30(3): 225–227, 2009.

63. C.C. Yang, J.K. Sheu, X.-W. Liang, M.-S. Huang, M.L. Lee, K.H. Chang, S.J. Tu, F.-W. Huang, and W.C. Lai. Enhancement of the conversion efficiency of GaN-based photovoltaic devices with Al-GaN/InGaN absorption layers. *Appl. Phys. Lett.*, 97(2): 021113-1–021113-3, 2010.

64. R.-H. Horng, S.-T. Lin, Y.-L. Tsai, M.-T. Chu, W.-Y. Liao, M.-H. Wu, R.-M. Lin, and Y.-C. Lu. Improved conversion efficiency of GaN/InGaN thin-film solar cells. *IEEE Electron Device Lett.*, 30(7): 724–726, 2009.

65. C.-L. Tsai, G.-S. Liua, G.-C. Fana, and Y.-S. Leea. Substrate-free large gap InGaN solar cells with bottom reflector. *Solid-State Electron.*, 54(5): 541–544, 2010.

66. Y.J. Lee, M.H. Lee, C.M. Cheng, and C.H. Yang. Enhanced conversion efficiency of InGaN multiple quantum well solar cells grown on a patterned sapphire substrate. *Appl. Phys. Lett.*, 98(26): 263504-1–263504-3, 2011.

67. Y. Kuwahara, T. Fujii, Y. Fujiyama, T. Sugiyama, M. Iwaya, T. Takeuchi, S. Kamiyama, I. Akasaki, and H. Amano. Realization of nitride-based solar cell on freestanding GaN substrate. *Appl. Phys. Express.*, 3(2): 111001-1–111001-3, 2010.

68. Y. Kuwahara, T. Fujii, T. Sugiyama, D. Iida, Y. Isobe, Y. Fujiyama, Y. Morita et al. GaInN-based solar cells using strained-layer GaInN/GaInN superlattice active layer on a freestanding GaN substrate. *Appl. Phys. Express.*, 4(2): 021001-1–021001-3, 2011.

69. B.W. Liou. Design and fabrication of $In_xGa_{1-x}N$/GaN solar cells with a multiple-quantum well structure on SiCN/Si(111) substrates. *Thin Solid Films*, 520(3): 1084–1090, 2011.

70. L. Hsu and W. Walukiewicz. Modeling of InGaN/Si tandem solar cells. *J. Appl. Phys.*, 10(2): 024507-1–024507-7, 2008.

71. J.W. Ager III, L.A. Reichertz, Y. Cui, Y.E. Romanyuk, D. Kreier, S.R. Leone, K.M. Yu, W.J. Schaff, and W. Walukiewicz. Electrical properties of InGaN-Si heterojunction. *Phys. Stat. Sol. (c)*, 6(S2): S413–S416, 2009.

72. D. Holec, Y. Zhang, D.V. Sridhara Rao, M.J. Kappers, C. McAleese, and C.J. Humphreys. Equilibrium critical thickness for misfit dislocations in III-nitrides. *J. Appl. Phys.*, 104(12): 123514-1–123514-7, 2008.

73. K. Pantzas, G. Patriarche, G. Orsal, S. Gautier, T. Moudakir, M. Abid, V. Gorge, Z. Djebbour, P.L. Voss, and A. Ougazzaden. Investigation of a relaxation mechanism specific to InGaN for improved MOVPE growth of nitride solar cell materials. *Phys. Status Solidi*, A 209(1): 25–28, 2012.

74. X.H. Wu, C.R. Elsass, A. Abare, M. Mack, S. Keller, P.M. Petroff, S.P. DenBaars, J.S. Speck, and S.J. Rosner. Structural origin of V-defects and correlation with localized excitonic centers in InGaN/GaN multiple quantum wells. *Appl. Phys. Lett.*, 72: 692–694, 1998.

75. Y.L. Hu, R.M. Farrell, C.J. Neufeld, M. Iza, S.C. Cruz, N. Pfaff, D. Simeonov et al. Effect of quantum well cap layer thickness on the microstructure and performance of InGaN/GaN solar cells. *Appl. Phys. Lett.*, 100: 161101-1–161101-4, 2012.

76. N.G. Young, R.M. Farrell, Y.L. Hu, Y. Terao, M. Iza, S. Keller, S.P. DenBaars, S. Nakamura, and J.S. Speck. High performance thin quantum barrier InGaNGaN solar cells on sapphire and bulk (0001) GaN substrates. *Appl. Phys. Lett.*, 103: 173903-1–173903-5, 2013.

77. F. Bernardini, V. Fiorentini, and D. Vanderbilt. Spontaneous polarization and piezoelectric constants of III-V nitrides. *Phys. Rev. B*, 56: 10024–10024, 1997.

78. Morkoç, R. Cingolani, and B. Gill. Polarization effects in nitride semiconductors and device structures. *Mat. Res. Innovat.*, 3: 97–106, 1999.

79. F. Bernardini and V. Fiorentini. Polarization fields in nitride nanostructures: 10 Points to think about. *App. Surface Science.*, 166: 23–29, 2000.

80. Q.Z. Li, M. Lestradet, Y.G. Xiao, and S.S. Li. Effects of polarization charge on the photovoltaic properties of InGaN solar cells. *Phys. Stat. Sol. A*, 208(4): 928–931, 2011.

81. J.J. Wierer Jr., A.J. Fischer, and D.D. Koleske. The impact of piezoelectric polarization and non-radiative recombination on the performance of (0001) face GaN/InGaN photovoltaic devices. *Appl. Phys. Lett.*, 96(5): 051107-1–051107-3, 2010.

82. Y.D. Lin, S. Yamamoto, C.Y. Huang, C.L. Hsiung, F. Wu, K. Fujito, H. Ohta, J.S. Speck, S.P. Denbaars, and S. Nakamura. High quality InGaN/AlGaN multiple quantum wells for semipolar InGaN green laser diodes. *Appl. Phys. Express.*, 3: 082001-1–082001-3, 2010.

83. J.W. Raring, M.C. Schmidt, C. Poblenz, Y.C. Chang, M.J. Mondry, B. Li, J. Iveland et al. High-efficiency blue and true-green-emitting laser diodes based on non-c-plane oriented GaN substrates. *Appl. Phys. Express*, 3: 112101-1–112101-3, 2010.

84. Y. Zhao, Q. Yan, C.Y. Huang, S.C. Huang, P.S. Hsu, S. Tanaka, C.C. Pan et al. Indium incorporation and emission properties of nonpolar and semipolar InGaN quantum wells. *Appl. Phys. Lett.*, 100: 201108-1–201108-4, 2012.

85. J.-O. Song, J.-S. Ha, and T.-Y. Seong. Ohmic contact technology for GaN-based LEDs: Role of p-type Contact. *IEEE Transactions Electron. Dev.*, 57(1): 42–59, 2010.

86. K. Kumakura, T. Makimoto, and N. Kobayashi. High hole concentrations in Mg-doped InGaN grown by MOVPE. *J. Cryst. Growth.*, 221(1): 267–270, 2000.

87. D. Yan, G. Weiling, Z. Yanxu, L. Jianpeng, and Y. Weiwei. Rapid thermal annealing effects on vacuum evaporated ITO for InGaN/GaN blue LEDs. *J. Semicond.*, 33(6): 066004-1–066004-4, 2012.

88. S. Krishnamoorthy, D.N. Nath, F. Akyol, P.S. Park, M. Esposto, and S. Rajan. Polarization-engineered GaN/InGaN/GaN tunnel diodes. *Appl. Phys. Lett.*, 97: 203502-1–s203502-3, 2010.

# 8 III-Nitride Semiconductors: New Infrared Intersubband Technologies

*Mark Beeler and Eva Monroy*

## CONTENTS

GaN/Al(Ga)N semiconductors have emerged within the last decade as promising materials for founding new intersubband (ISB) technologies relying on infrared (IR) optical transitions between quantum-confined electronic states in the conduction band of nanostructures (quantum wells [QWs], quantum dots [QDs], and nanowires). The large conduction-band offset (about 1.8 eV for GaN/AlN) and the sub-picosecond ISB relaxation time render III-nitrides suitable for the fabrication of ultrafast photonic devices for optical telecommunication networks in the near-IR range. Furthermore, the high energy of GaN longitudinal optical phonons (92 meV) opens prospects for high-temperature THz quantum cascade lasers (QCLs) and ISB devices covering the 5- to 10-THz band, inaccessible to As-based technologies due to phonon absorption.

Key words: nitride, infrared, intersubband, quantum well.

## 8.1 INTRODUCTION

The term "intersubband" is used to describe electronic transitions between confined states in either the conduction band or the valence band of semiconductor nanostructures. In such systems, ISB optical transitions at a desired operation wavelength can be obtained by the adequate choice of layer thicknesses, which is the base of the "band-structure engineering" concept. The first studies of ISB absorption date back to the 1970s [1,2] and refer to electronic transitions between confined levels in an accumulation layer on $n$-type Si (100), observed under far-IR illumination. The extrapolation of these principles to GaAs/AlGaAs QWs to shift the transition wavelength to mid-IR was first suggested by Esaki and Sakaki [3], and subsequent experimental [4,5] and theoretical studies [6,7] led to the first experimental measurement of strong ISB absorption in a series of multiple quantum wells (MQWs) performed by West and Eglash [8]. These results lead to the fabrication of the first quantum well infrared photodetector (QWIP), by Levine et al. [9,10]. In 1994, Faist et al. [11] presented a major breakthrough in ISB technology: the QCL. This was the beginning of tremendous development of the ISB technology, which resulted in commercially available devices operating in the mid- and far-IR. For a comprehensive introduction to ISB physics in QWs, we refer the readers to the work by Bastard [12] or Liu and Capasso [13].

Nowadays, ISB optoelectronic devices based on the III-As material system (GaAs/AlGaAs, InGaAs/AlInAs, or GaInAs/AlAsSb) can be tuned from the mid-IR to the THz spectral range. Operation at shorter wavelengths (<3 μm) is limited by the available conduction-band offset and by material transparency. III-Nitride semiconductors (GaN, AlN, InN, and their alloys), with their wide bandgap and large conduction-band offset (~1.8 eV for GaN/AlN [14–16]), are attracting much interest for ISB devices operating in the near-IR spectral range, particularly in the 1.3- to 1.55-μm wavelength window used for fiber-optic communications [17–20]. GaN is transparent in a large spectral region, notably for wavelengths longer than 360 nm (bandgap), except for the Restrahlen band (from 9.6 to 19 μm). Absorption in the range of 7.3–9 μm has been observed in bulk GaN substrates with carrier concentrations < $10^{16}$ cm$^{-3}$ [21–23] and was attributed to the second harmonic of the Restrahlen band. Although this second band might hinder the fabrication of waveguided devices in this spectral region, its effect in planar devices with micrometer-sized active regions is negligible, since the absorption coefficient related to two-phonon processes is much smaller than the one associated with ISB transitions [24,25]. On the other hand, III-nitrides do not present problems of inter-valley scattering, since the L and X points are much higher in energy (>2 eV) than the $\Gamma$ point.

There is also an interest to push the III-nitride ISB technology toward longer wavelengths, particularly to the THz frequency range [20]. The potential of this spectral region in applications like security screening, quality control, and medical diagnostics has driven extensive development of optoelectronic components. Because of the large LO-phonon energy of GaN (about three times that of GaAs), room temperature operation becomes feasible for ISB devices covering the IR band that was typically inaccessible to As-based semiconductors due to phonon absorption.

## 8.2 INTERSUBBAND ABSORPTION IN III-NITRIDE NANOSTRUCTURES

### 8.2.1 GaN/AlGaN Polar Quantum Wells

#### 8.2.1.1 Modeling

The optical properties of (0001)-oriented nitride QWs are strongly affected by the presence of internal electric fields arising from the piezoelectric and spontaneous polarization discontinuity between the various nitride compounds [26]. Figure 8.1a presents the band diagram of a GaN/AlN (2 nm/3 nm) superlattice calculated using the nextnano[3] 8-band k.p Schrödinger–Poisson solver [27] with the material parameters described by Kandaswamy et al. [28]. The electronic potential takes on a characteristic saw-tooth profile due to the internal electric field. The electron wave functions of the ground hole state $h_1$, ground electron state $e_1$, and excited electron states $e_2$ and $e_3$ are presented in Figure 8.1. In narrow QWs (~1 nm) the energy difference between $e_1$ and $e_2$ is mostly determined by the confinement in the QW, whereas for larger QWs (>2 nm) this difference is ruled by the internal electric field, since both electronic levels lie in the triangular part of the QW potential profile. The evolution of the $e_2$–$e_1$ and $e_3$–$e_1$ energy differences with the QW thickness and strain state is presented in Figure 8.1b and compared with experimental data from GaN/AlN MQWs. The increase in the $e_2$–$e_1$ ISB energy calculated when considering the MQW strained on GaN is related to the enhancement of the electric field in the QW, due to the larger piezoelectric coefficients of the AlN barrier in comparison with the GaN QW [28]. On the other hand, the band structure simulations show that certain QW thicknesses can result in a configuration where the $e_1 \rightarrow e_2$ transition has approximately the same energy as $e_2 \rightarrow e_3$ (or as $e_2 \rightarrow e_4$ for very thick barrier layers), which can result in the enhancement of second-order and third-order nonlinear effects such as two-photon absorption [29], second-harmonic generation [30], or saturable absorption [31].

In nitride heterostructures, charge distribution depends not only on the Si doping level in the QWs but also on the presence of non-intentional dopants and on the carrier redistribution due to the internal electric field. The polarization discontinuity between heterostructure layers leads to strong band bending, which typically results in the formation of a depletion layer on one side of the active region and an accumulation layer

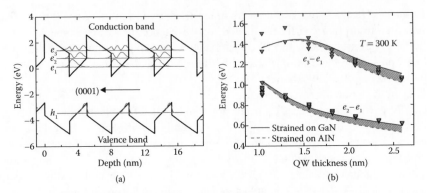

**FIGURE 8.1** (a) Band diagram of GaN/AlN quantum wells (QWs) in an infinite superlattice with 3-nm-thick AlN barriers and 2-nm-thick GaN QWs. The structure is considered strained on an AlN substrate. The electron wave functions of the ground hole state, $h_1$, ground electron state, $e_1$; and excited electron states, $e_2$ and $e_3$, are presented. (b) Variation of the $e_2-e_1$ and $e_3-e_1$ intersubband (ISB) transition energy as a function of the QW thickness in GaN/AlN MQW structures with 3-nm-thick barriers. Triangles indicate experimental data, and solid and dashed lines correspond to theoretical calculations, assuming that the structure fully strained on AlN and on GaN substrates, respectively. (From Kandaswamy et al., *J. Appl. Phys.*, 104(9), 093501, 2008 [28].)

on the other side (Figure 8.2a). Therefore, a realistic view of the charge distribution in a device is only achieved by extending the electronic modeling to the complete structure. As an illustration of this phenomena, Kandaswamy et al. [28] have studied the contribution of the internal electric field induced by a 50-nm-thick $Al_xGa_{1-x}N$ ($x = 0$, 0.25, 0.5, and 1) cap layer to the ISB absorption of 40-period non-intentionally doped GaN/AlN (1.5/1.5 nm) MQWs grown on AlN [28]. Measurements of ISB absorption in these samples confirm a monotonous increase and broadening of the absorption when increasing the Al mole fraction of the cap layer. This trend is consistent with the simulations of the electronic structure in Figure 8.2a and b, where the use of AlN as a cap layer lowers the conduction band of the first GaN QWs below the Fermi level (dash-dotted line at 0 eV in the figures), whereas the use of GaN as a cap layer results in the depletion of the MQW active region.

### 8.2.1.2 Growth and Defect Analysis

A main requirement for the growth of III-nitride nanostructures for ISB devices is the precise control of thickness and interfaces. Molecular beam epitaxy (MBE) seems to be the most suitable technique for this application thanks to its low growth temperature, which hinders GaN–AlN interdiffusion [32]. Plasma-assisted molecular beam epitaxy (PAMBE) was the first method to produce III-nitride nanostructures displaying ISB transitions at telecommunication wavelengths (1.3, 1.55 μm) [33–39]. The growth of (0001)-oriented GaN, AlN, and AlGaN by PAMBE is extensively discussed in the literature [40–45]. Deposition of two-dimensional (2D) III-nitride layers requires a precise control of the III/V flux ratio during the growth; it demands

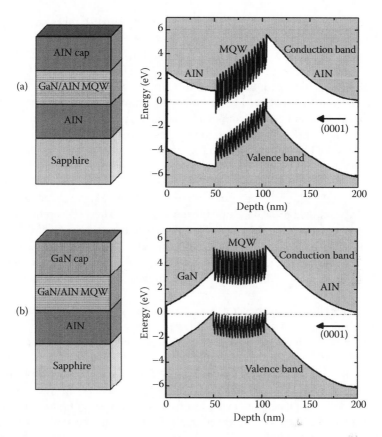

**FIGURE 8.2** Band diagram of non-intentionally doped GaN/AlN (1.5/1.5 nm) MQW structures with (a) GaN cap layer and (b) AlN cap layer. (From Kandaswamy et al., *J. Appl. Phys.*, 104(9), 093501, 2008 [28].)

slightly metal-rich conditions, and hence growth optimization requires the determination of stoichiometric flux conditions and precise control of the growth temperature. In the case of GaN, at a substrate temperature higher than 700°C there is a certain range of Ga fluxes for which the Ga excess remains in a situation of dynamic equilibrium on the growing front, that is, the Ga coverage is independent of the Ga exposure time. Smooth surfaces are generally achieved under moderate Ga-rich conditions [40,42,46,47], when the Ga excess arranges into a so-called laterally contracted Ga bilayer, which consists of two Ga layers adsorbed on the Ga-terminated (0001) GaN surface [40,41,47–49].

In the case of AlN, the deposition of layers with atomically flat surface morphology also requires metal-rich conditions [45]. However, Al does not desorb from the surface at the standard growth temperature for GaN. Therefore, to eliminate the Al excess at the surface it is necessary to perform periodic growth interruptions under nitrogen. An alternative approach to achieve 2D growth of AlN and low Al content (<50%) of Al(Ga)N layers is to use Ga as a surfactant, with the Al flux corresponding to the required Al mole fraction [28,43,50].

GaN/AlN QWs displaying ISB transitions in the near-IR can also be synthesized by metalorganic vapor phase epitaxy (MOVPE) [51–53]. In this case, a critical parameter to attain devices in the telecommunication spectral range (1.3 μm, 1.55 μm) is the reduction in growth temperature from the 1050°C–1100°C optimum range for GaN growth down to 900°C–950°C (or even to 770°C [54]), to minimize the GaN–AlN interdiffusion. Furthermore, deposition under compressive strain (e.g., using AlN substrates) is recommended at these growth temperatures to prevent the red shift of the ISB transition due to instabilities of the GaN/AlN interface [55].

Since GaN/AlN is a lattice-mismatched system (2.5% in-plane lattice mismatch), it is important to understand the effects of strain and misfit relaxation. The density of edge-type threading dislocations (TDs) should be kept to a minimum since they cause losses in the transmission of transverse-magnetic (TM)-polarized light, which adversely affects the performance of ISB devices [56]. These TDs propagate from the heteroepitaxial substrates (typical TD densities in commercial GaN-on-sapphire or AlN-on-sapphire templates are in the $10^8$ cm$^{-2}$ and $10^9$ cm$^{-2}$ range, respectively), but they are also generated during the growth due to the plastic strain relaxation, since misfit dislocations (MDs) often fold toward the growth direction, giving rise to edge-type TDs [57].

It is not clear how semiconductors with hexagonal symmetry, such as III-nitrides, relax the misfit stress. In the case of nitride heterostructures grown along the [0001] axis, the formation of regular networks of MDs is hindered since the most crystallographically favorable slip system, (0001) basal plane with <11-20> {0002} slip directions, lies parallel to the heterointerfaces. This means that the resolved misfit stress on the main slip plane is zero [58]. Thus, only secondary slip systems that are oblique to the basal plane can have a resolved misfit stress and may contribute to plastic relaxation. It has been observed that MDs following the secondary <11-23> {11-22} slip system can be generated at heterointerfaces when shear stress is intentionally or unintentionally induced by three-dimensional (3D) growth [59,60], by crack formation [57,61], or in close proximity to V-defects [62]. Therefore, the relaxation mechanism depends not only on the structure itself but also on the growth conditions. In general, GaN/AlGaN heterostructures grown under tensile stress on GaN substrates tend to crack along the <11-20> crystallographic direction at a certain critical thickness [61,63,64]. In the case of crack-free GaN/AlGaN superlattices deposited under compressive strain, the main relaxation mechanism is the tilt of the $a$-type TDs toward <1-100>, the inclination angle depending on the lattice misfit between the MQWs and the underlayer [65]. It has been proposed that the diagonal movement is due to a staircase-like movement of the dislocations through the stack, with a misfit segment at each well. Strain relaxation via TD inclination has also been observed in AlGaN layers deposited on mismatched AlGaN [66].

In PAMBE growth, the metal to N ratio and the growth temperature are key parameters that define the strain relaxation rate during growth [67]. Ga-rich conditions delay crack propagation and minimize strain relaxation [50]. They also allow good control of the layer thickness and Al incorporation in AlGaN alloys. In the case of GaN/AlN MQWs and in addition to the relaxation mechanisms described earlier, the periodic misfit relaxation appears to be associated with the formation of stacking fault loops that initiate at the beginning of the AlN barrier deposition, propagate through the barrier, and close within the following QW [50]. In contrast, transmission electron microscopy (TEM) images from GaN/AlGaN superlattices (ternary alloy barriers) do not reveal stacking faults or other periodic defects [68].

### 8.2.1.3  Intersubband Optical Characterization

Figure 8.3 shows the ISB absorption of Si-doped AlN/GaN MQWs with 3-nm-thick AlN barriers and QW thicknesses of 5, 6, 7, and 9 monolayers (MLs) [16,28]. The samples show a pronounced TM-polarized absorption, attributed to the transition from the first to the second electronic levels in the QW ($e_1 \rightarrow e_2$), whereas no absorption was observed for transverse-electric (TE)-polarized light within experimental sensitivity. For large QWs ($\geq$8 ML), the $e_1 \rightarrow e_3$ transition is observed, as indicated in Figure 8.3; this transition is forbidden in symmetric QWs [69] but is allowed in nitride QWs because the internal electric field in the well breaks the symmetry of the potential. The experimental values of $e_2 - e_1$ and $e_3 - e_1$ as a function of QW width are presented in Figure 8.1, showing a good fit with theoretical calculations.

The line width of the absorption remains in the 70- to 100-meV range for QWs doped at $5 \times 10^{19}$ cm$^{-2}$, and the ISB absorption efficiency per reflection attains 3%–5%. A record small line width of ~40 meV has been achieved in non-intentionally doped structures [70]. The spectra either present Lorentzian shape or is structured with two or three well-defined peaks in Lorentzian shape [16]. These multiple peaks correspond to the expected values of the $e_1 \rightarrow e_2$ line in QWs whose thickness is equal to an integer number of GaN monolayers. For very narrow QWs, a variation

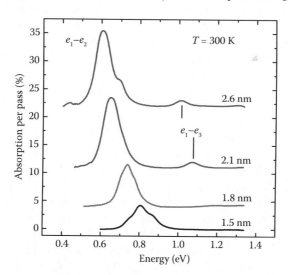

**FIGURE 8.3**  Room-temperature transverse-magnetic-polarized intersubband absorption spectra from 20-period Si-doped GaN/AlN multiple quantum well structures with 3-nm-thick AlN barriers and different GaN quantum well thickness. All the samples were grown on 1-μm-thick AlN-on-sapphire templates without a cap layer. The absorption peaks labeled $e_1 - e_2$ and $e_1 - e_3$ are assigned to the corresponding intra-conduction-band transitions as described in Figure 8.1a. The spectra are vertically shifted for clarity. (From Kandaswamy et al., *J. Appl. Phys.*, 104(9), 093501, 2008 [28].)

in the thickness by 1 ML implies an important shift in the ISB transitions (about 100 meV for QWs of 4 to 5 ML). This value is comparable to the full width at half maximum (FWHM) of the absorption lines and hence results in well-resolved absorption peaks. Thickness fluctuations might originate from a drift of the growth rate with time, resulting in a variation in the QW thickness. However, in situ measurements of the growth rate and TEM studies confirmed that structured absorption spectra appear in samples where no growth rate drift is detected. In these samples, cathodoluminescence studies confirmed the presence of in-plane thickness fluctuations, which appear to be associated with dislocations or extended defects [28]. Regarding the thermal stability of the ISB transition in GaN/AlN MQWs, it has been found that the ISB absorption energy decreases only by ~6 meV at 400°C relative to its room temperature value [71].

Using GaN/AlN QWs, the $e_1 \rightarrow e_2$ ISB transition can be tuned in the 1.0- to 3.5-μm wavelength range by changing the QW thickness from 1 to 7 nm [16,28,35,36,38,52,64,71,72] with AlN barrier thicknesses in the 1.5- to 5.1-nm range. For larger QWs (>5 nm), the first two electron-confined levels get trapped in the triangular section of the QW, which results in a saturation of the $e_2 - e_1$ value. Therefore, to shift the absorption toward longer wavelengths it is necessary to reduce the effect of the internal electric field in the QWs. A first approach consisted of using GaN/AlGaN MQWs, thereby reducing the Al mole fraction in the barriers. By changing the geometry and composition, the ISB absorption can be tailored to cover the near-IR range above 1.0 μm and the mid-IR region up to 5.3 μm [33,34,73–82]. To attain longer wavelengths, the requirement of substrate transparency imposes the replacement of sapphire-based templates by semi-insulating Si (111) as a substrate [24]. Using GaN-on-Si (111) templates, Kandaswamy et al. [24] have demonstrated the extension of the ISB absorption range of GaN/AlGaN QWs up to 10 μm, as illustrated in Figure 8.4a. A slight red shift of the ISB transition is observed when increasing the compressive strain in the QWs, as theoretically predicted [78].

### 8.2.1.4  Effect of Doping

To observe ISB absorption, it is necessary to control the carrier concentration in QWs to guarantee that the first electronic level is populated. High doping levels also have an effect on the targeted operating wavelength. The ISB absorption energy can blue shift markedly due to many-body effects [25,38,83], mostly due to exchange interaction and depolarization shift, as illustrated in Figure 8.4b and (c). On the other hand, studies of the effect of dopant location have shown a dramatic reduction of the ISB absorption line width by using a δ-doping technique with Si donors placed at the end of the QW [84]. This line width reduction is attributed to an improvement of interfacial roughness. It has also been theorized that doping in the wells should provide red-shifted [85] and stronger [86] ISB absorptions versus doping in the barriers.

When integrating III-nitride nanostructures into complete devices, it is necessary to keep in mind that the magnitude of the carrier distribution depends not only on the Si doping level in the QWs, but also on the presence of non-intentional dopants and on the carrier redistribution due to the internal electric field. The large polarization discontinuities in the III-N material system can result in a significant (even dominant) contribution to the IR absorption in GaN/AlN superlattices [28].

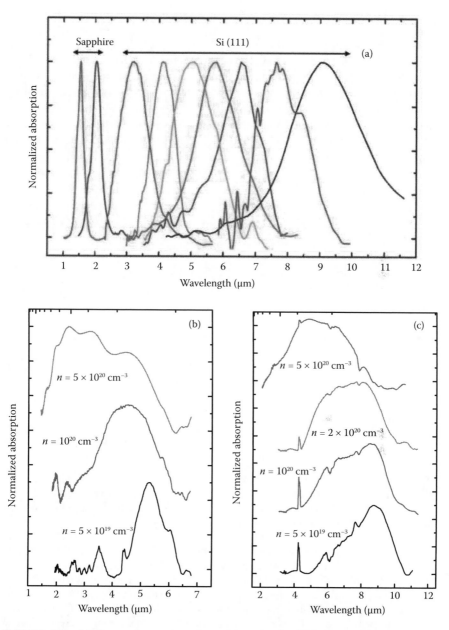

**FIGURE 8.4** (a) Room-temperature transverse-magnetic (TM)-polarized infrared (IR) photo-induced absorption spectra measured in GaN/AlGaN multiple quantum well (MQW) structures with different barrier Al contents and quantum well (QW) widths grown either on sapphire or on Si (111) templates [24]. (b) Infrared absorption spectra for TM-polarized light measured from GaN/Al$_{0.2}$Ga$_{0.8}$N (3 nm/3 nm) MQWs with different doping levels. Spectra are vertically shifted for clarity [25]. (c) IR absorption spectra for TM-polarized light measured from GaN/Al$_{0.1}$Ga$_{0.9}$N (7 nm/4 nm) MQWs with different doping levels. Spectra are vertically shifted for clarity. (From Kandaswamy et al., *Appl. Phys. Lett.*, 96(14), 141903, 2010 [25].)

## 8.2.2 FAR-INFRARED MULTILAYER ARCHITECTURES

AlGaN/GaN QWs displaying ISB transitions at 9.08 μm (136 meV) have been demonstrated [24]. However, in large QWs the internal electric field associated with the spontaneous and piezoelectric polarization discontinuities in the GaN/AlGaN system becomes the dominating characteristic for determining energy levels, which are no longer sensitive to changes in QW thickness. To further reduce the ISB transition energy, the first proposed architecture consisted of a three-layer well (step QW) with a virtually flat potential profile [87–89]. Figure 8.5a shows the conduction-band

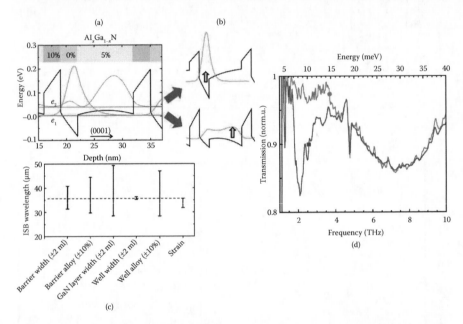

**FIGURE 8.5** (a) Conduction-band profile, and first ($e_1$) and second ($e_2$) electronic levels with their associated wave functions for an $Al_{0.05}Ga_{0.95}N/Al_{0.1}Ga_{0.9}N/GaN$ (10 nm/3 nm/3 nm) step quantum well (QW) [88]. (b) Shift of the wave function of the first electronic level associated with a variation of the Al concentration in the well layer. A higher step-well Al concentration (top) creates a more triangular well and increases confinement toward the GaN layer. A lower step-well Al concentration (down) creates a secondary confinement area at the opposite side of the well. To illustrate the effect clearly, the band profiles correspond to $Al_{0.07}Ga_{0.93}N/Al_{0.1}Ga_{0.9}N/GaN$ (10 nm/3 nm/3 nm) (top) and $Al_{0.03}Ga_{0.97}N/Al_{0.1}Ga_{0.9}N/GaN$ (10 nm/3 nm/3 nm) (bottom) [88]. (c) Illustration of the robustness of the step-QW system. The dashed line indicates the nominal transition wavelength for an $Al_{0.05}Ga_{0.95}N/Al_{0.1}Ga_{0.9}N/GaN$ (10 nm/3 nm/3 nm) step QW. The error bars represent the minimum and maximum values attributed to the uncertainties associated to growth. The barrier and GaN well thicknesses were changed from 3 nm to 2.5 nm and 3.5 nm. The barrier Al content was changed from 10% to 11% and 9%. The step-well alloy was changed from 5% to 4.5% and 5.5% (±10%). The step-well thickness was changed from 10 nm to 12 nm and 8 nm. The strain error bar illustrates the variation of the intersubband transition when evolving from a structure fully strained on GaN to a structure fully strained on $Al_{0.1}Ga_{0.9}N$ [88]. (d) Transmission spectra for transverse-magnetic- (square) and transverse-electric- (circle) polarized light at $T = 4.7$ K. (From Machhadani et al., *Appl. Phys. Lett.*, 97(19), 191101, 2010 [87].)

diagram of a step-QW design, in this example consisting of $Al_{0.1}Ga_{0.9}N/GaN/Al_{0.05}Ga_{0.95}N$ (3 nm/3 nm/10 nm). This three-layer structure is designed around the principle of polarization equivalency. The design can be broken effectively into two portions: The first is the "barrier," which comprises the high-Al-content $Al_xGa_{1-x}N$ layer and the GaN layer. The second portion is the "well," which is the low-Al-content $Al_xGa_{1-x}N$ layer. The design creates a semi-flat band in the well by having the barrier balanced at the same average Al percentage, that is, the average polarization in the barrier is approximately equal to the average polarization in the well. This configuration is associated with the minimum energy spacing between the ground electronic state and the first excited state, as described by Wu et al. [90]. Samples following the step-QW design in Figure 8.5a have been synthesized by PAMBE on GaN templates on float-zone Si (111) to evade problems of substrate transparency [24], and ISB absorption at ~2 THz (~70 µm) (illustrated in Figure 8.5d) and at ~13 THz (~22 µm) has been reported [87,89].

The weakness of the step-QW design lies in the fact that any deviation from this balance results in an internal electric field in the well, which shifts the wave function associated with the first electronic level toward the GaN layer (Figure 8.5b top, for higher Al content in the well) or toward the high-Al-content layer (Figure 8.5b down, for lower average Al content in the well). Thus, any imbalance in the structure has a drastic effect on ISB wavelength. The limitations of the step-QW configuration can be surmounted by the insertion of an additional AlGaN layer to separate the GaN layer from the low-Al-content $Al_xGa_{1-x}N$ well [88]. The "separation layer" is designed so that there is no confined state in the GaN layer. This architecture, described in Figure 8.6a, does not evade the quantum-confined Stark effect, but the GaN layer contributes to reduce the average spontaneous polarization of the complex barrier structure ($Al_{0.1}Ga_{0.9}N/GaN/Al_{0.07}Ga_{0.93}N$), which results in a lower electric field in the QW. The robustness of this design is analyzed in Figure 8.6c and shows much less variation of the ISB transition energy than the step-QW architecture with respect to the growth uncertainties.

The incorporation of the separation layer results in a geometry where the internal electric field is not fully compensated, that is, the QW keeps a triangular potential profile. As a consequence, the four-layer MQW system is more sensitive to changes in strain state versus the step-QW design. The strain error bar in Figure 8.6c illustrates the variation of the ISB transition when evolving from a structure fully strained on GaN to a structure fully strained on $Al_{0.1}Ga_{0.9}N$. These error bars are comparable to those generated by uncertainties in the structural parameters. However, the MQWs are expected to evolve toward a minimum elastic energy configuration independently of the substrate [68], so that the uncertainty in the strain state of the structure (neglecting the initial relaxation) is much smaller (<±0.025% variation of the in-plane lattice parameter) than the values simulated in Figure 8.5c (±0.12%).

AlGaN/GaN 40-period MQW structures following the four-layer MQW design in Figure 8.6b were deposited by PAMBE. Figure 8.6d shows the low-temperature ($T$ = 5–10 K) far-IR transmission spectra of samples with different doping concentrations. In the sample with a lower doping level ([Si] = $1.5 \times 10^{19}$ cm$^{-3}$), a TM-polarized absorption dip centered around 27–29 µm (~14 THz), which gets deeper and broader with increasing doping level, is observed. This absorption line is

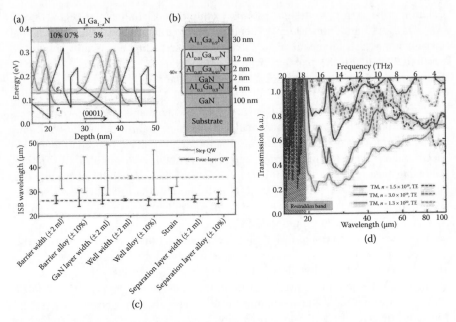

**FIGURE 8.6** (a) Conduction band profile, and first ($e_1$) and second ($e_2$) electronic levels with their associated wave functions for an $Al_{0.1}Ga_{0.9}N/GaN/Al_{0.07}Ga_{0.93}N/Al_{0.03}Ga_{0.097}N$ (4 nm/2 nm/2 nm/12 nm) four-layer quantum well (QW) design. (b) Schematic drawing of the structure synthesized by plasma-assisted molecular beam epitaxy (PAMBE). (c) Illustration of the robustness of the four-layer QW system compared with the step QW. The dashed lines indicate the nominal transition wavelengths for an $Al_{0.05}Ga_{0.95}N/A_{10.1}Ga_{0.9}N/GaN$ (10 nm/3 nm/3 nm) step QW and an $Al_{0.1}Ga_{0.9}N/GaN/Al_{0.07}Ga_{0.93}N/Al_{0.03}Ga_{0.97}N$ (4 nm/2 nm/2 nm/12 nm) four-layer QW. The error bar represents the minimum and maximum values attributed to the uncertainties associated with growth. In the four-layer QW, the barrier thickness was changed from 2 nm to 1.5 nm and 2.5 nm. The barrier Al content was changed from 10% to 11% and 9%. The GaN layer thickness was changed from 2 nm to 1.5 nm and 2.5 nm. The well thickness was changed from 12 nm to 11.5 nm and 12.5 nm, and its Al content was changed from 3% to 2.7% and 3.3% (±10%). The separation layer thickness was changed from 2 nm to 1.5 nm and 2.5 nm, and its Al content was changed from 7% to 6.3% and 7.7%. The strain error bar illustrates the variation of the ISB transition when evolving from a structure fully strained on GaN to a structure fully strained on $Al_{0.1}Ga_{0.9}N$. (d) Far-IR transmission measurement of four-layer multiple quantum wells with different doping levels for transverse-electric- and transverse-magnetic (TM)-polarized light. The spectra have been normalized by the response of a similar undoped superlattice, which exhibits no intersubband activity. The noise observed for wavelengths <10 μm is due to the GaN Restrahlen absorption. The dip in TM-polarized transmission at 27–30 μm is assigned to the transition between the first and the second electronic levels in the QWs. (From Beeler et al., *Appl. Phys. Lett.*, 103(9), 091108, 2013 [88].)

attributed to the transition from the first to the second electronic level in the QW, in good agreement with theoretical calculations.

The normalized absorption line width for the sample with a doping level [Si] = 1.5 × 10$^{19}$ cm$^{-3}$ is $\Delta f/f \sim 0.25$ [88], which is a significant improvement in comparison to results in step QWs ($\Delta f/f \sim 0.5$ in the study by Machhadani et al. [87]).

From progress on the four-layer design, robustness was deemed a key attribute to create these devices. However, the four-layer design used complex barrier systems to achieve this desired robustness. Such complex barrier structures inhibited tunneling transport, and therefore the incorporation of the architecture into ISB devices using the quantum cascade principle. This led Beeler et al. [91] to propose a pseudo-square QW, a four-layer architecture where the compensation of the polarization-induced internal electric field is obtained by creating a gradual increase in polarization field throughout the quantum "trough" generated by three low-Al-content layers (see schematic description in Figure 8.7). This design has single-layer barriers that can permit tunneling transport under bias. Experimentally, it is shown that the ISB wavelength can be varied from 100 to 75 µm by changing the size of the quantum trough, and from 150 to 50 µm by changing the doping level, as illustrated in Figure 8.7d.

### 8.2.3 Coupled Quantum Wells

The design of advanced ISB devices, like optically or electrically pumped ISB lasers, requires the exploitation of multilevel systems with finely tuned oscillator

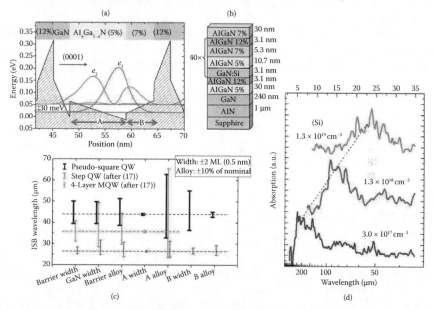

**FIGURE 8.7** (a) Band profile of a pseudo-square quantum well (QW) where the first and second confined electronic levels are indicated as $e_1$ and $e_2$, respectively. (b) Schematic description of the structure that was grown on an AlN-on-sapphire substrate. (c) Illustration of the robustness of the pseudo-square QWs. The dashed lined indicates the nominal transition wavelengths. The error bars represent the minimum and maximum values attributed to the uncertainties associated with growth (geometries changed by ±0.5 nm and alloys changed by ±10%). Similar data for step QWs and four-layer QWs are included for comparison [17]. (d) Spectral absorption of transverse-magnetic (TM)-polarized light in samples with different doping levels. Spectra are found from dividing transverse-electric (TE)- by TM-polarized light, then normalizing this division against reference spectra not showing absorption. (From Beeler et al., *Appl. Phys. Lett.*, 105(13), 131106, 2014. [91].)

strengths and lifetimes. Coupled QWs are the basic element for such systems. In III-nitrides, the realization of coupled QWs is complicated by the relatively heavy electron effective mass and the large conduction-band offset. As a result, very thin barriers (1–3 ML) are required to achieve strong interwell coupling in the GaN/AlN material system [92].

Coupling between GaN QWs was first experimentally investigated by Gmachl et al. [34,37] by using double GaN QWs coupled by AlGaN barriers with large (0.65 or 0.9) Al mole fractions. The coupling barrier thicknesses varied from 0.7 to 6 nm. Degenerate doping of the QWs ($10^{20}$ cm$^{-3}$) was used to establish a common reference energy at the Fermi level, which decreases the uncertainties related to intrinsic internal electric fields. The broadening and structuration of ISB absorption peaks were attributed to transitions toward excited states exhibiting anticrossing.

Coupled GaN/AlN QWs were first demonstrated by Tchernycheva et al. [93] using 0.5-nm-thick AlN coupling barriers. The ISB absorption spectra present two distinct peaks attributed to the transition: first, between the ground states of the two coupled wells and, second, between the ground state and the delocalized excited state between the two wells. As an alternative approach, Driscoll et al. have opted to decrease the Al content of the coupling barrier to 39%–53%, so that strong coupling is achieved with thicker barriers (~5 ML). In this fashion, the barrier's Al content can be used as a tunable parameter to control the coupling strength [94].

The influence of polarization-induced electric fields on ISB absorption and the associated variation of refractive index in AlN/GaN-coupled QWs has been theoretically analyzed by Cen et al. [95,96] for their application in ultrafast two-color optoelectronic devices and electro-optical modulators operating within the optical communication wavelength range.

### 8.2.4 In-Containing Superlattices: AlInN/GaN, AlInN/GaInN, and GaN/GaInN

The lattice mismatch between GaN and AlN can lead to high defect densities and the risk of cracking in GaN/AlN superlattices. An alternative material approach to overcome this problem is the use of AlInN alloys. AlInN with an In composition around 17%–18% is lattice matched to GaN and presents a refractive index contrast equivalent to AlGaN with 46% Al content (6% contrast with GaN at 1.55-μm wavelength). Therefore, AlInN is a promising material to form distributed Bragg reflectors and thick waveguide layers [97]. However, lattice-matched AlInN/GaN heterostructures still exhibit an electric field as large as 3 MV/cm, solely generated by the spontaneous polarization discontinuity.

The potential of AlInN/GaN lattice-matched systems for application in ISB technology has been explored [98,99]. However, this material system is not adapted to serve as the active region for telecommunication devices since the conduction-band offset is in the range of ~1 eV [100]. Nevertheless, ISB absorption in the near-IR spectral region has been reported at 2.3–2.9 μm in lattice-matched GaN/AlInN superlattices grown by MOVPE [98] and by MBE [101].

An alternative approach to manage the strain in the structure while retaining access to shorter wavelengths is possible by adding small concentrations of In (below 10%) both in the barrier and in the QW, forming an AlInN/GaInN superlattice [102,103]. This material combination reduces the probability of crack propagation in comparison with GaN/AlN, although it maintains a certain degree of strain. Despite the challenges to precisely control the In mole fraction, room-temperature ISB absorption in the 1.52- to 2.45-μm wavelength range has been theorized [104,105] and was demonstrated in AlInN/GaInN MQWs [102].

Regarding the polar GaN/GaInN system, only theoretical calculations of ISB transition energy have been published so far [106,107].

## 8.2.5 Quantum Dots

An alternative approach to QW structures for the fabrication of devices is based on optical transitions between bound states in the conduction band of QD superlattices [108,109]. Quantum dot infrared photodetectors (QDIPs) are expected to ultimately outperform QWIPs in terms of low dark current, high photoelectric gain, and sensitivity [110]. Furthermore, under certain conditions intraband bound to continuum transitions in QDs can be nearly independent of the polarization of excitation [111–113].

In the case of III-nitrides, GaN/AlN QD structures can be synthesized by PAMBE through GaN deposition under compressive strain and under N-rich conditions [70,114]. In this situation, 2D growth proceeds normally to create a 2-ML-thick wetting layer. Due to the lattice mismatch between AlN and GaN, further GaN deposition leads to the formation of 3D islands above this 2D wetting layer (Stranski–Krastanov growth mode). These GaN QDs are well-defined hexagonal truncated pyramids with {1-103} facets [115]. The QD size can be tuned by modifying the amount of GaN in the QDs, growth temperature, or growth interruption time after deposition of the QDs (Ostwald ripening). By adjusting the growth conditions, QDs with height (diameter) in the range of 1–1.5 nm (10–40 nm) and density between $10^{11}$ cm$^{-2}$ and $10^{12}$ cm$^{-2}$ can be synthesized (see atomic force microscopy image in Figure 8.8a as an example) [116]. To populate the first electronic level, silicon can be incorporated into the QDs without significant perturbation of the QD morphology.

Andreev et al. [117–119] have calculated the electronic structure of GaN/AlN QDs using the $k \cdot p$ model ($k$ = wavevector and $p$ = momentum operator) while taking the internal electric field into account. These calculations have been complemented by Ranjan et al. [120] through the use of the tight-binding theory and a self-consistent treatment to account for carrier screening of the electric field.

The models show that the polarization-related internal electric field localizes the electrons at the pyramid apex, whereas holes are rather located close to the wetting layer. In addition to the carrier separation along the growth axis, the electric field can provide a strong additional lateral confinement for carriers localized in the dot, which strongly modifies their electronic structure and optical properties [119].

From the experimental viewpoint, Si-doped QD superlattices have been reported to exhibit strong TM-polarized intraband absorption at room temperature, which can be tuned from 0.74 eV (1.68 μm) to 0.90 eV (1.38 μm) as a function of the QD size

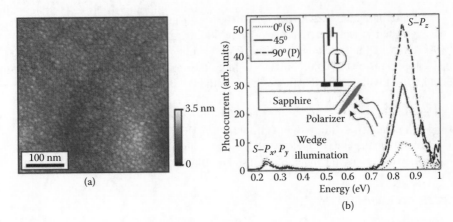

**FIGURE 8.8** (a) Atomic force microscopy image of a GaN quantum dot (QD) layer synthesized on AlN by plasma-assisted molecular beam epitaxy (PAMBE), showing a high density (~$10^{12}$ cm$^{-3}$) of small (height, ~1.3 nm; base diameter, ~11 nm) QDs. (b) Polarization-dependent spectral response (photocurrent) of a GaN/AlN QD stack measured in wedge illumination. The normalized response of both $s$–$p_z$ in the near infrared (IR) and $s$–$p_{x,y}$ in the mid-IR; the $s$–$p_z$ peak is about 10 times larger than the $s$–$p_{x,y}$ peak. (From Vardi et al., *Phys. Rev. B*, 80(15), 155439, 2009 [122].)

[116,121]. The broadening of the absorption peak can be as small as ~80 meV for the most homogeneous samples. This absorption line is attributed to transitions from the ground state of the conduction band, $s$, to the first excited electronic state confined along the growth axis, $p_z$. The lateral confinement in the QDs should give rise to additional transitions under TE-polarized excitation. However, taking into account the lateral dimension of the QDs, ~7 nm, the $s$–$p_{x,y}$ transitions should be masked by the sapphire absorption for $\lambda > 5$ μm. The optical signature associated with $s$–$p_{x,y}$ was first observed by Vardi et al., who studied near-IR and mid-IR intraband transitions in GaN/AlN QDs using in-plane electronic transport at low temperatures [122], as illustrated in Figure 8.8b. The measured $s$–$p_{x,y}$ energy separation (0.1–0.3 eV) was significantly larger than the equivalent transition energy in InGaAs\GaAs QDs. Their analysis shows that the appearance of large-energy $s$–$p_{x,y}$ in GaN\AlN QDs is due to the strong internal electric field in the QDs, which results in stronger confinement of the electrons at the QD top facet.

The homogeneous line width of the $s$–$p_z$ intraband transition at 1.55 μm in GaN/AlN QDs was assessed by means of nonlinear spectral hole-burning experiments [123]. These measurements demonstrated that electron–electron scattering plays a minor role in the coherence relaxation dynamics, since the homogeneous line width of 15 meV at 5 K does not depend on the incident pump power. This suggests the predominance of other dephasing mechanisms such as spectral diffusion.

TM-polarized IR absorption in the 1.6- to 2-μm wavelength range, attributed to $s$–$p_z$ intraband transitions, has also been reported in ternary Al$_x$Ga$_{1-x}$N/AlN QDs ($x = 0$–0.42) measured at room temperature. The $s$–$p_z$ transition red shifts for increasing Al mole fraction in the QDs as a result of the reduction in band offset, in good agreement with theoretical calculations [124,125].

## 8.2.6 Nanowire Heterostructures

Using the enhanced quantum-confined properties of QDs, a variety of structures can be created. However, QDs can only be grown under specific strain situations, limiting the degree to which these QD devices can be engineered. Nanowires do not have these limitations and due to the unique strain relaxation mechanisms can provide a much larger array of material pairings, beyond the limitations of planar and QD systems. Nanowire heterostructures in particular offer a unique situation for a variety of devices requiring low-defect-density active regions and large lattice-mismatched materials.

Following these benefits, nanowire heterostructures of Ge-doped GaN/AlN have been grown by PAMBE, with the result illustrated in the TEM images in Figure 8.9a and b [126]. The nanowire disk sizes were varied in the 2- to 8-nm range, and the doping level was varied over two orders of magnitude. The ISB energies were measured via Fourier transform infrared (FTIR), and Figure 8.9c displays the results in a series of samples with different doping levels. The TM-polarized absorption is assigned to the $s-p_z$ intraband transition in the Ge-doped GaN/AlN nanodisks. The $s-p_z$ absorption line experiences a blue shift with increasing Ge concentration and a red shift with increasing nanodisk thickness. Theoretical calculations in Figure 8.9d show that the intraband transitions are strongly blue shifted due to many-body effects, that is, the exchange interaction and depolarization shift.

## 8.2.7 Alternative Crystallographic Orientations

The presence of internal electric field in polar materials increases the design complexity of ISB devices. A simple approach to circumvent this problem consists of using nonpolar crystallographic orientations like $m$-plane {1-100} or $a$-plane {11-20} [127]. However, epitaxy for these orientations is an arduous task, due to strong anisotropy of surface properties, resulting in a high density of crystalline defects. An alternative approach is the growth on semipolar planes [127], which are those ($hkil$) planes with at least two nonzero $h$, $k$, or $i$ Miller indices and a nonzero $l$ Miller index. Semipolar planes allow considerable reduction in the internal electric field [128] while presenting a lower in-plane anisotropy than nonpolar surfaces [129,130].

Regarding nonpolar materials, ISB optical absorption at $\lambda \sim 2.1$ μm with FWHM = 120 meV has been reported in Si-doped 1.75-nm-thick GaN QWs with 5.1-nm-thick AlN barriers grown by PAMBE on $r$-plane sapphire and displaying pure $a$-plane orientation [131]. Furthermore, ISB absorption has been shown at THz frequencies in $m$-plane GaN/AlGaN MQW structures [132]. The ISB energy was tuned in the 15.6- to 26.1-meV range by changing the well sizes and alloy compositions of the barriers.

Near-IR ISB absorption has also been reported on semipolar (11-22)-oriented GaN/AlN MQWs grown by PAMBE [130,133], as illustrated in Figure 8.10a. In comparison with polar QWs, semipolar structures exhibit quasi-square potential band profiles with symmetric wave functions, due to the reduced electric field in the range of

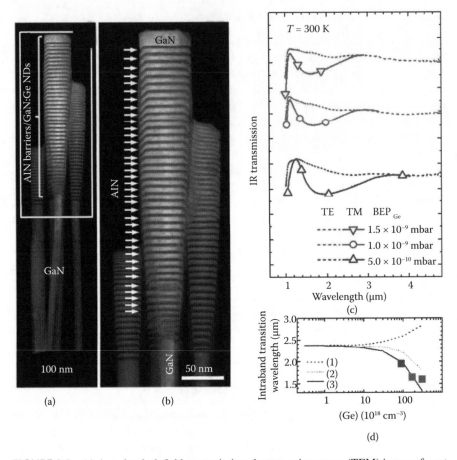

**FIGURE 8.9** (a) Annular dark field transmission electron microscopy (TEM) image of a set of GaN nanowires from a sample containing 40 periods of AlN/GaN nanodisks. (b) Zoom-in of the squared area in (a). (c) Room-temperature infrared transmission spectra for transverse-electric (TE)- (dashed) and transverse-magnetic (TM)-polarized (solid) light measured for Ge-doped GaN/AlN (4 nm/4 nm) heterostructured nanowires with different doping levels in the GaN nanodisks (NDs). The spectra are vertically shifted for clarity. (d) Variation of the intraband transition wavelength as a function of the estimated Ge concentration. Dots are experimental values from (c); the dashed line labeled (1) is a one-dimensional calculation of the intraband transition accounting for the screening of the internal electric field; the dotted line (2) incorporates corrections associated with both screening and exchange interaction; the solid line (3) accounts for screening, exchange interaction, and depolarization shift. (From Beeler et al., *Nano Lett.*, 14(3), 1665–1673, 2014 [126].)

0.5–0.6 MV/cm in the QWs. The evolution of the $e_2$–$e_1$ ISB transition energy with QW thickness is represented in Figure 8.10b, where symbols correspond to experimental measurements obtained from identical polar and semipolar samples consisting of 40 periods of GaN/AlN with 3-nm-thick AlN barriers. The absorption FWHM (~ 80–110 meV [130,133]) is comparable to the one measured in polar structures

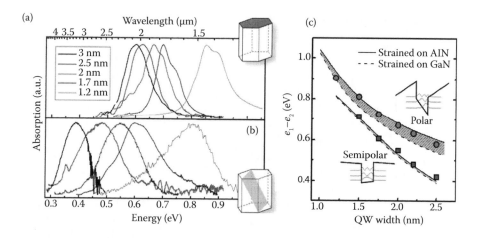

**FIGURE 8.10**  Intersubband (ISB) absorption spectra for (a) polar and (b) semipolar GaN/AlN quantum wells (QWs) with different well thicknesses. (c) Calculated and measured ISB absorption energy versus well thickness. (From Machhadani et al., *J. Appl. Phys.*, 113(14), 143109, 2013 [133].)

[16,28]. However, in semipolar structures the reduction in internal electric field results in a red shift of the ISB energy. Also, the spectral shape of semipolar absorption is Gaussian, in contrast to the Lorentzian shape described for polar GaN/AlN QWs. This is due to the spectral dispersion generated by thickness fluctuations and presence of stacking faults in semipolar material, both inducing carrier localization in the range of a few millielectron volts, much smaller than the FWHM of the ISB absorption line.

### 8.2.8  CUBIC III-NITRIDES

Another approach to eliminate the internal electrical field in III-nitride heterostructures is the use of III-nitride semiconductors crystallized in the zinc-blend crystallographic phase. The LO-phonon energy in cubic GaN is almost the same as in wurtzite GaN (92.7 meV [134]), but the effective mass is significantly smaller ($m^* = 0.11–0.17$ [135,136]) than in wurtzite GaN ($m^* = 0.2$), which should result in higher gain and lower threshold current in QCLs. It has also been theorized that these systems could be used to create QD photodetectors with high gain and sensitivity [137].

The cubic orientation can be selected by PAMBE using 3C–SiC substrates. However, due to their thermodynamically unstable nature cubic films present low structural quality with a high density of stacking faults. ISB absorption in the 1.40- to 4.0-μm spectral range has been reported in cubic GaN/AlN MQWs [138–140], in agreement with theoretical calculations assuming a conduction-band offset of 1.2 eV and an effective mass of $m^* = 0.11$. ISB THz absorption at 4.7 THz has also been observed in cubic GaN/Al$_{0.05}$Ga$_{0.95}$N (12 nm/15 nm) QWs [139,140]. Using slightly different material properties (a conduction-band offset of 1.6 eV and an effective mass $m^* = 0.11$), Radosavljevic et al. have theorized that mid-IR ISB absorption could be tuned by applying a bias perpendicular to the layers [141].

## 8.3  ALL-OPTICAL SWITCHES

The increasing bandwidth demand in optical communication networks impels the development of all-optical devices, particularly targeting the 1.55-μm transmission band of optical fibers. In particular, there is a need for all-optical switches capable of sustaining high repetition rates (sub-picosecond response time) with low switching energy and high modulations depths. These specifications led researchers to consider the use of resonant nonlinearities [142], to exploit phenomena like nonlinear absorption, self- and cross-phase modulation, self- and cross-gain modulation, and four-wave mixing. In this case, switching can be achieved through absorption saturation by an intense control pulse, as originally demonstrated in GaAs-based structures [143]. Using GaN/AlGaN, the switching is achieved by ISB absorption bleaching. Thanks to the ultrafast ISB recovery time (in the 140- to 400-fs range [37,144–148]) associated with the strong interaction of electrons with LO phonons, GaN/AlN QWs or QDs have been proposed as the active medium for all-optical switches (saturable absorbers) operating at terabit-per-second data rates and at telecommunication wavelengths.

The use of GaN/AlGaN QWs for all-optical modulators at telecommunication wavelengths was first proposed by Suzuki et al. [17,149]. Since then, all-optical switches at ~1.55 μm with sub-picosecond commutation times have been demonstrated by several groups [31,56,146,147,150–153]. In general, these devices consist of GaN/AlN MQWs embedded in a ridge waveguide. In such structures, a critical parameter to reduce transmission losses is the reduction of edge-type dislocations. These defects introduce acceptor centers where electrons can be captured and therefore can effectively act as a wire-grid polarizer, which leads to selective attenuation of the TM-polarized signal [154]. Control switching energies of 38 pJ for 10-dB modulation depth [151] and 25 pJ for 5-dB contrast [153] have been demonstrated using a waveguide with an AlN cladding below the active GaN/AlN QWs and GaN or $Si_xN_y$ as the upper cladding layer, respectively. Theoretical calculations predict a reduction of the switching energy by a factor of 30 by replacing the GaN/AlN QWs with properly designed AlN/GaN/AlGaN-coupled QWs [155].

From the material viewpoint, the parameter responsible for absorption saturation is the optical third-order susceptibility, $\chi^{(3)}$. Comparative studies using the forward degenerate four-wave mixing technique in a boxcars configuration point to an increase of $\chi^{(3)}$ by a factor of five in QDs compared to QWs [156]. From the experimental viewpoint, the intraband absorption saturation of GaN/AlN QDs was probed by Nevou et al. [31], obtaining values in the range of 15–137 MW/cm² (0.03–0.27 pJ/μm²). In spite of the large signal variation (a consequence of the focusing uncertainty in the sample), even the upper estimate of the saturation intensity for QDs is smaller than the corresponding value for GaN/AlN QWs (9.46 W/μm² [151]). Based on these results, Monteagudo-Lerma et al. made a comparison of the performance as saturable absorbers of 3 periods of GaN/AlN QWs and QDs inserted in a GaN-on-AlN waveguide structure [157]. In the case of 5-μm-wide QW-based waveguides, a 10-dB change in transmittance was achieved for input energies of ~24 pJ with 150-fs pulses. This value was improved by almost a factor of two by the replacement of QW by QDs as active elements. The reduction of the waveguide width to 2 μm (monomode waveguide) resulted in a further decrease in the required control pulse energy to ~8 pJ for 10-dB modulation (Figure 8.11).

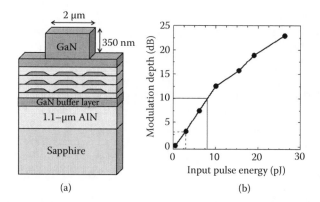

**FIGURE 8.11** Schematic description of a quantum dot (QD)-based AlN/GaN all-optical switch in a ridge-waveguide configuration. Transmittance increase versus control pulse energy for transverse-magnetic (TM)-polarized light measured in a 1.5-mm-long 2-µm-wide QD-based waveguide. (Data from Monteagudo-Lerma et al., *Opt. Express*, 21(23), 27578, 2013 [157].)

## 8.4 ELECTRO-OPTICAL MODULATORS

Electro-optical amplitude and phase modulators allow tuning the amplitude, phase, and/or polarization state of an optical beam as a function of the control voltage. State-of-the-art technologies in this field include modulators based on the quantum-confined Stark effect of interband transitions in InGaAsP QWs [158] and on the electro-optical effect in materials like $LiNbO_3$ within a Mach-Zehnder configuration [159]. These devices present several drawbacks, such as low saturation power and positive chirp in the former case and the need for high driving voltage and larger size in the latter case. Achieving a significant improvement in device performance requires a change in technological approach. Exploiting ISB transitions in QWs has been proposed as a means to reduce the driving voltage and increase the bandwidth [160–162]. Moreover, the ISB transitional oscillator strength is higher in comparison to interband transitions and should allow for enhanced miniaturization of the devices.

The first electro-absorption ISB modulation experiments on AlN/GaN QWs were based on the electrical depletion of a 5-period AlN/GaN (1.5 nm/1.5 nm) MQW structure grown on a thick GaN buffer [163]. The absorption spectrum of such a sample presents two distinct peaks related to ISB transitions in both the QWs and the 2D electron gas located at the interface of the lowest AlN barrier and the underlying GaN buffer. The ratio of these two absorption peaks can be adjusted by applying an external field, which modifies the charge distribution.

To increase the modulation depth, the interaction of light with the active medium should be enhanced, which can be achieved with a waveguide geometry [19]. Through the use of a 1-µm-thick $Al_{0.5}Ga_{0.5}N$ waveguiding layer on AlN, and with three active GaN/AlN QWs operating at $\lambda = 1.55$ µm, a modulation depth of 13.5 dB was observed for a –9 V/+7 V voltage swing (10 dB for a 5-V voltage swing).

The intrinsic speed limit can be greatly improved by emptying active QWs into a local reservoir, instead of transferring carriers over the whole active region. This is the principle of the coupled-QW modulator: The electro-modulation originates from electron tunneling between a wide well (reservoir) and a narrow well separated by an ultrathin (~1-nm) AlN barrier. Experiments on GaN QW coupling via AlN [93] or AlGaN [94] barriers have set the basis for the demonstration of room-temperature ISB electro-modulated absorption at telecommunication wavelengths in GaN/AlN-coupled QWs with AlGaN contact layers [164–166]. Such devices displayed a $BW_{-3}$ $_{dB}$ cutoff frequency is limited by the resistance × capacitance ($RC$) time constant to 3 GHz for 15 × 15 $\mu m^2$ mesas, but it could be further improved by reducing the access resistance of the AlGaN contact layers. According to Hölmstrom, the high-speed performance of such modulators will ultimately be determined by the ISB absorption line width $\Gamma$, since their capacitance depends on the line width as $C \sim \Gamma^3$ [160,161].

All the aforementioned electro-optical modulators rely on light amplitude modulation via ISB absorption. Based on Kramers–Kronig relations, the ISB absorption should also translate into a variation of the refractive index at wavelengths close to the transition, which can be used for phase modulation. This concept was experimentally verified at mid-IR (~10 μm) wavelengths using the Stark shift of ISB transitions in GaAs/AlGaAs step QWs [167]. The strongly nonlinear susceptibility observed in GaN/AlN QWs [156,168], which might be even enhanced in three-layer QW designs [169], has led to the first theoretical proposals of all-optical cross-phase modulators [95].

Using a depletion modulator consisting of three GaN/AlN QWs inserted in an $Al_{0.5}Ga_{0.5}N$/AlN ridge waveguide on sapphire, Lupu et al. [170] reported a variation of the refractive index around ~1.5 μm deduced from the shift of the beating interference maxima for different order modes. The change in refractive index was derived to be $\Delta n = -5 \times 10^{-3}$ as the population was changed from complete depletion to full population of the QWs. This result is in close agreement with the observation of a refractive index variation from $-5 \times 10^{-3}$ to $6 \times 10^{-3}$ in 100-period Si-doped GaN/AlN (1.5 nm/3 nm) MQWs using a free-space Mach-Zehnder interferometer configuration [171]. The values of $\Delta n$ are comparable to those obtained at the same wavelength in phase modulators based on interband transitions in InGaAsP/InP QWs using the quantum-confined Stark effect [172], and they are one order of magnitude higher than the index variation obtained in silicon [173]. These results open the way for the realization of ISB Mach-Zehnder interferometer phase modulators in the optical communication wavelength range.

## 8.5 INFRARED PHOTODETECTORS

### 8.5.1 QUANTUM WELL/QUANTUM DOT INFRARED PHOTODETECTORS

The main motivation for the development of III-nitride QWIPs is their potential application in optical communications, thanks to the possibility to tune ISB transitions in the 1.3- to 1.55-μm range with sub-picosecond carrier relaxation times. Photoconductive QWIPs based on hexagonal [174,175] and cubic [176] Si-doped

GaN/AlN QW superlattices have been reported; however, these devices operate at cryogenic temperatures and exhibit a non-explained photovoltaic effect. Lateral QDIPs have also been fabricated by depositing planar contacts on samples consisting of 20 periods of Si-doped GaN/AlN QDs, first operating at liquid nitrogen temperature [177] and then at room temperature [122,178]. These devices exhibit photocurrent for TM-polarized excitation in the 1.4- to 1.9-μm spectral range, which follows the intraband $s$–$p_z$ selections rules. At low temperatures ($T = 10$ K), mid-IR photoresponse to TE-polarized light is also observed and attributed to $s$–$p_{x,y}$ transitions. The appearance of photocurrent due to these bound to bound transitions is attributed to conductivity via lateral hopping [122]. Further studies have shown that deep levels in the AlN barriers may also contribute to the photocurrent, giving rise to negative photoconductivity effects [179].

In spite of these early demonstrations, photoconductive devices keep presenting a low yield due to the large dark current originating from the high density of structural defects in heteroepitaxial III-nitrides, particularly in highly mismatched GaN/AlN devices targeting near-IR wavelengths. An alternative to bypass the leakage problem consists of exploiting the device's photovoltaic response, where zero-bias operation guarantees a minimum dark current. Photoconductive QWIPs have already displayed a photovoltaic response [180], which was less sensitive to defects [175], in agreement with observations in photovoltaic versus photoconductive interband detectors [181].

The photovoltaic operation of GaN/AlN QWIPs at telecommunication wavelengths and at room temperature was first studied in detail by Hofstetter et al. [29,51,182–184], as illustrated in Figure 8.12. The working principle of these photovoltaic ISB detectors is based on resonant optical rectification processes [29,185]. In a GaN/AlN superlattice, due to the asymmetric potential profile in QWs, the excitation of an electron into the upper quantized level is accompanied

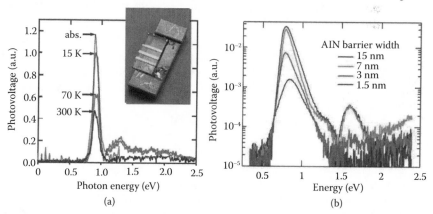

**FIGURE 8.12** (a) Photovoltage measurements on a sample consisting of a 1.5-nm GaN/1.5-nm AlN superlattice with a 50-nm-thick AlN, measured in transverse-magnetic polarization at 15, 70, and 300 K. Note the good agreement with the absorbance spectra shown as a dotted line. Inset: photograph of the mounted device [182]. (b) The measured spectral photovoltage response as a function of barrier thickness. The enhancement for larger barriers is associated with the higher internal electric field in the wells. (From Hofstetter, *Appl. Phys. Lett.*, 91(13), 131115, 2007 [29].)

by a charge displacement in the growth direction, so that an electrical dipole moment is created. For a high electron density and many QWs, these microscopic dipole moments add up to a macroscopic polarization of the crystal, which can be detected as an external photovoltage. This interpretation is consistent with the enhancement of the photovoltaic response observed in structures with larger barriers, where electron tunneling is not possible (Figure 8.12b). A strong performance enhancement (responsivity increase by a factor of 60) of these detectors has been achieved by using QDs instead of QWs in the active region [186]. The improvement is attributed to the longer electron lifetime in the upper QD states and the increased lateral electron displacement.

An interesting application of GaN-based photovoltaic ISB photodetectors is the so- called multispectral detectors, which operate in various wavelength ranges. Hofstetter et al. [187] have combined optical interband and ISB transitions with a monolithic integration of a photoconductive ultraviolet (UV) interband (solar-blind) detector based on an AlGaN thin film and a photovoltaic near-IR ISB detector based on an AlN/GaN superlattice, as illustrated in Figure 8.13. The two detectors exhibit spectrally narrow responsivity curves, thus enlarging the UV to visible rejection ratio in the case of the UV device and improving the noise behavior in the case of the IR detector at room temperature.

In the far-IR spectral region, the reduction of lattice mismatch in the structure makes it more accessible to fabricate photoconductive QWIPs. The first demonstration of a nitride-based THz ISB photodetector is reported by Sudradjat et al. [188] using a bound to quasi-bound configuration following the step-QW design [87], so that the first excited subband can be positioned at any desired energy relative to the top of the barriers by changing the QW thickness (Figure 8.14a). The fabricated devices present a photocurrent spectrum centered at 23-µm wavelength (13-THz frequency), well resolved from low temperature ($T = 20$ K in Figure 8.14b) up to $T = 50$ K, with a responsivity of ~7 mA/W [188].

More recently, Pesach et al. demonstrated InGaN/(Al)GaN QWIPs fabricated on freestanding nonpolar $m$-plane GaN substrates [189]. Devices consisting of 2.5-nm $In_{0.095}Ga_{0.905}N$/56.2 nm $Al_{0.07}Ga_{0.93}N$ and 3-nm $In_{0.1}Ga_{0.9}N$/50 nm GaN superlattices displayed photocurrent peaks at 7.5 µm and 9.3 µm, respectively, when characterized at a low temperature (14 K).

### 8.5.2 QUANTUM CASCADE DETECTORS

Quantum cascade detectors (QCDs) are photovoltaic devices consisting of several periods of an active QW coupled to a short-period superlattice, which serves as extractor [190,191]. Under illumination, electrons from the ground state, $e_1$, are excited to the upper state of the active QW, $e_2$, and then transferred to the extractor region where they experience multiple relaxations toward the next active QW. This results in a macroscopic photovoltage in an open circuit configuration. A major advantage is that their dark current is extremely low and the capacitance can be reduced by increasing the number of periods, which enables high frequency response.

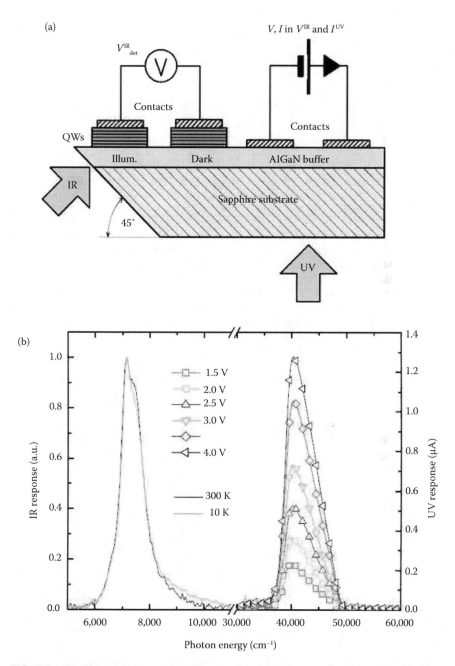

**FIGURE 8.13** (a) Schematic cross section through the sample showing the relative positions of the ultraviolet (UV) and the infrared (IR) detectors. Quantum wells (QWs) are used as the detection layer for the IR, whereas AlGaN buffer is the detection layer for the UV radiation. (b) Measured spectral responsivity curves for the UV (1.5–4.0 V in steps of 0.5 V at 300 K) and the IR detector (10 and 300 K). (From Hofstetter et al., *Electron. Lett.*, 44(16), 986, 2008 [187].)

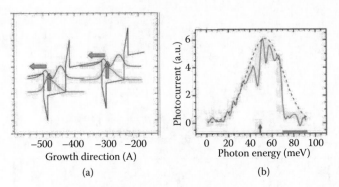

**FIGURE 8.14** (a) Conduction band profile of a far-infrared (IR) quantum well infrared photodetector (QWIP) under bias. Each repeat unit consists of an $Al_{0.16}Ga_{0.84}N/GaN/Al_{0.08}Ga_{0.92}N$ step quantum well (QW). The squared envelope functions of the ground state and first excited state of each QW are also shown, referenced to their respective energy levels. The vertical and horizontal arrows indicate, respectively, photon absorption and photoelectron escape into the continuum of unbound states over the barriers. (b) Photocurrent spectrum of a double-step-QW AlGaN THz QWIP measured at $T = 20$ K under 0.8-V bias (solid line) and Gaussian fit (dashed line). The gray band near the horizontal axis indicates the Restrahlen band of GaN. The vertical arrow marks the calculated transition energy. (From Sudradjat et al., *Appl. Phys. Lett.*, 100(24), 241113, 2012 [188].)

GaN/AlGaN QCDs operating in the near-IR have been reported [192,193], with their structure illustrated in Figure 8.15. These devices take advantage of the polarization-induced internal electric field to design an efficient AlGaN/AlN (or GaN/AlGaN) electron extractor where the energy levels are separated by approximately the LO-phonon energy (~90 meV), forming a phonon ladder. The peak responsivity of these GaN/AlGaN QCDs at room temperature was ~10 mA/W [192,194]. Detectors containing 40 periods of active region with a size of $17 \times 17$ μm² exhibit an $RC$-limited $BW_{-3\,dB}$ cutoff frequency at 19.7 GHz [195]. However, pump and probe measurements of these devices (Figure 8.15d) pointed to an ISB scattering time in the active QW of 0.1 ps and a transit time through the extractor of 1 ps [196]. With these data, the intrinsic frequency bandwidth is expected to be above 160 GHz, significantly higher than the theoretical predictions by Gryshchenko et al. [197]. Sakr et al. have shown improved performance by illuminating the side facet of the QCDs (illumination perpendicular to the growth axis), and by reducing the top contact resistance as well as the contact layer resistivity. They reached a responsivity of at least $9.5 \pm 2$ mA/W for $10 \times 10$ μm² devices at 1.5-μm peak detection wavelength at room temperature with a $BW_{-3\,dB}$ frequency response of ~40 GHz [194].

Based on the presence of the internal field in III-nitride QWs, symmetry breaking of the potential permits ISB transitions not only between the ground electronic state and the first excited state, $e_1 \rightarrow e_2$, but also between $e_1$ and the second excited state $e_3$, a transition forbidden in symmetric QWs [16]. This feature was exploited for the fabrication of a two-color GaN-based QCD operating at 1.7 and 1 μm at room temperature [198].

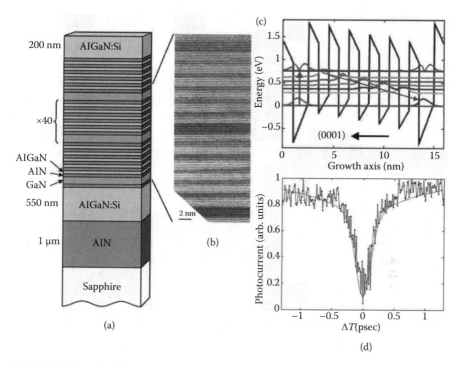

**FIGURE 8.15** (a) Schematic description of a GaN/AlN/AlGaN quantum cascade detector (QCD). (b) High-resolution TEM image of a period of the structure (active GaN quantum well followed by 5-period AlGaN/AlN extractor), viewed along the <11-20> axis [192]. (c) Band diagram and energy levels in one stage of the structure [192]. (d) QCD photocurrent as a function of pump–probe delay at room temperature under zero bias conditions. Full line: simulation fit based on rate equations and phonon scattering theory (From Vardi et al., *Appl. Phys. Lett.*, 99(20), 202111, 2011 [196].)

Finally, a simplified QCD design replacing the extractor superlattice by an AlGaN layer has been proposed [199]. The thickness and composition of the extractor alloy is chosen so that the energy separation between the ground state of the extractor and the ground state of the active QW is close to the LO-phonon energy. An alloy-extractor device presenting peak photovoltaic response at 1.9 μm has been demonstrated [199]. The photoresponse of such detectors at normal incidence can be increased by a factor of 30 by using a 2D nanohole Ti/Au array integrated on top of the detector [200].

## 8.6 TOWARD THE QUANTUM CASCADE LASER

### 8.6.1 LIGHT EMISSION IN SUPERLATTICES

ISB luminescence is an inefficient process due to the competition with nonradiative electron relaxation via interactions with LO phonons (sub picosecond), or electron–electron interactions and impurity scattering (tens of picoseconds). However, this does not hinder the realization of QCLs: In the population inversion regime, short radiative lifetime and high stimulated gain can be achieved thanks to the strong ISB transitional oscillator strength.

Despite the inefficiency of the process, room-temperature ISB luminescence in the 2- to 2.3-μm spectral range has been observed in GaN/AlN MQWs under optical pumping [201–203]. The QWs were designed to exhibit three bound states in the conduction band. The emission arises from the $e_3$–$e_2$ ISB transition. Photoluminescence excitation spectroscopy shows that the emission is only observed for TM-polarized excitation at wavelengths corresponding to the $e_1$–$e_3$ ISB transition. Further research has also provided mid-IR ISB electroluminescence measurements on chirped AlGaN/GaN MQW structures [204]. The emission line was shifted from 115 meV (FWHM = 38 meV) to 180 meV (FWHM = 58 meV) by changing the applied bias from 7 V to 14 V.

Room-temperature intraband emission has also been observed in optically pumped GaN/AlN QDs [205]. The $p_z$–s intraband luminescence was observed at $\lambda$ = 1.48 μm under optical excitation at $\lambda$ = 1.34 μm perpendicular to the [0001] growth axis.

The population of the $p$–$z$ state arises from Raman scattering by GaN $A_1$ longitudinal optical phonons. Based on the emission spectral shape, we estimate that the homogeneous line width of the $s$–$p_z$ intraband transition is less than 4 meV.

### 8.6.2 QUANTUM CASCADE LASER STRUCTURES

Quantum Cascade Laser Structures (QCLs) rely on transitions between quantized conduction-band states of a suitably designed semiconductor MQW structure [11]. Due to the polarization selection rules associated with ISB transitions, these devices are in-plane emitters, with their electric-field vector perpendicular to the plane of the layers. An electron injected into the *active QWs* first undergoes an ISB lasing transition and is rapidly extracted by a fast nonradiative transition, which maintains the population inversion. Then, the electron tunnels through the *injector region* toward the upper level of the next active QWs. By using several tens or even hundreds of periods of active region + injector in a series (a cascade), higher optical gains and multiple photons per electron are obtained. These complex structures require precise structure control and excellent homogeneity of the material, both in plane and along the multiple periods that compose the active region. Due to the large lattice mismatch and defect structure of the GaN/AlN system, the fabrication of GaN-based QCLs operating in the near-IR does not appear feasible, despite several theoretical proposals [18,206,207] and promising results in terms of waveguide fabrication [208]. However, there is increasing interest and research effort in the fabrication of the first GaN QCL in the far-IR, particularly in the so-called THz domain, spectral region where the lattice mismatch of the structure is reduced and where it should be possible to exploit the large LO phonon of III-nitrides to realize devices operating at room temperature.

Since the first demonstration of a GaAs-based THz QCL in 2001 [209], rapid progress has been made in terms of device performance. To date, QCLs have been demonstrated in the 0.85- to 5-THz range [210], with pulsed operation up to 186 K [211,212] and pulsed output powers of up to 250 mW [213]. The devices have evolved through different designs including the resonant-phonon, chirped superlattice, bound to continuum, and hybrid designs [211,214]. There are two major processes that cause the degradation of population inversion (and thus gain) in THz QCLs at high temperature: thermal backfilling and thermally activated phonon scattering. Backfilling of the lower radiative state with electrons from the heavily populated injector occurs either by thermal excitation (according to the Boltzmann distribution) or by reabsorption of nonequilibrium LO phonons (the hot-phonon effect) [215]. The other main degradation mechanism is the onset of thermally activated LO-phonon scattering, as electrons in the upper radiative state acquire sufficient in-plane kinetic energy to emit an LO phonon and relax nonradiatively to the lower radiative state. Both of these mechanisms greatly depend on the electron gas temperature, which is 50–100 K higher than the lattice temperature during device operation. The low LO-phonon energy in arsenide compounds constitutes a major bottleneck for operation at higher temperatures. Furthermore, the LO phonon of GaAs systems causes an unobtainable emission gap in lasing systems (Restrahlen band at 8 to 9 THz), which is an intrinsic property of the material system.

GaN has a LO-phonon energy of 92 meV, much higher than the ambient thermal energy. A number of designs for a GaN THz QCL have been proposed [186,216–223], all focusing on the resonant-phonon architecture first theorized in 2003 [214]. Figure 8.16 presents the basic device structure for polar III-nitrides. In the following paragraphs, we summarize the efforts of various groups working on this topic, who have introduced design improvements but keep the same underlying concept.

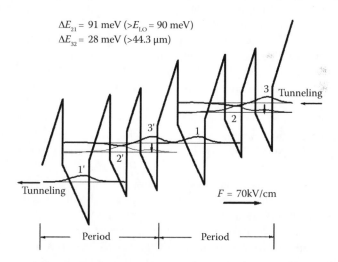

**FIGURE 8.16** Band structure, subband energy separations, and envelope wave functions of the active region of the proposed AlGaN THz quantum cascade laser (QCL) structure. Two periods are shown, each with three quantum wells (QWs) with layer thicknesses (in nanometers) 3, 4, 3, 2.5, 2, and 2.5. Wells are underlined, and barriers are plain. The transition from 3-2 is engineered through a coupled QW to be 28 meV, whereas the 2-1 transition is engineered to be on the order of the longitudinal optical phonon (90 meV). With an applied bias of 70 kV/cm, tunneling between states 1 and 3 occurs for the cascade effect and carrier recycling. (From Sun et al., *Superlattice. Microst.*, 37(2), 107–113, 2005 [217].)

Researchers from the University of Leeds, United Kingdom, have engineered one of the first designs for GaN-based QCLs using a fully consistent scattering rate equation model [216] and an energy balance method [224]. Both electron–LO phonon and electron–electron scattering mechanisms are taken into account. They have created a contour plot outlining the wavelengths of emission theorized with different well and barrier thicknesses within a superlattice, after appropriate strain balancing [206]. They have also proposed a 34-µm-wavelength QCL design in both *a* and *c* planes [216].

The group of Paiella and Moustakas at Boston University has proposed a QCL design emitting at 2 THz, designed using a Schrödinger-equation solver based on the effective-mass approximation [219]. They have also performed a rigorous comparison between a GaAs/AlGaAs and GaN/AlGaN THz QCLs emitting at the same wavelength using a microscopic model of carrier dynamics in QCL gain media based on a set of Boltzmann-like equations solved with a Monte Carlo technique [219,225]. Results show that the population inversion within GaN lasers is much less dependent on temperature than conventional GaAs designs. Furthermore, they have theoretically studied methods to create lattice-matched QCL structures using quaternary InAlGaN alloys [226]. From the experimental viewpoint, they have explored tunneling effects in cascade-like superlattices, their temperature dependence, and effect of bias for multiple device architectures [227].

Sun et al. [228] have modeled a QCL structure based on a three-well design that depopulates via the LO phonon and emits at 6.77 THz and have proposed the use of a spoof surface plasmon waveguide instead of a normal surface plasmon waveguide, which should result in an order of magnitude less losses in the guiding structure.

Mirzaei et al. [223] have proposed a dual-wavelength QCL to emit at both 33 and 52 µm with similar behavior of the output optical power for both wavelengths. The design is based on the LO-phonon resonance to extract electrons from the lower radiative levels and incorporates a miniband injector, theorized via rate equation analysis to operate properly up to 265 K.

Chou et al. [222] have modeled GaN-based resonant-phonon THz lasers using a transfer matrix method, paying particular attention to the effect of the strain state [222]. They predict higher THz power in GaN/AlGaN heterostructures compared to heterostructures incorporating In [229].

Yasuda et al. have used the nonequilibrium Green's function to model GaN THz QCL devices, that is, a four-well resonant-phonon InAlGaN/GaN structure on (0001)-oriented GaN [221], and a two-well nonpolar GaN/AlGaN structure [230].

Finally, Terashima and Hirayama (RIKEN, Japan) have presented THz QCL designs based on four-well resonant-phonon GaN/AlGaN structures [220,231,232]. The structures have been synthesized by PAMBE using a "droplet elimination by thermal annealing" technique [233], and they have been processed in a single-metal plasmon waveguide geometry [220,231,234]. Electroluminescence at 1.37 THz has been reported in a first structure [231] both grown on a GaN-on-sapphire template and grown on bulk GaN, and polarization-dependent electroluminescence at 2.82 THz, slightly tunable by changing the driving voltage in the 20- to 21-V range, has been reported using a second design [234] (band diagram and emission shown in Figure 8.17) grown on an AlN-on-sapphire template. More recently, this group has

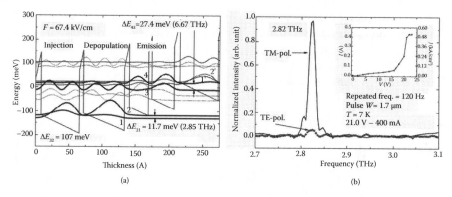

**FIGURE 8.17**   (a) Conduction band profile and square wave functions for a GaN/AlGaN quantum cascade laser (QCL) structure grown on an AlN template emitting at 6.67 and/or 2.85 THz under a biased external electric field of 67.4 kV/cm. (b) Electroluminescence spectra with each polarization direction for the THz QCL structure grown on an AlN template. The solid and dotted lines are spectra taken under transverse-magnetic- and transverse-electric-polarization directions, respectively. (From Hirayama, H., and Terashima, W., *Proc. SPIE*, 8993, 89930G, 2013 [234].)

reported electroluminescence at 6.97 THz in a double QW structure. In this case, the active region consisted of 200 repeats of 1.5 nm $Al_{0.15}Ga_{0.85}N$/4 nm GaN/1.5 nm $Al_{0.15}Ga_{0.85}N$/6 nm GaN grown on an AlN-on-sapphire template.

## 8.7   CONCLUSIONS AND PROSPECTS

In this chapter, we have reviewed recent research on III-nitride ISB optoelectronics. III-Nitride heterostructures are excellent candidates for high-speed ISB devices in the near-IR thanks to their large conduction-band offset (~1.8 eV for the GaN/AlN system) and sub-picosecond ISB scattering rates. However, bandgap engineering requires exquisite control of material growth and modeling, which is notoriously difficult in GaN/AlGaN. First prototypes of nitride-based ISB devices are room-temperature multi-Tbit/s all-optical switches operating at 1.5 μm, photovoltaic and photoconductive QWIPs, QDIPs, and ISB electro-optical modulators. Near-IR ISB luminescence from GaN/AlN QWs and QDs has been reported. The concept of quantum cascade applied to III-nitrides has been demonstrated by the development of QCDs operating in the 1.0- to 4.5-μm spectral range.

An emerging field for GaN-based ISB devices is the extension toward the far-IR spectral range, with several theoretical designs of GaN-based THz QCLs being recently reported. At far-IR wavelengths, the large GaN LO-phonon energy (92 meV) becomes a valuable property to achieve ISB operation at relatively high temperatures, and also to cover IR wavelengths that are not accessible by other III-V semiconductors due to Restrahlen absorption. Overall, many different designs of GaN-based QCLs have been presented, all focusing on the resonant-phonon architecture and predicting functionality at high temperatures. All the current research suggests that room-temperature THz QCL devices are feasible, but there are still numerous problems for device engineering, including unavailability of substrates;

difficult band engineering; weak current transport; as well as the problems from lattice mismatch, doping, and waveguide construction. The papers summarized in this chapter provide solutions toward managing the optical/electronic design and the lattice mismatch and fabrication methodology, but the fabrication of a high-efficiency laser device still remains a challenge.

## ACKNOWLEDGMENTS

This chapter is supported by the European Union ERC-StG "TeraGaN" (#278428) project.

## REFERENCES

1. Kamgar, A., Kneschaurek, P., Dorda, G., and Koch, J., 1974, "Resonance spectroscopy of electronic levels in a surface accumulation layer," *Physical Review Letters*, **32**(22), pp. 1251–1254.
2. Ando, T., Fowler, A. B., and Stern, F., 1982, "Electronic properties of two-dimensional systems," *Reviews of Modern Physics*, **54**(2), pp. 437–672.
3. Esaki, L., and Sakaki, H., 1977, "New photoconductor," *IBM Technical Disclosure Bulletin*, **20**, pp. 2456–2457.
4. Chiu, L. C., Smith, J. S., Margalit, S., Yariv, A., and Cho, A. Y., 1983, "Application of internal photoemission from quantum-well and heterojunction superlattices to infrared photodetectors," *Infrared Physics*, **23**(2), pp. 93–97.
5. Smith, J. S., Chiu, L. C., Margalit, S., Yariv, A., and Cho, A. Y., 1983, "A new infrared detector using electron emission from multiple quantum wells," *Journal of Vacuum Science & Technology B: Microelectronics and Nanometer Structures*, **1**(2), pp. 376–378.
6. Coon, D. D., and Karunasiri, R. P. G., 1984, "New mode of IR detection using quantum wells," *Applied Physics Letters*, **45**(6), pp. 649–651.
7. Coon, D. D., Karunasiri, R. P. G., and Liu, H. C., 1986, "Fast response quantum well photodetectors," *Journal of Applied Physics*, **60**(7), pp. 2636–2638.
8. West, L. C., and Eglash, S. J., 1985, "First observation of an extremely large- dipole infrared transition within the conduction band of a GaAs quantum well," *Applied Physics Letters*, **46**(12), pp. 1156–1158.
9. Levine, B. F., Choi, K. K., Bethea, C. G., Walker, J., and Malik, R. J., 1987, "New 10 µm infrared detector using intersubband absorption in resonant tunneling GaAlAs superlattices," *Applied Physics Letters*, **50**(16), pp. 1092–1094.
10. Levine, B. F., 1993, "Quantum-well infrared photodetectors," *Journal of Applied Physics*, **74**(8), pp. R1–R81.
11. Faist, J., Capasso, F., Sivco, D. L., Sirtori, C., Hutchinson, A. L., and Cho, A. Y., 1994, "Quantum cascade laser," *Science*, **264**(5158), pp. 553–556.
12. Bastard, G., 1988, *Wave Mechanics Applied to Semiconductor Heterostructures*, Les Editions de Physique; Halsted Press, Les Ulis, France; New York, NY.
13. Liu, H. C., and Capasso, F., 2000, *Intersubband Transitions in Quantum Wells: Physics and Device Applications I*, Academic Press, San Diego, CA.
14. Binggeli, N., Ferrara, P., and Baldereschi, A., 2001, "Band-offset trends in nitride heterojunctions," *Physical Review B*, **63**(24), p. 245306.
15. Cociorva, D., Aulbur, W. G., and Wilkins, J. W., 2002, "Quasiparticle calculations of band offsets at AlN–GaN interfaces," *Solid State Communications*, **124**(1–2), pp. 63–66.

16. Tchernycheva, M., Nevou, L., Doyennette, L., Julien, F., Warde, E., Guillot, F., Monroy, E., Bellet-Amalric, E., Remmele, T., and Albrecht, M., 2006, "Systematic experimental and theoretical investigation of intersubband absorption in GaN/AlN quantum wells," *Physical Review B*, **73**(12), p. 125347.

17. Suzuki, N., and Iizuka, N., 1997, "Feasibility study on ultrafast nonlinear optical properties of 1.55-μm intersubband transition in AlGaN/GaN quantum wells," *Japanese Journal of Applied Physics*, **36**(Part 2, No. 8A), pp. L1006–L1008.

18. Hofstetter, D., Baumann, E., Giorgetta, F. R., Théron, R., Wu, H., Schaff, W. J., Dawlaty, J., George, P. A., Eastman, L. F., Rana, F., Kandaswamy, P. K., Guillot, F., and Monroy, E., 2010, "Intersubband transition-based processes and devices in AlN/GaN-based heterostructures," *Proceedings of the IEEE*, **98**(7), pp. 1234–1248.

19. Machhadani, H., Kandaswamy, P., Sakr, S., Vardi, A., Wirtmüller, A., Nevou, L., Guillot, F., Pozzovivo, G., Tchernycheva, M., Lupu, A., Vivien, L., Crozat, P., Warde, E., Bougerol, C., Schacham, S., Strasser, G., Bahir, G., Monroy, E., and Julien, F. H., 2009, "GaN/AlGaN intersubband optoelectronic devices," *New Journal of Physics*, **11**(12), p. 125023.

20. Beeler, M., Trichas, E., and Monroy, E., 2013, "III-nitride semiconductors for intersubband optoelectronics: A review," *Semiconductor Science and Technology*, **28**(7), p. 074022.

21. Hao, M., Mahanty, S., Qhalid Fareed, R. S., Tottori, S., Nishino, K., and Sakai, S., 1999, "Infrared properties of bulk GaN," *Applied Physics Letters*, **74**(19), pp. 2788–2790.

22. Yang, J., Brown, G. J., Dutta, M., and Stroscio, M. A., 2005, "Photon absorption in the Restrahlen band of thin films of GaN and AlN: Two phonon effects," *Journal of Applied Physics*, **98**(4), p. 043517.

23. Welna, M., Kudrawiec, R., Motyka, M., Kucharski, R., Zając, M., Rudziński, M., Misiewicz, J., Doradziński, R., and Dwiliński, R., 2012, "Transparency of GaN substrates in the mid-infrared spectral range," *Crystal Research and Technology*, **47**(3), pp. 347–350.

24. Kandaswamy, P. K., Machhadani, H., Bougerol, C., Sakr, S., Tchernycheva, M., Julien, F. H., and Monroy, E., 2009, "Midinfrared intersubband absorption in GaN/AlGaN superlattices on Si (111) templates," *Applied Physics Letters*, **95**(14), p. 141911.

25. Kandaswamy, P. K., Machhadani, H., Kotsar, Y., Sakr, S., Das, A., Tchernycheva, M., Rapenne, L., Sarigiannidou, E., Julien, F. H., and Monroy, E., 2010, "Effect of doping on the mid-infrared intersubband absorption in GaN/AlGaN superlattices grown on Si (111) templates," *Applied Physics Letters*, **96**(14), p. 141903.

26. Bernardini, F., Fiorentini, V., and Vanderbilt, D., 1997, "Spontaneous polarization and piezoelectric constants of III-V nitrides," *Physical Review B*, **56**(16), pp. R10024–R10027.

27. Birner, S., Zibold, T., Andlauer, T., Kubis, T., Sabathil, M., Trellakis, A., and Vogl, P., 2007, "nextnano: General purpose 3-D simulations," *IEEE Transactions on Electron Devices*, **54**(9), pp. 2137–2142.

28. Kandaswamy, P. K., Guillot, F., Bellet-Amalric, E., Monroy, E., Nevou, L., Tchernycheva, M., Michon, A., Julien, F. H., Baumann, E., Giorgetta, F. R., Hofstetter, D., Remmele, T., Albrecht, M., Birner, S., and Dang, L. S., 2008, "GaN/AlN short-period superlattices for intersubband optoelectronics: A systematic study of their epitaxial growth, design, and performance," *Journal of Applied Physics*, **104**(9), p. 093501.

29. Hofstetter, D., Baumann, E., Giorgetta, F. R., Guillot, F., Leconte, S., and Monroy, E., 2007, "Optically nonlinear effects in intersubband transitions of GaN/AlN-based superlattice structures," *Applied Physics Letters*, **91**(13), p. 131115.

30. Nevou, L., Tchernycheva, M., Julien, F., Raybaut, M., Godard, A., Rosencher, E., Guillot, F., and Monroy, E., 2006, "Intersubband resonant enhancement of second-harmonic generation in GaN/AlN quantum wells," *Applied Physics Letters*, **89**(15), p. 151101.

31. Nevou, L., Mangeney, J., Tchernycheva, M., Julien, F. H., Guillot, F., and Monroy, E., 2009, "Ultrafast relaxation and optical saturation of intraband absorption of GaN/AlN quantum dots," *Applied Physics Letters*, **94**(13), p. 132104.

32. Sarigiannidou, E., Monroy, E., Gogneau, N., Radtke, G., Bayle-Guillemaud, P., Bellet-Amalric, E., Daudin, B., and Rouvière, J. L., 2006, "Comparison of the structural quality in Ga-face and N-face polarity GaN/AlN multiple-quantum-well structures," *Semiconductor Science and Technology*, **21**(5), pp. 612–618.

33. Gmachl, C., Ng, H. M., George Chu, S.-N., and Cho, A. Y., 2000, "Intersubband absorption at λ ~1.55μm in well- and modulation-doped GaN/AlGaN multiple quantum wells with superlattice barriers," *Applied Physics Letters*, **77**(23), pp. 3722–3724.

34. Gmachl, C., Ng, H. M., and Cho, A. Y., 2001, "Intersubband absorption in degenerately doped GaN/Al[sub x]Ga[sub 1–x]N coupled double quantum wells," *Applied Physics Letters*, **79**(11), pp. 1590–1592.

35. Kishino, K., Kikuchi, A., Kanazawa, H., and Tachibana, T., 2002, "Intersubband transition in (GaN)[sub m]/(AlN)[sub n] superlattices in the wavelength range from 1.08 to 1.61 μm," *Applied Physics Letters*, **81**(7), pp. 1234–1236.

36. Iizuka, N., Kaneko, K., and Suzuki, N., 2002, "Near-infrared intersubband absorption in GaN/AlN quantum wells grown by molecular beam epitaxy," *Applied Physics Letters*, **81**(10), pp. 1803–1805.

37. Heber, J. D., Gmachl, C., Ng, H. M., and Cho, A. Y., 2002, "Comparative study of ultrafast intersubband electron scattering times at ~1.55 μm wavelength in GaN/AlGaN heterostructures," *Applied Physics Letters*, **81**(7), pp. 1237–1239.

38. Helman, A., Tchernycheva, M., Lusson, A., Warde, E., Julien, F. H., Moumanis, K., Fishman, G., Monroy, E., Daudin, B., Le Si Dang, D., Bellet-Amalric, E., and Jalabert, D., 2003, "Intersubband spectroscopy of doped and undoped GaN/AlN quantum wells grown by molecular-beam epitaxy," *Applied Physics Letters*, **83**(25), pp. 5196–5198.

39. Zhou, Q., Chen, J., Pattada, B., Manasreh, M. O., Xiu, F., Puntigan, S., He, L., Ramaiah, K. S., and Morkoç, H., 2003, "Infrared optical absorbance of intersubband transitions in GaN/AlGaN multiple quantum well structures," *Journal of Applied Physics*, **93**(12), pp. 10140–10142.

40. Adelmann, C., Brault, J., Mula, G., Daudin, B., Lymperakis, L., and Neugebauer, J., 2003, "Gallium adsorption on (0001) GaN surfaces," *Physical Review B*, **67**(16), p. 165419.

41. Neugebauer, J., Zywietz, T., Scheffler, M., Northrup, J., Chen, H., and Feenstra, R., 2003, "Adatom kinetics on and below the surface: The existence of a new diffusion channel," *Physical Review Letters*, **90**(5), p. 056101.

42. Heying, B., Averbeck, R., Chen, L. F., Haus, E., Riechert, H., and Speck, J. S., 2000, "Control of GaN surface morphologies using plasma-assisted molecular beam epitaxy," *Journal of Applied Physics*, **88**(4), pp. 1855–1860.

43. Iliopoulos, E., and Moustakas, T. D., 2002, "Growth kinetics of AlGaN films by plasma-assisted molecular-beam epitaxy," *Applied Physics Letters*, **81**(2), pp. 295–297.

44. Monroy, E., Daudin, B., Bellet-Amalric, E., Gogneau, N., Jalabert, D., Enjalbert, F., Brault, J., Barjon, J., and Dang, L. S., 2003, "Surfactant effect of In for AlGaN growth by plasma-assisted molecular beam epitaxy," *Journal of Applied Physics*, **93**(3), pp. 1550–1556.

45. Koblmueller, G., Averbeck, R., Geelhaar, L., Riechert, H., Hösler, W., and Pongratz, P., 2003, "Growth diagram and morphologies of AlN thin films grown by molecular beam epitaxy," *Journal of Applied Physics*, **93**(12), pp. 9591–9597.

46. Mula, G., Adelmann, C., Moehl, S., Oullier, J., and Daudin, B., 2001, "Surfactant effect of gallium during molecular-beam epitaxy of GaN on AlN (0001)," *Physical Review B*, **64**(19), p. 195406.

47. Brown, J. S., Koblmüller, G., Wu, F., Averbeck, R., Riechert, H., and Speck, J. S., 2006, "Ga adsorbate on (0001) GaN: In situ characterization with quadrupole mass spectrometry and reflection high-energy electron diffraction," *Journal of Applied Physics*, **99**(7), p. 074902.

48. Northrup, J., Neugebauer, J., Feenstra, R., and Smith, A., 2000, "Structure of GaN (0001): The laterally contracted Ga bilayer model," *Physical Review B*, **61**(15), pp. 9932–9935.

49. Koblmüller, G., Brown, J., Averbeck, R., Riechert, H., Pongratz, P., and Speck, J. S., 2005, "Continuous evolution of Ga adlayer coverages during plasma-assisted molecular-beam epitaxy of (0001) GaN," *Applied Physics Letters*, **86**(4), p. 041908.

50. Kandaswamy, P. K., Bougerol, C., Jalabert, D., Ruterana, P., and Monroy, E., 2009, "Strain relaxation in short-period polar GaN/AlN superlattices," *Journal of Applied Physics*, **106**(1), p. 013526.

51. Baumann, E., Giorgetta, F. R., Hofstetter, D., Golka, S., Schrenk, W., Strasser, G., Kirste, L., Nicolay, S., Feltin, E., Carlin, J. F., and Grandjean, N., 2006, "Near infrared absorption and room temperature photovoltaic response in AlN/GaN superlattices grown by metal-organic vapor-phase epitaxy," *Applied Physics Letters*, **89**(4), p. 041106.

52. Bayram, C., Péré-laperne, N., and Razeghi, M., 2009, "Effects of well width and growth temperature on optical and structural characteristics of AlN/GaN superlattices grown by metal-organic chemical vapor deposition," *Applied Physics Letters*, **95**(20), p. 201906.

53. Sodabanlu, H., Yang, J. -S., Sugiyama, M., Shimogaki, Y., and Nakano, Y., 2009, "Strain effects on the intersubband transitions in GaN/AlN multiple quantum wells grown by low-temperature metal organic vapor phase epitaxy with AlGaN interlayer," *Applied Physics Letters*, **95**(16), p. 161908.

54. Yang, J. -S., Sodabanlu, H., Sugiyama, M., Nakano, Y., and Shimogaki, Y., 2009, "Blueshift of intersubband transition wavelength in AlN/GaN multiple quantum wells by low temperature metal organic vapor phase epitaxy using pulse injection method," *Applied Physics Letters*, **95**(16), p. 162111.

55. Nicolay, S., Feltin, E., Carlin, J. -F., Grandjean, N., Nevou, L., Julien, F. H., Schmidbauer, M., Remmele, T., and Albrecht, M., 2007, "Strain-induced interface instability in GaN/AlN multiple quantum wells," *Applied Physics Letters*, **91**(6), p. 061927.

56. Iizuka, N., Kaneko, K., and Suzuki, N., 2006, "All-optical switch utilizing intersubband transition in GaN quantum wells," *IEEE Journal of Quantum Electronics*, **42**(8), pp. 765–771.

57. Floro, J. A., Follstaedt, D. M., Provencio, P., Hearne, S. J., and Lee, S. R., 2004, "Misfit dislocation formation in the AlGaN/GaN heterointerface," *Journal of Applied Physics*, **96**(12), pp. 7087–7094.

58. Ponce, F. A., 1997, "Detects and interfaces in GaN epitaxy," *MRS Bulletin*, **22**(2), pp. 51–57.

59. Moran, B., Wu, F., Romanov, A. E., Mishra, U. K., Denbaars, S. P., and Speck, J. S., 2004, "Structural and morphological evolution of GaN grown by metalorganic chemical vapor deposition on SiC substrates using an AlN initial layer," *Journal of Crystal Growth*, **273**(1–2), pp. 38–47.

60. Kehagias, T., Delimitis, A., Komninou, P., Iliopoulos, E., Dimakis, E., Georgakilas, A., and Nouet, G., 2005, "Misfit accommodation of compact and columnar InN epilayers grown on Ga-face GaN (0001) by molecular-beam epitaxy," *Applied Physics Letters*, **86**(15), p. 151905.

61. Hearne, S. J., Han, J., Lee, S. R., Floro, J. A., Follstaedt, D. M., Chason, E., and Tsong, I. S. T., 2000, "Brittle-ductile relaxation kinetics of strained AlGaN/GaN heterostructures," *Applied Physics Letters*, **76**(12), pp. 1534–1536.

62. Liu, R., Mei, J., Srinivasan, S., Ponce, F. A., Omiya, H., Narukawa, Y., and Mukai, T., 2006, "Generation of misfit dislocations by basal-plane slip in InGaN/GaN heterostructures," *Applied Physics Letters*, **89**(20), p. 201911.
63. Einfeldt, S., Heinke, H., Kirchner, V., and Hommel, D., 2001, "Strain relaxation in AlGaN/GaN superlattices grown on GaN," *Journal of Applied Physics*, **89**(4), pp. 2160–2167.
64. Andersson, T. G., Liu, X. Y., Aggerstam, T., Holmström, P., Lourdudoss, S., Thylen, L., Chen, Y. L., Hsieh, C. H., and Lo, I., 2009, "Macroscopic defects in GaN/AlN multiple quantum well structures grown by MBE on GaN templates," *Microelectronics Journal*, **40**(2), pp. 360–362.
65. Cherns, P. D., McAleese, C., Kappers, M. J., and Humphreys, C. J., 2007, "Strain relaxation in an AlGaN/GaN quantum well system," *Microscopy of Semiconducting Materials*, A.G. Cullis and P.A. Midgley, eds., Springer, Dordrecht, Netherlands, pp. 25–28.
66. Cantu, P., Wu, F., Waltereit, P., Keller, S., Romanov, A. E., DenBaars, S. P., and Speck, J. S., 2005, "Role of inclined threading dislocations in stress relaxation in mismatched layers," *Journal of Applied Physics*, **97**(10), p. 103534.
67. Bellet-Amalric, E., Adelmann, C., Sarigiannidou, E., Rouvière, J. L., Feuillet, G., Monroy, E., and Daudin, B., 2004, "Plastic strain relaxation of nitride heterostructures," *Journal of Applied Physics*, **95**(3), pp. 1127–1133.
68. Kotsar, Y., Doisneau, B., Bellet-Amalric, E., Das, A., Sarigiannidou, E., and Monroy, E., 2011, "Strain relaxation in GaN/Al$_x$Ga$_1$N superlattices grown by plasma-assisted molecular-beam epitaxy," *Journal of Applied Physics*, **110**(3), p. 033501.
69. Yang, R., Xu, J., and Sweeny, M., 1994, "Selection rules of intersubband transitions in conduction-band quantum wells," *Physical Review B*, **50**(11), pp. 7474–7482.
70. Guillot, F., Amstatt, B., Bellet-Amalric, E., Monroy, E., Nevou, L., Doyennette, L., Julien, F. H., and Dang, L. S., 2006, "Effect of Si doping on GaN/AlN multiple-quantum-well structures for intersubband optoelectronics at telecommunication wavelengths," *Superlattices and Microstructures*, **40**(4–6), pp. 306–312.
71. Berland, K., Stattin, M., Farivar, R., Sultan, D. M. S., Hyldgaard, P., Larsson, A., Wang, S. M., and Andersson, T. G., 2010, "Temperature stability of intersubband transitions in AlN/GaN quantum wells," *Applied Physics Letters*, **97**(4), p. 043507.
72. Liu, X. Y., Holmström, P., Jänes, P., Thylén, L., and Andersson, T. G., 2007, "Intersubband absorption at 1.5–3.5 µm in GaN/AlN multiple quantum wells grown by molecular beam epitaxy on sapphire," *Physica Status Solidi (b)*, **244**(8), pp. 2892–2905.
73. Suzuki, N., and Iizuka, N., 1999, "Effect of polarization field on intersubband transition in AlGaN/GaN quantum wells," *Japanese Journal of Applied Physics*, **38**(Part 2, No. 4A), pp. L363–L365.
74. Ng, H. M., Gmachl, C., Siegrist, T., Chu, S. N. G., and Cho, A. Y., 2001, "Growth and characterization of GaN/AlGaN superlattices for near-infrared intersubband transitions," *Physica Status Solidi (a)*, **188**(2), pp. 825–831.
75. Ng, H. M., Gmachl, C., Heber, J. D., Hsu, J. W. P., Chu, S. N. G., and Cho, A. Y., 2002, "Recent progress in GaN-based superlattices for near-infrared intersubband transitions," *Physica Status Solidi (b)*, **234**(3), pp. 817–821.
76. Sherliker, B., Halsall, M., Kasalynas, I., Seliuta, D., Valusis, G., Vengris, M., Barkauskas, M., Sirutkaitis, V., Harrison, P., Jovanovic, V. D., Indjin, D., Ikonic, Z., Parbrook, P. J., Wang, T., and Buckle, P. D., 2007, "Room temperature operation of AlGaN/GaN quantum well infrared photodetectors at a 3–4 µm wavelength range," *Semiconductor Science and Technology*, **22**(11), pp. 1240–1244.
77. Péré-Laperne, N., Bayram, C., Nguyen-The, L., McClintock, R., and Razeghi, M., 2009, "Tunability of intersubband absorption from 4.5 to 5.3 µm in a GaN/Al$_{0.2}$Ga$_{0.8}$N superlattices grown by metalorganic chemical vapor deposition," *Applied Physics Letters*, **95**(13), p. 131109.

78. Tian, W., Yan, W. Y., Hui, X., Li, S. L., Ding, Y. Y., Li, Y., Tian, Y., Dai, J. N., Fang, Y. Y., Wu, Z. H., Yu, C. H., and Chen, C. Q., 2012, "Tunability of intersubband transition wavelength in the atmospheric window in AlGaN/GaN multi-quantum wells grown on different AlGaN templates by metalorganic chemical vapor deposition," *Journal of Applied Physics*, **112**(6), p. 063526.

79. Bayram, C., 2012, "High-quality AlGaN/GaN superlattices for near- and mid- infrared intersubband transitions," *Journal of Applied Physics*, **111**(1), p. 013514.

80. Huang, C. C., Xu, F. J., Yan, X. D., Song, J., Xu, Z. Y., Cen, L. B., Wang, Y., Pan, J. H., Wang, X. Q., Yang, Z. J., Shen, B., Zhang, B. S., Chen, X. S., and Lu, W., 2011, "Intersubband transitions at atmospheric window in Al[sub x]Ga[sub 1−x]N/GaN multiple quantum wells grown on GaN/sapphire templates adopting AlN/GaN superlattices interlayer," *Applied Physics Letters*, **98**(13), p. 132105.

81. Edmunds, C., Tang, L., Li, D., Cervantes, M., Gardner, G., Paskova, T., Manfra, M. J., and Malis, O., 2012, "Near-infrared absorption in lattice-matched AlInN/GaN and strained AlGaN/GaN heterostructures grown by MBE on low- defect GaN substrates," *Journal of Electronic Materials*, **41**(5), pp. 881–886.

82. Chen, G., Li, Z. L., Wang, X. Q., Huang, C. C., Rong, X., Sang, L. W., Xu, F. J., Tang, N., Qin, Z. X., Sumiya, M., Chen, Y. H., Ge, W. K., and Shen, B., 2013, "Effect of polarization on intersubband transition in AlGaN/GaN multiple quantum wells," *Applied Physics Letters*, **102**(19), p. 192109.

83. Liu, D. -F., Jiang, J. -G., Cheng, Y., and He, J. -F., 2013, "Effect of delta doping on mid-infrared intersubband absorption in AlGaN/GaN step quantum well structures," *Physica E: Low-Dimensional Systems and Nanostructures*, **54**, pp. 253–256.

84. Edmunds, C., Tang, L., Shao, J., Li, D., Cervantes, M., Gardner, G., Zakharov, D. N., Manfra, M. J., and Malis, O., 2012, "Improvement of near-infrared absorption linewidth in AlGaN/GaN superlattices by optimization of delta-doping location," *Applied Physics Letters*, **101**(10), p. 102104.

85. Wu, T., Wei-Yi, Y., Hui, X., Jian-Nan, D., Yan-Yan, F., Zhi-Hao, W., Chen-Hui, Y., and Chang-Qin, C., 2013, "Effects of polarization on intersubband transitions of AlxGa1−xN/GaN multi-quantum wells," *Chinese Physics B*, **22**(5), p. 057302.

86. Zhuo, X., Ni, J., Li, J., Lin, W., Cai, D., Li, S., and Kang, J., 2014, "Band engineering of GaN/AlN quantum wells by Si dopants," *Journal of Applied Physics*, **115**(12), p. 124305.

87. Machhadani, H., Kotsar, Y., Sakr, S., Tchernycheva, M., Colombelli, R., Mangeney, J., Bellet-Amalric, E., Sarigiannidou, E., Monroy, E., and Julien, F. H., 2010, "Terahertz intersubband absorption in GaN/AlGaN step quantum wells," *Applied Physics Letters*, **97**(19), p. 191101.

88. Beeler, M., Bougerol, C., Bellet-Amalric, E., and Monroy, E., 2013, "Terahertz absorbing AlGaN/GaN multi-quantum-wells: Demonstration of a robust 4-layer design," *Applied Physics Letters*, **103**(9), p. 091108.

89. Beeler, M., Bougerol, C., Bellet-Amalaric, E., and Monroy, E., 2014, "THz intersubband transitions in AlGaN/GaN multi-quantum-wells: THz intersubband transitions in AlGaN/GaN MQWs," *Physica Status Solidi (a)*, **211**(4), pp. 761–764.

90. Wu, F., Tian, W., Yan, W. Y., Zhang, J., Sun, S. C., Dai, J. N., Fang, Y. Y., Wu, Z. H., and Chen, C. Q., 2013, "Terahertz intersubband transition in GaN/AlGaN step quantum well," *Journal of Applied Physics*, **113**(15), p. 154505.

91. Beeler, M., Bougerol, C., Bellet-Amalric, E., and Monroy, E., 2014, "Pseudo-square AlGaN/GaN quantum wells for terahertz absorption," *Applied Physics Letters*, **105**(13), p. 131106.

92. Suzuki, N., Iizuka, N., and Kaneko, K., 2003, "Calculation of near-infrared intersubband absorption spectra in GaN/AlN quantum wells," *Japanese Journal of Applied Physics*, **42**(Part 1, No. 1), pp. 132–139.

93. Tchernycheva, M., Nevou, L., Doyennette, L., Julien, F. H., Guillot, F., Monroy, E., Remmele, T., and Albrecht, M., 2006, "Electron confinement in strongly coupled GaN/AlN quantum wells," *Applied Physics Letters*, **88**(15), p. 153113.
94. Driscoll, K., Bhattacharyya, A., Moustakas, T. D., Paiella, R., Zhou, L., and Smith, D. J., 2007, "Intersubband absorption in AlN/GaN/AlGaN coupled quantum wells," *Applied Physics Letters*, **91**(14), p. 141104.
95. Cen, L. B., Shen, B., Qin, Z. X., and Zhang, G. Y., 2009, "Influence of polarization induced electric fields on the wavelength and the refractive index of intersubband transitions in AlN/GaN coupled double quantum wells," *Journal of Applied Physics*, **105**(9), p. 093109.
96. Cen, L. B., Shen, B., Qin, Z. X., and Zhang, G. Y., 2009, "Near-infrared two-color intersubband transitions in AlN/GaN coupled double quantum wells," *Journal of Applied Physics*, **105**(5), p. 053106.
97. Lupu, A., Julien, F. H., Golka, S., Pozzovivo, G., Strasser, G., Baumann, E., Giorgetta, F., Hofstetter, D., Nicolay, S., Mosca, M., Feltin, E., Carlin, J. -F., and Grandjean, N., 2008, "Lattice-Matched GaN-InAlN Waveguides at $\lambda$ = 1.55 µm Grown by Metal-Organic Vapor Phase Epitaxy," *IEEE Photonics Technology Letters*, **20**(2), pp. 102–104.
98. Nicolay, S., Carlin, J. -F., Feltin, E., Butté, R., Mosca, M., Grandjean, N., Ilegems, M., Tchernycheva, M., Nevou, L., and Julien, F. H., 2005, "Midinfrared intersubband absorption in lattice-matched AlInN/GaN multiple quantum wells," *Applied Physics Letters*, **87**(11), p. 111106.
99. Nicolay, S., Feltin, E., Carlin, J. -F., Mosca, M., Nevou, L., Tchernycheva, M., Julien, F. H., Ilegems, M., and Grandjean, N., 2006, "Indium surfactant effect on AlN/GaN heterostructures grown by metal-organic vapor-phase epitaxy: Applications to intersubband transitions," *Applied Physics Letters*, **88**(15), p. 151902.
100. Gonschorek, M., Carlin, J. -F., Feltin, E., Py, M. A., Grandjean, N., Darakchieva, V., Monemar, B., Lorenz, M., and Ramm, G., 2008, "Two-dimensional electron gas density in $Al_{1-x}In_xN/AlN/GaN$ heterostructures ($0.03 \leq x \leq 0.23$)," *Journal of Applied Physics*, **103**(9), p. 093714.
101. Malis, O., Edmunds, C., Manfra, M. J., and Sivco, D. L., 2009, "Near-infrared intersubband absorption in molecular-beam epitaxy-grown lattice-matched InAlN/GaN superlattices," *Applied Physics Letters*, **94**(16), p. 161111.
102. Cywiński, G., Skierbiszewski, C., Fedunieiwcz-Żmuda, A., Siekacz, M., Nevou, L., Doyennette, L., Tchernycheva, M., Julien, F. H., Prystawko, P., Kryśko, M., Grzanka, S., Grzegory, I., Presz, A., Domagała, J. Z., Smalc, J., Albrecht, M., Remmele, T., and Porowski, S., 2006, "Growth of thin AlInN/GaInN quantum wells for applications to high-speed intersubband devices at telecommunication wavelengths," *Journal of Vacuum Science & Technology B: Microelectronics and Nanometer Structures*, **24**(3), pp. 1505–1509.
103. Kudrawiec, R., Motyka, M., Cywin'ski, G., Siekacz, M., Skierbiszewski, C., Nevou, L., Doyennette, L., Tchernycheva, M., Julien, F. H., and Misiewicz, J., 2008, "Contactless electroreflectance spectroscopy of inter- and intersub-band transitions in AlInN/GaInN quantum wells," *Physica Status Solidi (c)*, **5**(2), pp. 503–507.
104. Dakhlaoui, H., 2013, "Influence of doping layer concentration on the electronic transitions in symmetric $Al_xGa_{(1-x)}N/GaN$ double quantum wells," *Optik—International Journal for Light and Electron Optics*, **124**(18), pp. 3726–3729.
105. Akabli, H., Almaggoussi, A., Abounadi, A., Rajira, A., Berland, K., and Andersson, T. G., 2012, "Intersubband energies in $Al_{1-y}In_yN/Ga_{1-x}In_xN$ heterostructures with lattice constant close to aGaN," *Superlattices and Microstructures*, **52**(1), pp. 70–77.
106. Zhu, J., Ban, S. -L., and Ha, S. -H., 2012, "Phonon-assisted intersubband transitions in wurtzite $GaN/In_xGa_{1-x}N$ quantum wells," *Chinese Physics B*, **21**(9), p. 097301.

107. Yıldırım, H., and Aslan, B., 2014, "Intersubband transitions in $In_xGa_{1-x}N/In_yGa_{1-y}N/$ GaN staggered quantum wells," *Journal of Applied Physics*, **115**(16), p. 164306.

108. Berryman, K. W., Lyon, S. A., and Segev, M., 1997, "Mid-infrared photoconductivity in InAs quantum dots," *Applied Physics Letters*, **70**(14), pp. 1861–1863.

109. Phillips, J., Kamath, K., and Bhattacharya, P., 1998, "Far-infrared photoconductivity in self-organized InAs quantum dots," *Applied Physics Letters*, **72**(16), pp. 2020–2022.

110. Ryzhii, V., 1996, "The theory of quantum-dot infrared phototransistors," *Semiconductor Science and Technology*, **11**(5), pp. 759–765.

111. Pan, D., Towe, E., and Kennerly, S., 1998, "Normal-incidence intersubband (In, Ga) As/GaAs quantum dot infrared photodetectors," *Applied Physics Letters*, **73**(14), pp. 1937–1939.

112. Chen, Z., Baklenov, O., Kim, E. T., Mukhametzhanov, I., Tie, J., Madhukar, A., Ye, Z., and Campbell, J. C., 2001, "Normal incidence $InAs/Al_xGa_{1-x}As$ quantum dot infrared photodetectors with undoped active region," *Journal of Applied Physics*, **89**(8), pp. 4558–4563.

113. Chu, L., Zrenner, A., Bichler, M., and Abstreiter, G., 2001, "Quantum-dot infrared photodetector with lateral carrier transport," *Applied Physics Letters*, **79**(14), pp. 2249–2251.

114. Daudin, B., Widmann, F., Feuillet, G., Samson, Y., Arlery, M., and Rouvière, J., 1997, "Stranski-Krastanov growth mode during the molecular beam epitaxy of highly strained GaN," *Physical Review B*, **56**(12), pp. R7069–R7072.

115. Chamard, V., Schülli, T., Sztucki, M., Metzger, T., Sarigiannidou, E., Rouvière, J. -L., Tolan, M., and Adelmann, C., 2004, "Strain distribution in nitride quantum dot multilayers," *Physical Review B*, **69**(12), p. 125327.

116. Guillot, F., Tchernycheva, M., Nevou, L., Doyennette, L., Monroy, E., Julien, F. H., Dang, L. S., Remmele, T., Albrecht, M., Shibata, T., and Tanaka, M., 2006, "Si-doped GaN/AlN quantum dot superlattices for optoelectronics at telecommunication wavelengths," *Physica Status Solidi (a)*, **203**(7), pp. 1754–1758.

117. Andreev, A., and O'Reilly, E., 2000, "Theory of the electronic structure of GaN/AlN hexagonal quantum dots," *Physical Review B*, **62**(23), pp. 15851–15870.

118. Andreev, A. D., and O'Reilly, E. P., 2001, "Optical transitions and radiative lifetime in GaN/AlN self-organized quantum dots," *Applied Physics Letters*, **79**(4), pp. 521–523.

119. Williams, D., Andreev, A., O'Reilly, E., and Faux, D., 2005, "Derivation of built- in polarization potentials in nitride-based semiconductor quantum dots," *Physical Review B*, **72**(23), p. 235318.

120. Ranjan, V., Allan, G., Priester, C., and Delerue, C., 2003, "Self-consistent calculations of the optical properties of GaN quantum dots," *Physical Review B*, **68**(11), p. 115303.

121. Tchernycheva, M., Nevou, L., Doyennette, L., Helman, A., Colombelli, R., Julien, F. H., Guillot, F., Monroy, E., Shibata, T., and Tanaka, M., 2005, "Intraband absorption of doped GaN/AlN quantum dots at telecommunication wavelengths," *Applied Physics Letters*, **87**(10), p. 101912.

122. Vardi, A., Bahir, G., Schacham, S. E., Kandaswamy, P. K., and Monroy, E., 2009, "Photocurrent spectroscopy of bound-to-bound intraband transitions in GaN/AlN quantum dots," *Physical Review B*, **80**(15), p. 155439.

123. Nguyen, D. T., Wüster, W., Roussignol, P., Voisin, C., Cassabois, G., Tchernycheva, M., Julien, F. H., Guillot, F., and Monroy, E., 2010, "Homogeneous linewidth of the intraband transition at 1.55 µm in GaN/AlN quantum dots," *Applied Physics Letters*, **97**(6), p. 061903.

124. Himwas, C., Songmuang, R., Le Si Dang, Bleuse, J., Rapenne, L., Sarigiannidou, E., and Monroy, E., 2012, "Thermal stability of the deep ultraviolet emission from AlGaN/AlN Stranski-Krastanov quantum dots," *Applied Physics Letters*, **101**(24), p. 241914.

125. Himwas, C., den Hertog, M., Bellet-Amalric, E., Songmuang, R., Donatini, F., Si Dang, L., and Monroy, E., 2014, "Enhanced room-temperature mid-ultraviolet emission from AlGaN/AlN Stranski-Krastanov quantum dots," *Journal of Applied Physics*, **116**(2), p. 023502.

126. Beeler, M., Hille, P., Schörmann, J., Teubert, J., de la Mata, M., Arbiol, J., Eickhoff, M., and Monroy, E., 2014, "Intraband absorption in self-assembled Ge-doped GaN/AlN nanowire heterostructures," *Nano Letters*, **14**(3), pp. 1665–1673.

127. Speck, J. S., and Chichibu, S. F., 2011, "Nonpolar and semipolar group III nitride-based materials," *MRS Bulletin*, **34**(05), pp. 304–312.

128. Romanov, A. E., Baker, T. J., Nakamura, S., Speck, J. S., and ERATO/JST UCSB Group, 2006, "Strain-induced polarization in wurtzite III-nitride semipolar layers," *Journal of Applied Physics*, **100**(2), p. 023522.

129. Lahourcade, L., Bellet-Amalric, E., Monroy, E., Abouzaid, M., and Ruterana, P., 2007, "Plasma-assisted molecular-beam epitaxy of AlN (11-22) on m sapphire," *Applied Physics Letters*, **90**(13), p. 131909.

130. Lahourcade, L., Kandaswamy, P. K., Renard, J., Ruterana, P., Machhadani, H., Tchernycheva, M., Julien, F. H., Gayral, B., and Monroy, E., 2008, "Interband and intersubband optical characterization of semipolar (11-22)-oriented GaN/AlN multiple-quantum-well structures," *Applied Physics Letters*, **93**(11), p. 111906.

131. Gmachl, C., and Ng, H. M., 2003, "Intersubband absorption at ~2.1 μm in A-plane GaN/AlN multiple quantum wells," *Electronics Letters*, **39**(6), pp. 567–569.

132. Edmunds, C., Shao, J., Shirazi-HD, M., Manfra, M. J., and Malis, O., 2014, "Terahertz intersubband absorption in non-polar *m*-plane AlGaN/GaN quantum wells," *Applied Physics Letters*, **105**(2), p. 021109.

133. Machhadani, H., Beeler, M., Sakr, S., Warde, E., Kotsar, Y., Tchernycheva, M., Chauvat, M. P., Ruterana, P., Nataf, G., De Mierry, P., Monroy, E., and Julien, F. H., 2013, "Systematic study of near-infrared intersubband absorption of polar and semipolar GaN/AlN quantum wells," *Journal of Applied Physics*, **113**(14), p. 143109.

134. Brazis, R., and Raguotis, R., 2006, "Monte Carlo modeling of phonon-assisted carrier transport in cubic and hexagonal gallium nitride," *Optical and Quantum Electronics*, **38**(4–6), pp. 339–347.

135. Pugh, S. K., Dugdale, D. J., Brand, S., and Abram, R. A., 1999, "Electronic structure calculations on nitride semiconductors," *Semiconductor Science and Technology*, **14**(1), pp. 23–31.

136. Suzuki, M., and Uenoyama, T., 1996, "Optical gain and crystal symmetry in III–V nitride lasers," *Applied Physics Letters*, **69**(22), pp. 3378–3380.

137. Ghasemi, F., and Razi, S., 2013, "Cuboid GaN/AlGaN quantum dot infrared photodetector; photoconductive gain and capture probability," *Optik—International Journal for Light and Electron Optics*, **124**(9), pp. 859–863.

138. DeCuir, E. A., Fred, E., Manasreh, M. O., Schörmann, J., As, D. J., and Lischka, K., 2007, "Near-infrared intersubband absorption in nonpolar cubic GaN/AlN superlattices," *Applied Physics Letters*, **91**(4), p. 041911.

139. Machhadani, H., Tchernycheva, M., Sakr, S., Rigutti, L., Colombelli, R., Warde, E., Mietze, C., As, D. J., and Julien, F. H., 2011, "Intersubband absorption of cubic GaN/Al (Ga)N quantum wells in the near-infrared to terahertz spectral range," *Physical Review B*, **83**(7), p. 075313.

140. As, D. J., and Mietze, C., 2013, "MBE growth and applications of cubic AlN/GaN quantum wells," *Physica Status Solidi A*, **210**(3), pp. 474–479.

141. Radosavljević, A., Radovanović, J., and Milanović, V., 2014, "Optimization of cubic GaN/AlGaN quantum well-based structures for intersubband absorption in the infrared spectral range," *Solid State Communications*, **182**, pp. 38–42.

142. Wada, O., 2004, "Femtosecond all-optical devices for ultrafast communication and signal processing," *New Journal of Physics*, **6**, pp. 183–183.

143. Noda, S., Yamashita, T., Ohya, M., Muromoto, Y., and Sasaki, A., 1993, "All- optical modulation for semiconductor lasers by using three energy levels in n-doped quantum wells," *IEEE Journal of Quantum Electronics*, **29**(6), pp. 1640–1647.

144. Iizuka, N., Kaneko, K., Suzuki, N., Asano, T., Noda, S., and Wada, O., 2000, "Ultrafast intersubband relaxation (≤150 fs) in AlGaN/GaN multiple quantum wells," *Applied Physics Letters*, **77**(5), pp. 648–650.

145. Gmachl, C., Frolov, S. V., Ng, H. M., Chu, S. -N. G., and Cho, A. Y., 2001, "Sub-picosecond electron scattering time for ≃ 1.55 µm intersubband transitions in GaN/AlGaN multiple quantum wells," *Electronics Letters*, **37**(6), p. 378.

146. Rapaport, R., Chen, G., Mitrofanov, O., Gmachl, C., Ng, H. M., and Chu, S. N. G., 2003, "Resonant optical nonlinearities from intersubband transitions in GaN/AlN quantum wells," *Applied Physics Letters*, **83**(2), p. 263.

147. Iizuka, N., Kaneko, K., and Suzuki, N., 2005, "Sub-picosecond all-optical gate utilizing GaN intersubband transition," *Optics Express*, **13**(10), pp. 3835–3840.

148. Hamazaki, J., Kunugita, H., Ema, K., Kikuchi, A., and Kishino, K., 2005, "Intersubband relaxation dynamics in GaN/AlN multiple quantum wells studied by two-color pump-probe experiments," *Physical Review B*, **71**(16), p. 165334.

149. Suzuki, N., Iizuka, N., and Kaneko, K., 2000, "Intersubband transition in AlGaN-GaN quantum wells for ultrafast all-optical switching at communication wavelength," *Proceedings of SPIE*, **3940**, pp. 127–138.

150. Iizuka, N., Kaneko, K., and Suzuki, N., 2004, "Sub-picosecond modulation by inter-subband transition in ridge waveguide with GaN/AlN quantum wells," *Electronics Letters*, **40**(15), p. 962963.

151. Li, Y., Bhattacharyya, A., Thomidis, C., Moustakas, T. D., and Paiella, R., 2007, "Ultrafast all-optical switching with low saturation energy via intersubband transitions in GaN/AlN quantum-well waveguides," *Optics Express*, **15**(26), pp. 17922–17927.

152. Sodabanlu, H., Yang, J. -S., Tanemura, T., Sugiyama, M., Shimogaki, Y., and Nakano, Y., 2011, "Intersubband absorption saturation in AlN-based waveguide with GaN/AlN multiple quantum wells grown by metalorganic vapor phase epitaxy," *Applied Physics Letters*, **99**(15), p. 151102.

153. Iizuka, N., Yoshida, H., Managaki, N., Shimizu, T., Hassanet, S., Cumtornkittikul, C., Sugiyama, M., and Nakano, Y., 2009, "Integration of GaN/AlN all-optical switch with SiN/AlN waveguide utilizing spot-size conversion," *Optics Express*, **17**(25), pp. 23247–23253.

154. Iizuka, N., Kaneko, K., and Suzuki, N., 2006, "Polarization dependent loss in III- nitride optical waveguides for telecommunication devices," *Journal of Applied Physics*, **99**(9), p. 093107.

155. Li, Y., and Paiella, R., 2006, "Intersubband all-optical switching based on Coulomb-induced optical nonlinearities in GaN/AlGaN coupled quantum wells," *Semiconductor Science and Technology*, **21**(8), pp. 1105–1110.

156. Valdueza-Felip, S., Naranjo, F. B., Gonzalez-Herraez, M., Fernandez, H., Solis, J., Guillot, F., Monroy, E., Nevou, L., Tchernycheva, M., and Julien, F. H., 2008, "Characterization of the resonant third-order nonlinear susceptibility of Si-doped GaN-AlN quantum wells and quantum dots at 1.5 µm," *IEEE Photonics Technology Letters*, **20**(16), pp. 1366–1368.

157. Monteagudo-Lerma, L., Valdueza-Felip, S., Naranjo, F. B., Corredera, P., Rapenne, L., Sarigiannidou, E., Strasser, G., Monroy, E., and González-Herráez, M., 2013, "Waveguide saturable absorbers at 1.55 µm based on intraband transitions in GaN/AlN QDs," *Optics Express*, **21**(23), p. 27578.

158. Lewen, R., Irmscher, S., Westergren, U., Thylen, L., and Eriksson, U., 2004, "Segmented transmission-line electroabsorption modulators," *Journal of Lightwave Technology*, **22**(1), pp. 172–179.
159. Jungo Kondo, Aoki, K., Kondo, A., Ejiri, T., Iwata, Y., Hamajima, A., Mori, T., Mizuno, Y., Imaeda, M., Kozuka, Y., Mitomi, O., and Minakata, M., 2005, "High-speed and low-driving-voltage thin-sheet X-cut LiNbO₃ modulator with laminated low-dielectric-constant adhesive," *IEEE Photonics Technology Letters*, **17**(10), pp. 2077–2079.
160. Holmstrom, P., 2001, "High-speed mid-IR modulator using Stark shift in step quantum wells," *IEEE Journal of Quantum Electronics*, **37**(10), pp. 1273–1282.
161. Holmstrom, P., 2006, "Electroabsorption modulator using intersubband transitions in GaN–AlGaN–AlN step quantum wells," *IEEE Journal of Quantum Electronics*, **42**(8), pp. 810–819.
162. Holmström, P., Liu, X. Y., Uchida, H., Aggerstam, T., Kikuchi, A., Kishino, K., Lourdudoss, S., Andersson, T. G., and Thylén, L., 2007, "Intersubband photonic devices by group-III nitrides," *Proceedings of SPIE*, **6782**, p. 67821N.
163. Baumann, E., Giorgetta, F. R., Hofstetter, D., Leconte, S., Guillot, F., Bellet- Amalric, E., and Monroy, E., 2006, "Electrically adjustable intersubband absorption of a GaN/AlN superlattice grown on a transistorlike structure," *Applied Physics Letters*, **89**(10), p. 101121.
164. Nevou, L., Kheirodin, N., Tchernycheva, M., Meignien, L., Crozat, P., Lupu, A., Warde, E., Julien, F. H., Pozzovivo, G., Golka, S., Strasser, G., Guillot, F., Monroy, E., Remmele, T., and Albrecht, M., 2007, "Short-wavelength intersubband electroabsorption modulation based on electron tunneling between GaN/AlN coupled quantum wells," *Applied Physics Letters*, **90**(22), p. 223511.
165. Kheirodin, N., Nevou, L., Machhadani, H., Crozat, P., Vivien, L., Tchernycheva, M., Lupu, A., Julien, F. H., Pozzovivo, G., Golka, S., Strasser, G., Guillot, F., and Monroy, E., 2008, "Electrooptical modulator at telecommunication wavelengths based on GaN/AlN coupled quantum wells," *IEEE Photonics Technology Letters*, **20**(9), pp. 724–726.
166. Dussaigne, A., Nicolay, S., Martin, D., Castiglia, A., Grandjean, N., Nevou, L., Machhadani, H., Tchernycheva, M., Vivien, L., Julien, F. H., Remmele, T., and Albrecht, M., 2010, "Growth of intersubband GaN/AlGaN heterostructures," *Proceedings of SPIE*, **7608**, p. 76080H.
167. Dupont, E. B., Delacourt, D., and Papuchon, M., 1993, "Mid-infrared phase modulation via Stark effect on intersubband transitions in GaAs/GaAlAs quantum wells," *IEEE Journal of Quantum Electronics*, **29**(8), pp. 2313–2318.
168. Li, Y., Bhattacharyya, A., Thomidis, C., Liao, Y., Moustakas, T. D., and Paiella, R., 2008, "Refractive-index nonlinearities of intersubband transitions in GaN/AlN quantum-well waveguides," *Journal of Applied Physics*, **104**(8), p. 083101.
169. Wu, F., Tian, W., Zhang, J., Wang, S., Wan, Q. X., Dai, J. N., Wu, Z. H., Xu, J. T., Li, X. Y., Fang, Y. Y., and Chen, C. Q., 2014, "Double-resonance enhanced intersubband second-order nonlinear optical susceptibilities in GaN/AlGaN step quantum wells," *Optics Express*, **22**(12), p. 14212.
170. Lupu, A., Tchernycheva, M., Kotsar, Y., Monroy, E., and Julien, F. H., 2012, "Electroabsorption and refractive index modulation induced by intersubband transitions in GaN/AlN multiple quantum wells," *Optics Express*, **20**(11), p. 12541.
171. Gross, E., Nevet, A., Pesach, A., Monroy, E., Schacham, S. E., Orenstein, M., Segev, M., and Bahir, G., 2013, "Measuring the refractive index around intersubband transition resonance in GaN/AlN multi-quantum wells," *Optics Express*, **21**(3), pp. 3800–3808.
172. Zucker, J. E., Bar-Joseph, I., Miller, B. I., Koren, U., and Chemla, D. S., 1989, "Quaternary quantum wells for electro-optic intensity and phase modulation at 1.3 and 1.55 μm," *Applied Physics Letters*, **54**(1), pp. 10–12.

173. Soref, R., and Bennett, B., 1987, "Electrooptical effects in silicon," *IEEE Journal of Quantum Electronics*, **23**(1), pp. 123–129.
174. Hofstetter, D., Schad, S. -S., Wu, H., Schaff, W. J., and Eastman, L. F., 2003, "GaN/AlN-based quantum-well infrared photodetector for 1.55 µm," *Applied Physics Letters*, **83**(3), pp. 572–574.
175. Baumann, E., Giorgetta, F. R., Hofstetter, D., Lu, H., Chen, X., Schaff, W. J., Eastman, L. F., Golka, S., Schrenk, W., and Strasser, G., 2005, "Intersubband photoconductivity at 1.6 µm using a strain-compensated AlN/GaN superlattice," *Applied Physics Letters*, **87**(19), p. 191102.
176. DeCuir, E. A., Manasreh, M. O., Tschumak, E., Schörmann, J., As, D. J., and Lischka, K., 2008, "Cubic GaN/AlN multiple quantum well photodetector," *Applied Physics Letters*, **92**(20), p. 201910.
177. Doyennette, L., Nevou, L., Tchernycheva, M., Lupu, A., Guillot, F., Monroy, E., Colombelli, R., and Julien, F. H., 2005, "GaN-based quantum dot infrared photodetector operating at 1.38 µm," *Electronics Letters*, **41**(19), pp. 1077–1078.
178. Vardi, A., Akopian, N., Bahir, G., Doyennette, L., Tchernycheva, M., Nevou, L., Julien, F. H., Guillot, F., and Monroy, E., 2006, "Room temperature demonstration of GaN/AlN quantum dot intraband infrared photodetector at fiber-optics communication wavelength," *Applied Physics Letters*, **88**(14), p. 143101.
179. Vardi, A., Bahir, G., Schacham, S. E., Kandaswamy, P. K., and Monroy, E., 2010, "Negative photoconductivity due to intraband transitions in GaN/AlN quantum dots," *Journal of Applied Physics*, **108**(10), p. 104512.
180. Baumann, E., Giorgetta, F. R., Hofstetter, D., Wu, H., Schaff, W. J., Eastman, L. F., and Kirste, L., 2005, "Tunneling effects and intersubband absorption in AlN/GaN superlattices," *Applied Physics Letters*, **86**(3), p. 032110.
181. Monroy, E., Omn s, F., and Calle, F., 2003, "Wide-bandgap semiconductor ultraviolet photodetectors," *Semiconductor Science and Technology*, **18**(4), pp. R33–R51.
182. Hofstetter, D., Baumann, E., Giorgetta, F. R., Graf, M., Maier, M., Guillot, F., Bellet-Amalric, E., and Monroy, E., 2006, "High-quality AlN/GaN-superlattice structures for the fabrication of narrow-band 1.4 µm photovoltaic intersubband detectors," *Applied Physics Letters*, **88**(12), p. 121112.
183. Giorgetta, F. R., Baumann, E., Guillot, F., Monroy, E., and Hofstetter, D., 2007, "High frequency (f = 2.37 GHz) room temperature operation of 1.55 µm AlN/GaN- based intersubband detector," *Electronics Letters*, **43**(3), pp. 185–187.
184. Hofstetter, D., Baumann, E., Giorgetta, F. R., Théron, R., Wu, H., Schaff, W. J., Dawlaty, J., George, P. A., Eastman, L. F., Rana, F., Kandaswamy, P. K., Leconte, S., and Monroy, E., 2009, "Photodetectors based on intersubband transitions using III-nitride superlattice structures," *Journal of Physics: Condensed Matter*, **21**(17), p. 174208.
185. Rosencher, E., and Bois, P., 1991, "Model system for optical nonlinearities: Asymmetric quantum wells," *Physical Review B*, **44**(20), pp. 11315–11327.
186. Hofstetter, D., Di Francesco, J., Kandaswamy, P. K., Das, A., Valdueza-Felip, S., and Monroy, E., 2010, "Performance improvement of AlN/GaN-based intersubband detectors by using quantum dots," *IEEE Photonics Technology Letters*, **22**(15), pp. 1087–1089.
187. Hofstetter, D., Theron, R., Baumann, E., Giorgetta, F. R., Golka, S., Strasser, G., Guillot, F., and Monroy, E., 2008, "Monolithically integrated AlGaN/GaN/AlN-based solar-blind ultraviolet and near-infrared detectors," *Electronics Letters*, **44**(16), p. 986.
188. Sudradjat, F. F., Zhang, W., Woodward, J., Durmaz, H., Moustakas, T. D., and Paiella, R., 2012, "Far-infrared intersubband photodetectors based on double-step III-nitride quantum wells," *Applied Physics Letters*, **100**(24), p. 241113.

189. Pesach, A., Gross, E., Huang, C. -Y., Lin, Y. -D., Vardi, A., Schacham, S. E., Nakamura, S., and Bahir, G., 2013, "Non-polar *m*-plane intersubband based InGaN/(Al)GaN quantum well infrared photodetectors," *Applied Physics Letters*, **103**(2), p. 022110.

190. Gendron, L., Carras, M., Huynh, A., Ortiz, V., Koeniguer, C., and Berger, V., 2004, "Quantum cascade photodetector," *Applied Physics Letters*, **85**(14), p. 2824.

191. Giorgetta, F. R., Baumann, E., Graf, M., Yang, Q., Manz, C., Kohler, K., Beere, H. E., Ritchie, D. A., Linfield, E., Davies, A. G., Fedoryshyn, Y., Jackel, H., Fischer, M., Faist, J., and Hofstetter, D., 2009, "Quantum cascade detectors," *IEEE Journal of Quantum Electronics*, **45**(8), pp. 1039–1052.

192. Vardi, A., Bahir, G., Guillot, F., Bougerol, C., Monroy, E., Schacham, S. E., Tchernycheva, M., and Julien, F. H., 2008, "Near infrared quantum cascade detector in GaN/AlGaN/AlN heterostructures," *Applied Physics Letters*, **92**(1), p. 011112.

193. Sakr, S., Kotsar, Y., Haddadi, S., Tchernycheva, M., Vivien, L., Sarigiannidou, I., Isac, N., Monroy, E., and Julien, F. H., 2010, "GaN-based quantum cascade photodetector with 1.5 μm peak detection wavelength," *Electronics Letters*, **46**(25), pp. 1685–1686.

194. Sakr, S., Crozat, P., Gacemi, D., Kotsar, Y., Pesach, A., Quach, P., Isac, N., Tchernycheva, M., Vivien, L., Bahir, G., Monroy, E., and Julien, F. H., 2013, "GaN/AlGaN waveguide quantum cascade photodetectors at λ ≈ 1.55 μm with enhanced responsivity and ~40 GHz frequency bandwidth," *Applied Physics Letters*, **102**(1), p. 011135.

195. Vardi, A., Kheirodin, N., Nevou, L., Machhadani, H., Vivien, L., Crozat, P., Tchernycheva, M., Colombelli, R., Julien, F. H., Guillot, F., Bougerol, C., Monroy, E., Schacham, S., and Bahir, G., 2008, "High-speed operation of GaN/AlGaN quantum cascade detectors at λ ≈ 1.55 μm," *Applied Physics Letters*, **93**(19), p. 193509.

196. Vardi, A., Sakr, S., Mangeney, J., Kandaswamy, P. K., Monroy, E., Tchernycheva, M., Schacham, S. E., Julien, F. H., and Bahir, G., 2011, "Femto-second electron transit time characterization in GaN/AlGaN quantum cascade detector at 1.5 micron," *Applied Physics Letters*, **99**(20), p. 202111.

197. Gryshchenko, S. V., Klymenko, M. V., Shulika, O. V., Sukhoivanov, I. A., and Lysak, V. V., 2012, "Temperature dependence of electron transport in GaN/AlGaN quantum cascade detectors," *Superlattices and Microstructures*, **52**(4), pp. 894–900.

198. Sakr, S., Giraud, E., Dussaigne, A., Tchernycheva, M., Grandjean, N., and Julien, F. H., 2012, "Two-color GaN/AlGaN quantum cascade detector at short infrared wavelengths of 1 and 1.7 μm," *Applied Physics Letters*, **100**(18), p. 181103.

199. Sakr, S., Giraud, E., Tchernycheva, M., Isac, N., Quach, P., Warde, E., Grandjean, N., and Julien, F. H., 2012, "A simplified GaN/AlGaN quantum cascade detector with an alloy extractor," *Applied Physics Letters*, **101**(25), p. 251101.

200. Pesach, A., Sakr, S., Giraud, E., Sorias, O., Gal, L., Tchernycheva, M. , Orenstein, M., Grandjean, N., Julien, F. H., and Bahir, G., 2014, "First demonstration of plasmonic GaN quantum cascade detectors with enhanced efficiency at normal incidence," *Optics Express*, **22**(17), pp. 21069–21078.

201. Nevou, L., Julien, F. H., Colombelli, R., Guillot, F., and Monroy, E., 2006, "Room-temperature intersubband emission of GaN/AlN quantum wells at λ = 2.3 μm," *Electronics Letters*, **42**(22), pp. 1308–1309.

202. Nevou, L., Tchernycheva, M., Julien, F. H., Guillot, F., and Monroy, E., 2007, "Short wavelength (λ=2.13 μm) intersubband luminescence from GaN/AlN quantum wells at room temperature," *Applied Physics Letters*, **90**(12), p. 121106.

203. Driscoll, K., Liao, Y., Bhattacharyya, A., Zhou, L., Smith, D. J., Moustakas, T. D., and Paiella, R., 2009, "Optically pumped intersubband emission of short-wave infrared radiation with GaN/AlN quantum wells," *Applied Physics Letters*, **94**(8), p. 081120.

204. Hofstetter, D., Bour, D. P., and Kirste, L., 2014, "Mid-infrared electro-luminescence and absorption from AlGaN/GaN-based multi-quantum well inter-subband structures," *Applied Physics Letters*, **104**(24), p. 241107.

205. Nevou, L., Julien, F. H., Tchernycheva, M., Guillot, F., Monroy, E., and Sarigiannidou, E., 2008, "Intraband emission at λ≈1.48 µm from GaN/AlN quantum dots at room temperature," *Applied Physics Letters*, **92**(16), p. 161105.
206. Jovanović, V. D., Ikonić, Z., Indjin, D., Harrison, P., Milanović, V., and Soref, R. A., 2003, "Designing strain-balanced GaN/AlGaN quantum well structures: Application to intersubband devices at 1.3 and 1.55 µm wavelengths," *Journal of Applied Physics*, **93**(6), pp. 3194–3197.
207. Stattin, M., Berland, K., Hyldgaard, P., Larsson, A., and Andersson, T. G., 2011, "Waveguides for nitride based quantum cascade lasers," *Physica Status Solidi (c)*, **8**(7–8), pp. 2357–2359.
208. Ive, T., Berland, K., Stattin, M., Fälth, F., Hyldgaard, P., Larsson, A., and Andersson, T. G., 2012, "Design and fabrication of AlN/GaN heterostructures for intersubband technology," *Japanese Journal of Applied Physics*, **51**(1), p. 01AG07.
209. Köhler, R., Tredicucci, A., Beltram, F., Beere, H. E., Linfield, E. H., Davies, A. G., Ritchie, D. A., Iotti, R. C., and Rossi, F., 2002, "Terahertz semiconductor- heterostructure laser," *Nature*, **417**(6885), pp. 156–159.
210. Scalari, G., Walther, C., Fischer, M., Terazzi, R., Beere, H., Ritchie, D., and Faist, J., 2009, "THz and sub-THz quantum cascade lasers," *Laser & Photonics Review*, **3**(1–2), pp. 45–66.
211. Williams, B. S., 2007, "Terahertz quantum-cascade lasers," *Nature Photonics*, **1**(9), pp. 517–525.
212. Kumar, S., Hu, Q., and Reno, J. L., 2009, "186 K operation of terahertz quantum- cascade lasers based on a diagonal design," *Applied Physics Letters*, **94**(13), p. 131105.
213. Williams, B. S., Callebaut, H., Hu, Q., and Reno, J. L., 2001, "Magnetotunneling spectroscopy of resonant anticrossing in terahertz intersubband emitters," *Applied Physics Letters*, **79**(26), p. 4444.
214. Williams, B. S., Callebaut, H., Kumar, S., Hu, Q., and Reno, J. L., 2003, "3.4-THz quantum cascade laser based on longitudinal-optical-phonon scattering for depopulation," *Applied Physics Letters*, **82**(7), pp. 1015–1017.
215. Lü, J. T., and Cao, J. C., 2006, "Monte Carlo simulation of hot phonon effects in resonant-phonon-assisted terahertz quantum-cascade lasers," *Applied Physics Letters*, **88**(6), p. 061119.
216. Jovanović, V. D., Indjin, D., Ikonić, Z., and Harrison, P., 2004, "Simulation and design of GaN/AlGaN far-infrared (λ ~ 34 µm) quantum-cascade laser," *Applied Physics Letters*, **84**(16), pp. 2995–2997.
217. Sun, G., Soref, R. A., and Khurgin, J. B., 2005, "Active region design of a terahertz GaN/Al0.15Ga0.85N quantum cascade laser," *Superlattices and Microstructures*, **37**(2), pp. 107–113.
218. Vukmirović, N., Jovanović, V. D., Indjin, D., Ikonić, Z., Harrison, P., and Milanović, V., 2005, "Optically pumped terahertz laser based on intersubband transitions in a GaN/AlGaN double quantum well," *Journal of Applied Physics*, **97**(10), p. 103106.
219. Bellotti, E., Driscoll, K., Moustakas, T. D., and Paiella, R., 2008, "Monte Carlo study of GaN versus GaAs terahertz quantum cascade structures," *Applied Physics Letters*, **92**(10), p. 101112.
220. Terashima, W., and Hirayama, H., 2009, "Design and fabrication of terahertz quantum cascade laser structure based on III-nitride semiconductors," *Physica Status Solidi (c)*, **6**(S2), pp. S615–S618.
221. Yasuda, H., Kubis, T., and Hirakawa, K., 2011, "Non-equilibrium Green's function calculation for GaN-based terahertz quantum cascade laser structures," 36th International Conference on Infrared, Millimeter and Terahertz Waves (IRMMW-THz).
222. Chou, H., Manzur, T., and Anwar, M., 2011, "Active layer design of THz GaN quantum cascade lasers," *Proceedings of SPIE*, **8023**, p. 802309.

223. Mirzaei, B., Rostami, A., and Baghban, H., 2012, "Terahertz dual-wavelength quantum cascade laser based on GaN active region," *Optics & Laser Technology*, **44**(2), pp. 378–383.

224. Harrison, P., Indjin, D., Jovanović, V. D., Mirčetić, A., Ikonić, Z., Kelsall, R. W., McTavish, J., Savić, I., Vukmirović, N., and Milanović, V., 2005, "A physical model of quantum cascade lasers: Application to GaAs, GaN and SiGe devices," *Physica Status Solidi (a)*, **202**(6), pp. 980–986.

225. Bellotti, E., Driscoll, K., Moustakas, T. D., and Paiella, R., 2009, "Monte Carlo simulation of terahertz quantum cascade laser structures based on wide-bandgap semiconductors," *Journal of Applied Physics*, **105**(11), p. 113103.

226. Shishehchi, S., Paiella, R., and Bellotti, E., 2014, "Numerical simulation of III- nitride lattice-matched structures for quantum cascade lasers," B. Witzigmann, M. Osinski, F. Henneberger, and Y. Arakawa, eds., *Proceedings of SPIE*, **8980**, p. 89800T.

227. Sudradjat, F., Zhang, W., Driscoll, K., Liao, Y., Bhattacharyya, A., Thomidis, C., Zhou, L., Smith, D. J., Moustakas, T. D., and Paiella, R., 2010, "Sequential tunneling transport characteristics of GaN/AlGaN coupled-quantum-well structures," *Journal of Applied Physics*, **108**(10), p. 103704.

228. Sun, G., Khurgin, J. B., and Tsai, D. P., 2013, "Spoof plasmon waveguide enabled ultrathin room temperature THz GaN quantum cascade laser: A feasibility study," *Optics Express*, **21**(23), p. 28054.

229. Chou, H., Anwar, M., and Manzur, T., 2012, "Active layer design and power calculation of nitride-based THz quantum cascade lasers," *Proceedings of SPIE*, **8268**, p. 82680O.

230. Yasuda, H., Hosako, I., and Hirakawa, K., 2012, "Designs of GaN-based terahertz quantum cascade lasers for higher temperature operations," 2012 Conference on Lasers and Electro-Optics (CLEO).

231. Terashima, W., and Hirayama, H., 2011, "Spontaneous emission from GaN/AlGaN terahertz quantum cascade laser grown on GaN substrate," *Physica Status Solidi (c)*, **8**(7–8), pp. 2302–2304.

232. Terashima, W., and Hirayama, H., 2011, "Terahertz intersubband electroluminescence from GaN/AlGaN quantum cascade laser structure on AlGaN template," 2011 36th International Conference on Infrared, Millimeter and Terahertz Waves (IRMMW-THz).

233. Terashima, W., and Hirayama, H., 2010, "The utility of droplet elimination by thermal annealing technique for fabrication of GaN/AlGaN terahertz quantum cascade structure by radio frequency molecular beam epitaxy," *Applied Physics Express*, **3**(12), p. 125501.

234. Hirayama, H., and Terashima, W., 2013, "Recent progress toward realizing GaN- based THz quantum cascade laser," *Proceedings of SPIE*, **8993**, p. 89930G.

# 9 Gallium Nitride-Based Interband Tunnel Junctions

*Siddharth Rajan, Sriram Krishnamoorthy, and Fatih Akyol*

## CONTENTS

## 9.1 INTRODUCTION

Efficient interband tunneling offers interesting opportunities in a wide bandgap material system such as gallium nitride. An efficient tunnel junction (TJ) can act as a carrier conversion center, converting electrons into holes and vice versa. In a reverse biased TJ, electrons in the valence band of the p-type material can tunnel into empty states in the conduction band, leaving behind a hole in the p-type material. In a forward biased TJ, electrons in the conduction band tunnel into the empty states available in the valence band of a heavily doped p-type material. Interband TJs have been investigated in different material systems such as Ge [1], Si [2], SiGe [3], and GaAs [4] for a variety of device applications such as light-emitting diodes (LEDs), laser diodes, multijunction solar cells, and tunnel field effect transistors (TFETs) [4–6]. In the III-nitride material system, the large bandgap makes it difficult to realize TJs using the standard approach of degenerately doping the p-side and the n-side. In this chapter, efforts on III-nitride-based interband TJ devices are reviewed.

Current challenges in nitrides-based optoelectronics include efficiency droop [7] in visible LEDs and the low wall plug efficiency in ultraviolet (UV) and deep ultraviolet (DUV) LEDs [8]. Efficiency droop refers to the reduction of efficiency of an LED device with higher current that is necessary to achieve high brightness. We propose the use of a novel design, in which multiple LED stacks can be connected in series using interband TJ interconnects (Figure 9.1). Cascaded LED structure can be operated at low current density with a higher operating voltage as compared to the conventional approach of running a single LED at high current density for high brightness. Such a design would enable high overall light output power (combined from all the subcells) while still operating each subcell at a current density level at which the efficiency droop effect does not kick in. This leads to an overall high wall plug efficiency and enables operation of the LED stack at a higher input power and hence, high brightness [9]. Epitaxial cascading of active regions leads to lowered semiconductor material cost, and reduced joule heating [9]. TJs are necessary for connecting multiple active regions in series, such as in multicolor LEDs [10], and multijunction solar cells [4]. This is especially attractive in the III-nitride material system with a large range of bandgaps accessible, enabling monolithic integration for applications such as white LEDs. Figure 9.1 shows a top emitting LED structure with a TJ to enable an n-type top contact layer. This device structure enables device designers to replace the relatively more challenging p-contact with low resistance n-type, which could be very important for larger bandgap AlGaN. In the case of p-type top contacts, full metal coverage is required to make contact to the semiconductor as p-type GaN has high sheet resistance. In the case of the tunnel junction LEDs, the n-type GaN top contact layer has low sheet resistance with excellent current spreading. Hence a smaller metal footprint would be sufficient on the top surface, thereby enhancing the light extraction efficiency. In case of DUV LEDs [11], TJs can eliminate the need for thick p-GaN that is otherwise required for hole injection

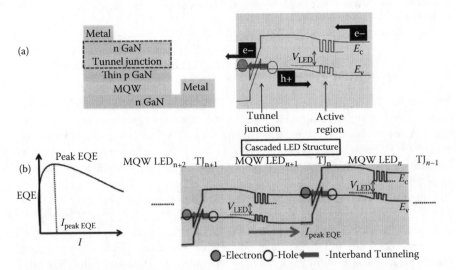

**FIGURE 9.1**   (a) Epitaxial structure and schematic band diagram showing hole injection in p-GaN using a reverse biased tunnel junction. (b) Left: Efficiency droop in a single light-emitting diode (LED) operated at high current density, with a peak external quantum efficiency at $I_{peak\ EQE}$. Right: Proposed cascaded LED structure operated at $I_{peak\ EQE}$, in which carriers are regenerated at the tunnel junction sites and multiple photons are emitted for every injected electron–hole pair.

into p-AlGaN. As the low bandgap layer required is thinner than the thick p-GaN cap, the associated absorption loss will also be reduced. The nonequilibrium tunnel-injected holes could also be used to substitute thermally ionized holes completely, enabling bipolar devices without any p-type doping [12]. In comparison to visible LEDs, the UV and DUV emitters are less mature. The external quantum efficiency of DUV LEDs is about 10% [13], with hole injection and light extraction being major bottlenecks. Hence, UV TJs could be very attractive for improving the efficiency of UV emitters. Efficient TJs could also help realize new device topologies. Emitters with reversed polarization are expected to improve carrier injection and confinement in the quantum well LEDs [14–16] and solar cells [17]. The p-down structure necessary to realize these reverse polarization structures along the +c orientation, has until now been challenging due to current crowding and poor ohmic p-contact on etched surfaces. A TJ-based structure would overcome both p-region spreading and contact issues and enable inverted polarity LEDs for potentially higher efficiency. The two important requirements for such a TJ to be incorporated in a LED or solar cell would be that (1) the voltage drop across the TJ during the device operation is negligible, with appreciable tunneling close to zero bias, and (2) the specific resistivity of the TJ is minimal.

The standard method of making an interband TJ is to heavily dope the p-type and n-type regions of a PN junction. High doping reduces the space charge region width that reduces the tunnel barrier width. Tunnel barrier height is determined by the built-in voltage of the $P^+$-$N^+$ junction. For a wide bandgap material such as GaN, the tunnel barrier height is large. There is an additional difficulty in achieving degenerate hole concentration. The large potential barrier due to the higher bandgap leads to high tunneling resistance and low current density. Hence, the standard approach of using degenerately doped PN junctions does not result in efficient interband TJs in the GaN material system. TJs in III-nitrides were first examined way back in 2001. Jeon et al [18] and Takeuchi et al. [19] demonstrated $p^+/n^+GaN$ and $p^+InGaN/n^+GaN$ TJ LEDs, respectively. These structures enable n-type GaN as a top contact layer instead of p-type GaN, eliminating the need for a semitransparent top contact required for current spreading. Ozden et al. [20] designed and implemented a monolithic dual wavelength blue (470 nm)/green (535 nm) LED using a $p^+/n^+GaN$ TJ, with independent electrical control of each LED. Diagne et al. [21] reported a vertical cavity violet LED with an in-situ grown dielectric distributed bragg reflector (DBR) and intracavity lateral current spreading layer using TJs. Jeon et al. [22] reported a GaN TJ as a current aperture in a blue surface–emitting LED. An InGaN/GaN photonic crystal LED with increased extraction efficiency was demonstrated by utilizing current spreading enabled by TJs [23]. Although these TJs showed promise for a variety of applications, a large voltage drop across the TJ was the major bottleneck.

## 9.2 POLARIZATION ENGINEERED TUNNEL JUNCTIONS

### 9.2.1 POLARIZATION-ASSISTED BAND BENDING

The high spontaneous and piezoelectric polarization charge [24] along the c-axis of III-nitrides and other highly polar semiconductors provides a new design approach for tunneling structures. The high density of sheet charge at polar heterointerface between GaN and InGaN can be exploited to bend the energy bands over a small distance, thereby, aligning the conduction and valence band of GaN with both low tunneling barrier width and height. Because of the deviation of the crystal structure in III-nitrides from an ideal wurtzite structure and the polar nature of the III-N bonds, each unit cell has a net spontaneous polarization along the c-axis. For the Ga-terminated surface (Ga-polar/Ga-face orientation), this results in a positive sheet charge at the bottom of the unit cell and a negative sheet charge at the top. Hence, the net polarization vector points along the [000-1]. For the N-terminated surface (N-polar or N-face orientation), the polarization vector direction is reversed along [0001]. However, in the bulk of a given III-N material, these sheet charges cancel out, resulting in uncompensated spontaneous charges at the surfaces. This spontaneous polarization charge can either get screened or affect the electrostatics in the device structures. In addition to the spontaneous polarization, III-nitrides have a large piezoelectric coefficient, which results in piezoelectric polarization charge at strained heterointerfaces. When AlGaN is grown pseudomorphically on GaN, the piezoelectric polarization and spontaneous polarization vectors point in the same direction in the AlGaN layer. In contrast, when InGaN is grown pseudomorphically on GaN, the piezoelectric polarization and spontaneous polarization vectors are along opposite directions. At a given polar heterointerface, the net polarization charge is the sum of the difference of the spontaneous polarization charge due to two different materials and the piezoelectric charge from the strained layer. In the case of InGaN/GaN heterostructures, the piezoelectric polarization is the dominant contribution to polarization charge whereas spontaneous polarization dominates in AlGaN/GaN heterostructures. Excellent reviews on polarization are available in references [25–29]. These polarization effects form the basis for the origin of the two-dimensional electron gas (2DEG) in AlGaN/GaN high electron mobility transistors (HEMTs), eliminating the need for modulation doping in such structures [30–35]. Polarization has been used to create electrostatic barriers to reduce base-collector leakage in III-nitride tunneling hot electron transistors (THETA) [36,37]. Although polarization is beneficial in providing charge in the channel for HEMTs, it creates electrostatic field in InGaN/GaN quantum wells (QW), leading to separation of the electron and hole wave function, causing Stark shift in InGaN/GaN LEDs [38]. In this work, the polarization charge at GaN/InGaN heterointerface is exploited to engineer the electrostatics of an interband junction for efficient tunneling.

The principle of polarization-induced TJ [10,12,39–42] was demonstrated using AlN as the barrier material for interband tunneling in GaN (Figure 9.2b). Even with a 2.6-nm-thin AlN barrier layer, the field in the AlN layer because of the polarization dipole is large enough to align the bands. However, the large bandgap of AlN reduces tunneling probability and the GaN/AlN/GaN devices have low current density and high tunneling resistance. As the orientation of the crystal dictates the direction of polarization, the polarization dipole has a fixed field direction for a given crystal orientation. In the case of GaN/AlN/GaN structure along the Ga-face orientation, the net positive and negative polarization charges are found at the bottom and top GaN/AlN heterointerfaces, respectively. Hence, to align the conduction and valence bands of the GaN layers, the p+GaN layer should be the top layer. The situation is reversed in the case of GaN/InGaN/GaN structure. The positive polarization charge is found

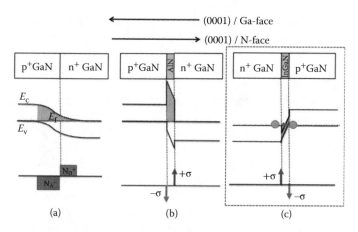

**FIGURE 9.2** Charge and band diagram of a (a) conventional degenerately doped tunnel junction, (b) polarization engineered GaN/AlN tunnel junctions, and (c) GaN/InGaN tunnel junctions.

at the top of the GaN/InGaN heterointerface, and hence, to use InGaN barrier for tunneling, the top layer has to be n⁺GaN. Tunneling probability across a potential barrier is determined by the barrier height and the thickness of the barrier material. Hence, it is expected that for the same tunneling width, a lower bandgap material will be a more efficient tunnel barrier between wide bandgap materials. It may, therefore, be expected that $In_xGa_{1-x}N$ with its low bandgap and a large piezoelectrically induced polarization when coherently strained to GaN, is an ideal candidate for efficient interband tunneling.

## 9.2.2 Design of GaN/InGaN/GaN Tunnel Junctions

For interband tunneling to occur at a very low reverse bias, the InGaN composition and thickness are chosen such that the polarization-induced band bending aligns the conduction and valence band on either side of the sandwiched InGaN layer [43]. GaN is degenerately doped to reduce the depletion region thickness at the GaN/InGaN interface, and hence, the major barrier to tunneling is the InGaN layer.

The equilibrium band bending $\Phi$ due to $In_xGa_{1-x}N$ layer of thickness, $t$, is given by the following:

$$\Phi(x) = \frac{q\sigma(x)t}{\varepsilon(x)} \tag{1}$$

where $\varepsilon(x)$ is the permittivity of $In_xGa_{1-x}N$ and $\sigma(x)$ is the fixed polarization-induced charge density at the GaN/$In_xGa_{1-x}N$ interface. For a critical InGaN layer thickness, $t_{cr}$, the equilibrium potential drop in the InGaN layer equals the bandgap of the InGaN layer, $E_{g,InGaN}$. If $t < t_{cr}$, the tunneling probability is very low as conduction and valence band extrema on either side are not aligned. With $t > t_{cr}$, the tunneling probability reduces due to the increased thickness of the barrier. With $t > t_{cr}$, degenerate carrier gases accumulate at the GaN/InGaN interface reducing the electric field in the InGaN layer if the InGaN layer thickness is further increased. Equilibrium band diagrams of the GaN/InGaN/GaN structure with increasing thicknesses of InGaN layer are shown in Figure 9.3.

Under reverse bias, the electrons in the valence band of p-GaN tunnel across the p-depletion region (intra-valence band), InGaN (interband), and n-depletion region (intra-conduction band), entering the conduction band of n-GaN (Figure 9.4a). Figure 9.4b shows the variation of the critical thickness $t_{cr}$ with the InN mole fraction in InGaN. The critical thickness, $t_{cr}$, decreases with increasing In composition due to the combined effects of increasing polarization charge and decreasing bandgap. The critical design thickness should not be confused with the critical thickness for relaxation of a strained epilayer. The spontaneous ($P_{Sp,InGaN}$), piezoelectric ($P_{Pz,InGaN}$), and the net polarization ($P_{Tot,InGaN/GaN}$) at InGaN/GaN interface as a function of InN mole fraction in InGaN, elastic constants, and bowing parameters used in the calculation can be found in reference [44]. For the critical thickness, the potential barrier ($V$[t]) seen by an electron is a triangular barrier of width $t_{cr}$ and maximum height $E_{g,InGaN}/q$. The depletion region width on the n-side and p-side are denoted as $x_n$, and $x_p$, respectively, with a barrier of $\Delta E_c$ and $\Delta E_v$, respectively. We estimate the tunneling probability ($T$) due to the three barriers by multiplying the individual probabilities calculated using a Wentzel-Kramers-Brillouin (WKB) approximation. The probability due to the intraband tunneling in n-type ($T_n$) and p-type GaN ($T_p$) is related to conduction band ($\Delta E_c$) and valence band discontinuity ($\Delta E_v$), respectively, and therefore, decreases as the indium composition is increased. The interband tunneling through the InGaN barrier ($T_{InGaN}$) increases with indium composition, as discussed earlier. For interband tunneling probability estimation, the Kane model was used to calculate the wave vector within the barrier [45,46]. The tunneling probability estimation assumes negligible reverse bias across the TJ required for tunneling that does not change the band diagram significantly from the equilibrium. The overall tunneling probability ($T_{net}$) of the device structure for a given InN mole fraction is the product of three tunneling probabilities, $T_n$, $T_p$, and $T_{InGaN}$. More details are available in reference [43]. These calculations provide some guidelines for the design of GaN/InGaN/GaN TJs. The result of this calculation is shown in Figure 9.4b, which shows that for the doping density assumed ($N_A = 1 \times 10^{19}$ cm$^{-3}$, $N_D = 5 \times 10^{19}$ cm$^{-3}$), indium

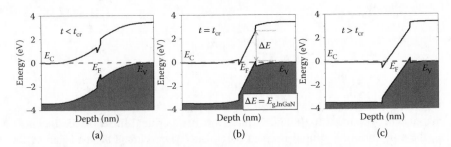

**FIGURE 9.3** Effect of InGaN thickness on band alignment in the case of (a) thin InGaN layer ($t < t_{cr}$), (b) design thickness ($t = t_{cr}$), and (c) thick InGaN layer ($t > t_{cr}$).

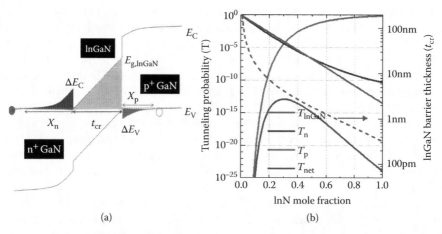

(a)  (b)

**FIGURE 9.4** (a) Equilibrium band diagram of a polarization engineered GaN/InGaN/GaN tunnel junction showing interband tunneling barrier due to InGaN, and intraband tunnel barriers due to band discontinuities ($\Delta E_c$ and $\Delta E_v$). (b) Tunneling probability estimation for the three different tunneling components, namely the interband tunneling through InGaN barrier, intraband tunneling through depletion regions in GaN ($N_A = 1 \times 10^{19}$ cm$^{-3}$, $N_D = 5 \times 10^{19}$ cm$^{-3}$). Overall tunnel probability is higher for InN mole fraction in the range of 25%–40%. Thickness of InGaN barrier ($t_{cr}$) as a function of InN mole fraction is plotted in the right hand ordinate axis. (Reproduced with permission from Krishnamoorthy, S. et al., Low resistance GaN/InGaN/GaN tunnel junctions, *Appl. Phys. Lett.*, 102, 113503, 2013. Copyright 2013, AIP Publishing LLC [43].)

composition of approximately 25%–30% results in the highest tunneling probability. Our calculation shows that there is an "optimal" composition that depends on doping density—higher doping density would enable a higher composition and higher probability. In addition, as the conduction band discontinuity is higher than the valence band discontinuity, increasing n-type doping density would have a much more important effect on tunneling probability. The doping density is limited by considerations of surface morphology, which can degrade at high doping density as well as intrinsic solubility. The effect of doping on tunneling current has been experimentally verified in a recently reported work [47], confirming our hypothesis regarding the effect of band offsets.

### 9.2.3 DEVICE CHARACTERISTICS OF GaN/InGaN/GaN TUNNEL JUNCTIONS

To study the stand-alone TJs, it is essential to make contacts to p$^+$GaN and n$^+$GaN layers separately. In such a case, the N-polar design with p$^+$GaN at the surface (Figure 9.2c) is an ideal structure for the following reasons. Along the N-face orientation, the indium uptake in the InGaN layers is generally found to be higher compared to the Ga-face orientation [48,49]. Hence, a large range of indium composition in the InGaN layer can be explored. Ga-face orientation has limitations in terms of indium uptake due to miscibility gap in the high indium content InGaN alloys due to phase separation [50–52]. Second, heavy doping of magnesium to achieve p$^+$GaN material does not invert the polarity [53], unlike the Ga-polar orientation that is crucial for the

TJ to work [54]. Third, for investigating the InGaN/GaN TJs along the Ga-polar orientation, the p-GaN layer has to be at the bottom. Dry etching that is typically used in nitride device fabrication for device isolation and reaching the bottom contact creates donor-like states on the etched surface, which makes ohmic contact formation very difficult [55]. Hence, the N-polar p-GaN up structure was the ideal one to investigate stand-alone GaN/InGaN/GaN TJs [56–58]. Nevertheless, similar results were obtained in Ga-polar p-down TJs [59].

N-polar InGaN was grown on 100-nm-thick Si-doped GaN ($N_D$ ~5 × $10^{18}$ cm$^{-3}$) and capped with 100 nm of p-GaN ($N_A$ ~1 × $10^{19}$ cm$^{-3}$). A reference sample with no InGaN layer was grown for comparison. Figure 9.5 shows the I–V characteristic of the InGaN TJ device (30 μm × 30 μm) and the reference PN junction. In the reverse direction, the PN junction shows reverse leakage current that is lower than the forward current. The TJ shows much higher reverse current than forward current, which is typical behavior for a backward diode. In addition, the reverse tunneling current in the TJ was several orders of magnitude higher than that in the PN junction. Two regimes were observed in the reverse I–V characteristics for the GaN/InGaN/GaN TJ. In the low reverse bias regime, a sharp increase in reverse current was observed (~70–130 mV/decade) from zero bias (inset to Figure 9.5) indicating the onset of tunneling in the device. Such a low turn-on voltage would be an ideal candidate to connect devices in series, especially in the case of multiple active region emitters. At a reverse voltage of 1 V, a current density of 118 A/cm$^2$ was obtained. In comparison, the reference PN junction sample had very low reverse current (five orders of magnitude lower at −1 V) as expected due to the thicker barrier in the absence of polarization induced field [56]. At higher current density levels (>100 A/cm$^2$), the differential resistance increased significantly. The maximum observed current density in reverse bias, 9.2 kA/cm$^2$, is the highest current density reported for III-nitride tunnel diodes. Although the maximum current density value is limited due to the characterization

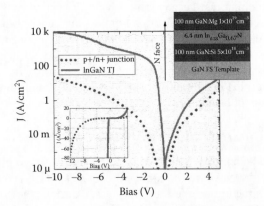

**FIGURE 9.5** Log J–V characteristics of the GaN/In$_{0.33}$Ga$_{0.67}$N/GaN tunnel junction (TJ) (solid line) and standard p+/n+ junction (dotted line). Inset (bottom left): Linear J–V characteristics of the GaN/In$_{0.33}$Ga$_{0.67}$N/GaN TJ (solid line) and reference p+/n+ junction (dotted line). Inset (top right): epitaxial stack of N-polar p$^+$ GaN/In$_{0.33}$Ga$_{0.67}$N/n$^+$ GaN TJ. (Reproduced with permission from Krishnamoorthy, S. et al., Polarization-engineered GaN/InGaN/GaN tunnel diodes, *Appl. Phys. Lett.*, 97, 203502, 2010. Copyright 2010, AIP Publishing LLC [56].)

equipment and joule heating, this demonstrates the higher current–carrying capability of the TJ. Pulsed I–V measurements were performed using Accent DIVA D265 and the device characteristics were measured with pulsing from zero bias, using a 200 ns pulse width and a period of 100 ms. At a reverse bias of 0.9 V and 9 V, a high current density of 141 A/cm² and 9.3 kA/cm², respectively, was measured, compared to 63.8 A/cm² and 6.1 kA/cm², respectively in the case of the DC measurement. This indicates that self-heating limits the current density in the device, mainly due to the resistive p-type contacts. The epitaxial structure was not capped with heavily doped p-type contact layer, and the contacts were not annealed, resulting in the formation of a Ni/p-GaN Schottky barrier. When the TJ is in forward bias, the Schottky contact is reverse biased. Hence, the actual forward characteristics of the TJ, which could include negative differential resistance, cannot be measured. The net current in the device would be limited by the reverse leakage current of the top Schottky contact. However, when the TJ device is reverse biased, the Schottky gets forward biased, and conducts with a voltage drop across it. Hence, the actual reverse characteristics of the TJ are reflected in the measured reverse bias characteristics of the device, with an additional voltage drop across the Schottky contact. At large reverse current density, the device is limited by the p-type contact resistance. Hence, electrical characteristics of such a stand-alone p-GaN/InGaN/n-GaN TJ will be limited by the ohmic nature and the on-resistance of the p-type contact. GaN/AlGaN/GaN backward tunnel diode with 3 A/cm² at −1 V has been reported recently [60].

When thickness of the InGaN barrier layer is increased beyond $t_{cr}$, degenerate carrier gases start to accumulate at the GaN/InGaN interfaces. At a low forward bias, there is resonant tunneling between the two-dimensional degenerate carrier gases resulting in interband tunneling in GaN. With further forward bias, the degenerate carrier gas density increases, but the conduction and the valence band edges are out of alignment, resulting in a sharp decrease in the tunneling current. This is manifested as the negative differential resistance regime. Beyond this, interband tunneling ceases and diffusion current across the GaN PN junction becomes the dominant current. Thus, an InGaN barrier with thickness $t > t_{cr}$ can be utilized for enhancing the forward tunneling across degenerately doped GaN PN junction, at the cost of reduced reverse tunneling current density owing to the increased thickness of InGaN layer beyond $t_{cr}$.

The epitaxial structure of an InGaN TJ structure with $t > t_{cr}$ is shown in Figure 9.6a. Up to 7-nm-thick 40% InGaN layer (confirmed by x-ray diffractometer) was grown on n⁺GaN, followed by 70 nm of p-GaN ($N_A$ ~5 × 10¹⁹ cm⁻³). The structure was capped with 30 nm of highly Mg-doped GaN cap layer to achieve ohmic contacts to the device. A current density of 15 A/cm² is obtained at −1 V, and there is a sharp increase in current density close to zero bias. Two regimes were observed in the forward bias of the sample A shown in Figure 9.6b. At a very low forward bias, there is a sharp increase in the current density, which is in agreement with the design for forward tunneling. A current density of 153 mA/cm² is achieved even at a low forward bias of 10 mV Hence, this device would be an efficient TJ for device applications requiring <100 mA/cm² current density during forward bias of the TJ. At higher forward bias, negative differential resistance regime is observed. The peak current density of 17.7 A/cm² is achieved at a forward bias of 0.8 V, followed by a

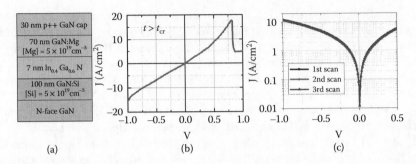

**FIGURE 9.6**   Epitaxial stack of (a) GaN/InGaN/GaN interband tunnel junction showing forward tunneling characteristics ($t > t_{cr}$), (b) linear J–V characteristics showing NDR in forward bias. (Reproduced with permission from Krishnamoorthy, S. et al., Demonstration of forward interband tunneling in GaN by polarization engineering, *Appl. Phys. Lett.*, 99, 233504, 2011. Copyright 2011, AIP Publishing LLC [57].) (c) Multiple scans showing no hysteresis in the device when the peak forward current is not reached.

sharp reduction in current density resulting in a peak-to-valley current ratio (PVCR) of 4 at room temperature. The sharp reduction in current density is because of the step function nature of the two-dimensional density of states. The higher peak current voltage can be attributed to the high ohmic contact resistance to p-GaN. As the voltage is increased beyond the negative differential resistance (NDR) onset voltage, the current is dominated by the diffusion current and excess current.

The device I–V was repeatable in both sweep directions (forward to reverse bias and vice versa) as long as the peak current voltage was not exceeded. When the device was biased beyond the NDR onset voltage, it showed hysteresis in current with respect to sweep direction [57]. During the reverse sweep (higher voltage to lower voltage sweep after a lower voltage to higher voltage sweep), NDR was not observed, and the current density was lowered. The current density was higher again at higher reverse bias and successive scans (low voltage to high voltage sweep) in forward direction show NDR. We believe that the hysteresis observed here can be attributed to trapping effects related to a donor-like hole trap found at positive polarization charge interfaces close to the valence band edge that has also been found previously in N-face HEMTs [61,62], and AlN/GaN TJs [12]. We note that similar hysteresis with respect to sweep direction was observed in AlN/GaN and AlGaN/GaN double barrier resonant tunnel diodes [63]. Hysteresis occurs when the Fermi level moves below this trap. To avoid this, the device needs to be operated such that the Fermi level never moves below the trap level, which translates to device operating under low forward bias below the voltage at which the peak current is observed. In fact, as expected from these experiments, the TJ "shows repeatable high current density" close to zero bias without any hysteresis (Figure 9.6c is the regime in which a typical forward biased TJ for a multijunction solar cell is operated). NDR with a PVCR of 147 was measured at low temperature (7 K) [64], confirming the interband tunneling mechanism.

Under forward bias, electrons from the 2DEG tunnel into states in the two-dimensional hole gas (2DHG) conserving both energy and lateral momentum. This

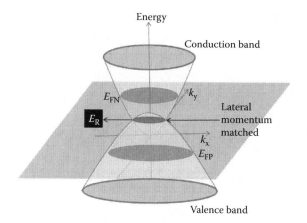

**FIGURE 9.7** Overlap of the first sub-band structure of two-dimensional electron gas and two-dimensional hole gas under forward bias showing lateral momentum conservation and energy conservation at a resonant energy $E_R$. Relaxing the momentum matching constraint increases the total current.

holds good only for a resonant energy $E_R$, where the two parabolic bands of the 2DEG and 2DHG first sub-bands intersect as illustrated in Figure 9.7. However, the uncertainty principle allows a range of states for tunneling where momentum is conserved, but the energy is not conserved. In reality, the current density is expected to be in between the extreme case of momentum conservation and the unconstrained one. The various scattering mechanisms that can relax the stringent 2D-2D tunneling momentum conservation are as follows. Delta doping at the interface would increase the scattering with dopants, and this could help in redistribution of momentum to enable momentum matching. Scattering with phonons (absorption and emission of optical and acoustic phonons) can redistribute the lateral momentum. If the scattering rate is comparable or higher than the attempt rate for tunneling, such processes are highly probable and would lead to increased current density. Other possible scattering processes are remote ionized impurity scattering and alloy scattering. Alloy scattering can be significantly higher at low temperatures. Delta doping was used in the GaN/InGaN/GaN structure, resulting in a very high peak current density ~40 kA/cm$^2$ [58]. Negative differential resistance has also been recently reported in GaN/AlGaN/GaN TJs [65].

### 9.2.4 Hole Injection Using Tunnel Junctions

In this section, GaN/InGaN TJ is used as a tunneling p-contact for hole injection in a p-down N-polar GaN PN junction. The epitaxial structure and the equilibrium band diagram of the PN junction TJ structure is shown in Figure 9.8. The structure consists of a GaN PN junction on top of a p-GaN/4-nm $In_{0.25}Ga_{0.75}N$/n-GaN TJ, so that tunneling is used to inject holes into the p-type layer of the PN junction. No p-type contact formation is necessary as the holes are injected by the TJ. Electrical characteristics of this p-contactless PN junction device (50 μm × 50 μm) are shown

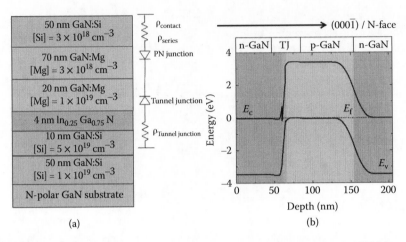

(a)                                              (b)

**FIGURE 9.8**  (a) Epitaxial stack and (b) equilibrium band diagram of a GaN PN junction with a GaN/InGaN/GaN tunneling contact layer to p-GaN. Circuit model showing two back-to-back connected diodes and the various resistance components. (Reproduced with permission from Krishnamoorthy, S. et al., Low resistance GaN/InGaN/GaN tunnel junctions, *Appl. Phys. Lett.*, 102, 113503, 2013. Copyright 2013, AIP Publishing LLC [43].)

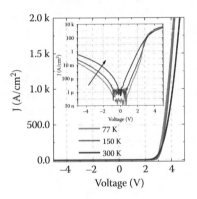

**FIGURE 9.9**  Temperature dependent I–V characteristics of the device showing efficient hole tunnel injection even at low temperatures. Inset: Temperature-dependent log J–V characteristics of the device. (Reproduced with permission from Krishnamoorthy, S. et al., Low resistance GaN/InGaN/GaN tunnel junctions, *Appl. Phys. Lett.*, 102, 113503, 2013. Copyright 2013, AIP Publishing LLC [43].)

in Figure 9.9. The device clearly shows rectification and a low resistance under forward bias. The total series resistance of the device in forward bias was found by fitting the linear region of the forward bias characteristics to be $4.7 \times 10^{-4}$ $\Omega \cdot cm^2$. This total resistance of the device is the sum of the contact resistance to n-type GaN, series resistance in the p-GaN and n-GaN regions, and the resistance of the TJ. The contact resistance of the top n-type contact was estimated to be $3.5 \times 10^{-4}$ $\Omega \cdot cm^2$ using transfer length measurement patterns on the n-type region. The specific resistivity of the TJ is therefore, lower than $1.2 \times 10^{-4}$ $\Omega \cdot cm^2$, which is the lowest observed resistance for a III-nitride TJ. At a forward current density of 100 A/cm², the voltage drop across the PN junction is 3.05 V, and the voltage drop across the TJ is 12 mV. To further investigate tunnel injection of holes, low temperature I–V measurements were performed on the PN junction sample, and the results are summarized in Figure 9.9. As expected, the reverse leakage of the PN junction was reduced by lowering the temperature as can be clearly seen in the log J–V plot. Under forward bias, low series resistance was observed even at low temperatures where hole freeze-out is expected. With polarization engineering demonstrated in this work, the tunneling resistivity for GaN can overcome the bandgap limits, bringing the resistance down to ~$10^{-4}$ Ω·cm². Although this device demonstration was performed on N-face orientation with the p-type layer down structure, similar characteristics would be obtained from a Ga-face n-GaN/InGaN TJ on a +c-plane oriented p-up PN junction or LED structure. This implies that the low resistance TJs demonstrated in this work can be directly extended to commercial c-plane LEDs. In summary, low resistance ($1.2 \times 10^{-4}$ Ω·cm²) GaN/InGaN-based TJs were demonstrated in a GaN PN junction along the N-face orientation. Along the Ga-face orientation, a resistance of $5 \times 10^{-4}$ Ω·cm² has been demonstrated recently [59] and calculations indicate that this can be further lowered [66].

## 9.3  MID-GAP STATES-ASSISTED TUNNELING

The main enabler for the polarization-assisted TJs is that the polarization field can be aligned to that of the depletion field and band bending required for interband tunneling can be achieved over a small distance. However, as polarization field direction is directly related to the crystal orientation, this approach cannot be used in all cases. For instance, GaN/InGaN TJ can result in undesirable band bending in the case of a p-up P-GaN/InGaN/n-GaN TJ along the Ga-polar orientation, compared to the polarization-assisted band bending observed in the p-down p-GaN/InGaN/n-GaN TJ structure. A p-up Ga-polar GaN/InGaN TJ would be attractive for realizing p-down emitters. However, the GaN/InGaN TJ would not work for such a structure and hence, there is a need for an approach to make low resistance TJs with no dependence on polarity of the substrate. Mid-gap states–assisted tunneling could provide a pathway to achieve polarity-independent low resistance TJs.

Mid-gap states created by intentional defects can be used to overcome tunneling resistance limitations imposed by constraints such as the energy gap and dopant solubility by providing intermediate states that can reduce the tunneling width between p and n regions in a tunnel diode. In the GaAs system, this idea was successfully used to improve tunneling using annealed low temperature (LT) molecular beam epitaxy (MBE) grown GaAs [67], where As precipitates provided the mid-gap states, and epitaxial semimetallic ErAs [68–70] and MnAs [71] nanoparticles in GaAs, where the nanoscale islands provide intermediate states to assist in tunneling. In this work, we show that rare earth nitride (GdN) nanoislands can create mid-gap states in a PN junction that enhance interband tunneling by several orders of magnitude [72]. Under forward bias, the holes in the valence band of p-GaN and electrons in the conduction band of n-GaN tunnel into states in GdN and recombine. As the forward bias current does not rely on direct band to band tunneling (from n-GaN to p-GaN), we do not expect band-crossing-related negative differential resistance (as expected in a normal heterojunction Esaki diodes). In reverse bias, electrons in the valence band of p-GaN tunnel into conduction band of n-GaN via states in GdN.

Bulk GdN [73,74] is a cubic rock salt semiconductor with a theoretical indirect bandgap of 0.7–0.85 eV and (111)-oriented GdN has a 9.4% lattice mismatch with the basal plane of wurtzite GaN. It should be noted that the band structure of GdN [75–79] is still debated. GdN nanoislands have successfully been demonstrated on c-plane GaN [80], in addition, it was shown that single crystalline GaN layers could be overgrown on these nanoislands [80–82]. The formation of GdN as nanoparticles is important because GaN overgrown on planar (111) GdN [83] was found to be polycrystalline due to broken rotational symmetry along the c-axis. The ability to grow high-quality crystalline GaN on top of GdN enables us to embed nanoislands within a matrix of single crystal GaN, while maintaining high crystalline quality in the GaN. Such "nanoheterostructures" extend the functionality of III-nitrides by enabling energy band profiles and mid-gap states that would not be achievable within the (In,Ga,Al)N alloy system.

GaN p⁺-GaN/GdN nanoisland/n⁺-GaN junctions as shown in Figure 9.10a were grown along N-face (000-1) and Ga-face (0001) orientations using $N_2$ plasma-assisted molecular beam epitaxy (PAMBE). GdN coverage was incomplete and deposition of 2.4 mL of GdN resulted in a partial coverage with 3-nm-tall islands oriented along the [111] direction on wurtzite [0001] GaN [80]. Details about island formation dynamics and transmission electron microscope images of the islands are described in the references [80–82]. Figure 9.10a shows the room temperature

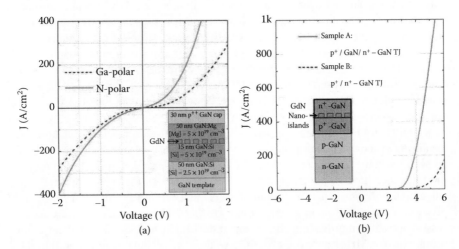

**FIGURE 9.10**   (a) I–V characteristics of Ga-polar and N-polar GaN/GdN/GaN tunnel junction (TJ). Reverse characteristics of the Ga-polar and N-polar TJ shows similar current density indicating a weak polarization dependence on tunneling resistance. (b) I–V characteristics of the p-contactless PN junction with and without GaN/GdN TJ. The sample with GaN/GdN TJ (sample A) has negligible voltage drop across the tunnel junction, while the reference sample (sample B) has a significant resistive drop. (Reprinted with permission from Krishnamoorthy, S. et al., GdN nanoisland-based GaN tunnel junctions, *Nano Letters*, 13(6), 2570–2575, 2013. Copyright 2013 American Chemical Society [72].)

electrical characteristics of the Ga-polar and N-polar TJ devices. At a reverse bias of 1 V, the Ga-polar GaN/GdN/GaN device had a current density of 78 A/cm², whereas the same epitaxial layer grown along the N-polar orientation had a comparable current density of 99 A/cm². This indicates that GaN/GdN/GaN TJs have relatively weak dependence on crystal orientation and polarization in the reverse direction. However, in the forward bias, as shown in Figure 9.10a, the N-polar TJ shows higher current density than the Ga-polar TJ. This effect is in fact quite similar to reduced turn-on voltage reported in previous work on N-polar green LEDs [84]. Preliminary study on GdN/GaN heterojunction band line-up suggests that the p-type doping is more crucial compared to n-type doping for GdN/GaN TJs [85].

To confirm tunneling injection of holes, we designed an n-GaN/GdN/p-GaN/n-GaN structure (sample A) as shown in Figure 9.10b. This structure uses an n-GaN/GdN/p-GaN TJ to contact the p-GaN layer of a GaN PN junction, thereby eliminating the need for a metal p-contact. The structure with the GaN/GdN TJ contact layer (sample A) was compared with a structure with a reference p⁺-n⁺GaN TJ contact layer (sample B) that has the same structure as sample A, but without the GdN nanoisland layer. The series resistance of the device with GaN/GdN TJ (sample A) in forward bias was calculated from a fit of the linear portion of the forward bias characteristics to be $1.3 \times 10^{-3}$ $\Omega \cdot$cm². This includes the contact resistance, series resistance in the p-GaN and n-GaN regions, and the TJ resistance. The top contact resistance was measured using transfer length method (TLM) method, and was found to be $1.3 \times 10^{-5}$ $\Omega \cdot$cm². The estimated series resistance of the thin p-GaN (~$10^{-5}$–$10^{-6}$ $\Omega \cdot$cm²) and n-GaN layers (~$10^{-7}$ $\Omega \cdot$cm²) used in this study are negligible compared to the measured total forward bias resistance. Hence, the specific TJ resistivity can be extracted to be $1.3 \times 10^{-3}$ $\Omega \cdot$cm². The forward turn-on voltage of this device with the incorporation of a TJ is the sum of the voltage drop across the forward biased PN junction and the voltage drop across the reverse biased TJ. At a reference current density of 20 A/cm², the p contactless PN junction with GaN/GdN TJ (sample A) had a voltage drop of 3.05 V. At a typical current density of 20 A/cm², the voltage drop across the GaN/GdN TJ would be 26 mV, which is negligible compared to the turn-on voltage of the GaN PN junction. A voltage drop of 4.5 V was measured at a current density of 20 A/cm² in the reference p⁺-n⁺GaN TJ device (sample B). This indicates an additional voltage drop of 1.45 V across the p⁺-n⁺GaN TJ, which is similar to past reports where a GaN-based tunnel junction light-emitting diodes (TJLEDs) led to an extra voltage drop ranging from 0.6 V to 1.5 V [86,87]. These results demonstrate that GdN nanoislands can provide very efficient hole tunneling into p-GaN layers. We observed blue-UV electroluminescence at low temperature indicating efficient hole injection due to the TJ even in the absence of any thermally ionized holes. At 20 K, the band to band (3.4 eV) and band to acceptor (3.2 eV) emission in GaN were observed.

## 9.4  TUNNEL JUNCTION LIGHT-EMITTING DIODES

Earlier reports on TJLEDs have demonstrated the advantage in using a low-spreading resistance n-type top contact layer. These TJs were primarily based on degenerately doped PN junctions (p⁺/n⁺ TJs) [19,86], current spreading by 2DEG [88], strained layer superlattices [89], and transparent conducting ZnO nanoelectrodes [90]. Although a higher output power was measured in all the aforementioned devices, voltage drop across the TJ was high, and led to high losses in LEDs. In this work, we demonstrate MBE-grown TJs on commercial LED wafers with the lowest voltage drop across the TJ ever reported [59].

InGaN/GaN TJs were grown by MBE on top of a metal-organic chemical vapor deposition (MOCVD)-grown blue LED (TJLED). In the case of the InGaN TJLED sample, magnesium delta doping was carried out to partially compensate the positive charge (O, C, Si impurities) at the regrowth interface marked as a dotted red line in Figure 9.11a. A reference LED (REFLED) with semitransparent Ni/Au contacts was also fabricated for comparison with the TJLED [59]. Electroluminescence measurements of the InGaN TJLED, and REFLED samples (pulsed measurement with a duty cycle of 0.1% to avoid joule heating effects) was carried out. As the current increased from 20 mA to 100 mA, a blue shift in the peak emission wavelength from 443.8 nm to 442.4 nm was measured in the REFLED. Electroluminescence characteristics of the InGaN TJLED were found to be similar to that of the REFLED, with a blue shift in the peak wavelength from 447.4 nm to 444.9 nm, corresponding to an increase in the current from 20 mA to 100 mA. It should be noted that the bandgap of the InGaN layer used in the TJLED would suggest absorption of blue photons. However, the photo-generated electron–hole pair would experience a very high electric field in the InGaN layer. As a result, the generated hole is swept back to the p-GaN layer. The photo-generated electron is swept to the top contact, and is injected back into the

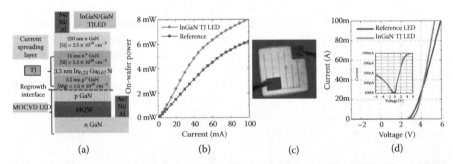

(a)                    (b)                    (c)                    (d)

**FIGURE 9.11**   (a) Epitaxial structure of InGaN tunnel junction light-emitting diode (LED) grown on metal-organic chemical vapor deposition (MOCVD)-LED with Mg and heavily doped p-GaN layer at the regrowth interface. (b) On-wafer output power of the tunnel junction light-emitting diode (TJLED) and the reference light-emitting diode (REFLED), showing higher output from TJLED due to less metal footprint on the top contact surface. (c) Optical micrograph showing uniform light emission at high current density (100 mA) due to the low-spreading resistance n-type top layer enabled by the tunnel junction. (d) Linear I–V characteristics of the TJLED and the REFLED (350 μm × 350 μm). Forward resistance of the TJLED was measured to be $2 \times 10^{-2}$ Ω·cm$^2$. TJLED shows lowest reported voltage drop of 5.3 V at 100 mA current drive. Inset: Log I–V characteristics of TJLED. (Reproduced with permission from Krishnamoorthy, S. et al., InGaN/GaN tunnel junctions for hole injection in GaN light emitting diodes, *Appl. Phys. Lett.*, 105, 141104, 2014. Copyright 2014, AIP Publishing LLC [59].)

n-GaN layer through the power supply. Thus, if the internal quantum efficiency of the device is high, which is the case for blue LEDs, the photo-generated carriers are reinjected into the active region and absorption losses due to the low bandgap InGaN tunnel barrier is not expected to be an appreciable loss mechanism. Hence, in spite of the low bandgap of the InGaN barrier, the electroluminescence spectrum of the TJLED does not show additional peaks compared to the REFLED, indicating that there is no measurable absorption/reemission due to the TJ.

On-wafer pulsed output power measurements (0.1% duty cycle) show an increased power output from the TJLEDs compared to the REFLED (Figure 9.11b). This is simply because of less electrode coverage on the top surface of the TJLED. The optical micrograph (Figure 9.11c) shows excellent current spreading in the n-GaN layer. These results are consistent with the earlier reports of higher output power in TJLEDs.

The main metric for a TJLED is the electrical loss introduced by the TJ. Current–voltage characteristics of the TJLED device and the REFLED (350 µm × 350 µm) are shown in Figure 9.11d. Forward voltage measured in the InGaN TJLED at 100 µA (500 µA) current is 2.625 V (2.9 V), which is slightly larger compared to 2.5 V (2.675 V) in the REFLED. However, after the device turn-on, the forward resistance of the InGaN TJLED is comparable to that of the REFLED. Up to 20 mA (100 mA) current drive, relevant for LED device operation, is achieved at 3.9 V (5.35 V) in InGaN/GaN TJLED compared to 3.775 V (5.975 V) in the REFLED. The additional forward voltage in the TJLED devices is attributed to the barrier for hole injection due to the regrowth interface depletion as no such voltage drop was observed when the entire TJ/PN device was grown by MBE without a regrowth interface [59]. The regrowth interface is generally found to contain oxygen, carbon, and silicon impurities, which act as n-type dopants on the surface. Although chemical treatment can reduce the amount of impurities, complete removal has been a challenge, as reported earlier for HEMT [91]. The positive impurity sheet charge concentration at the regrowth interface can lead to a parasitic PN junction in series with the TJ and the LED. With higher impurity sheet charge density, the resultant undesirable band bending at the regrowth interface can be large, and can cause additional voltage offset for the TJLED device. Even with the detrimental effect of the regrowth interface, the values reported here represent the lowest reported operating voltage for TJLEDs. Forward resistance of the TJLED device was extracted to be $2 \times 10^{-2} \, \Omega \cdot cm^2$, which is the lowest reported value for any III-nitride TJLED.

The regrowth interface is therefore, a major limitation in utilizing low resistance TJs for light emitters. Growing the entire structure without growth interrupts, such as in an all-MOCVD process would naturally eliminate the regrowth impurities. Surface treatment strategies earlier demonstrated for electronic devices [92] could be adapted for regrown TJs. Mg out-diffusion and activation of buried p-type layers are also a challenge for integration of MOCVD TJs, and solutions to these are under active investigation.

In summary, InGaN/GaN TJs were incorporated in 450-nm LEDs to realize a p-contact free LED design with a low-spreading resistance n-GaN layer as the top contact layer [59]. The reported operating voltage of 5.3 V at 20 mA is the lowest for GaN LEDs with TJ-based p-contacts. Higher output power was achieved in TJLEDs on account of the low metal footprint on the top surface and low-spreading resistance of the top n-type GaN contact layer. The elimination of regrowth interface impurities through an all-MOCVD process, or surface treatment procedures will allow LEDs to harness the full potential of polarization-engineered TJs for n-type tunneling contacts and multi-active region structures.

## 9.5  CASCADED ACTIVE REGIONS USING TUNNEL JUNCTIONS

Efficiency droop impacts the efficiency and affordability of solid state lighting, preventing large-scale adoption of LED lighting technology. The origin of efficiency droop is still being debated, and the phenomenon is attributed to various causes including Auger recombination processes, electron overflow, and inefficient hole injection and transport mechanisms [93–98]. Although many elegant solutions have been proposed to overcome efficiency droop, none has been completely successful until now, and it may be that the inherent properties of the III-nitride material system will preclude complete elimination of this effect. The underlying cause of efficiency droop is that for higher operating power density, LEDs are driven at higher currents. We show that it is possible to cascade multiple PN junctions using visible transparent TJs with low resistance and negligible voltage drop [9]. On the basis of performance parameters of current commercial LEDs, we show that the efficiency droop problem can be almost completely circumvented by using tunneling-based carrier regeneration in cascaded multiple active region LED structures. Higher luminous output power can be achieved by increasing the operating voltage, rather than the current, leading to unprecedented power density and efficiency for III-nitride LEDs.

To demonstrate the feasibility of cascading GaN emitters, we designed an experiment using multiple (1, 2, and 4) PN junctions connected epitaxially in series (Figure 9.12) with recently demonstrated GdN-based visible transparent TJs [72]. The

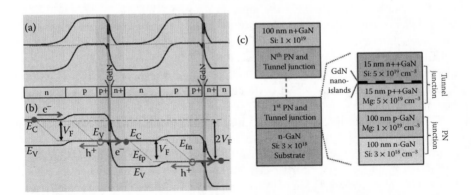

**FIGURE 9.12**  Energy band diagram of a cascaded PN junction under (a) equilibrium, (b) forward bias showing tunneling of electrons from p⁺GaN to n⁺GaN through GdN nanoislands, (c) epitaxial design of the cascaded PN junctions. (Reproduced with permission from Akyol, F. et al., Tunneling-based carrier regeneration in cascaded GaN light emitting diodes to overcome efficiency droop, Appl. Phys. Lett., 103, 081107, 2013. Copyright 2013, AIP Publishing LLC [9].)

energy band diagrams of this structure at equilibrium and forward bias are shown in Figure 9.12a and b, respectively. As the PN junction is forward biased, the TJ gets reverse biased, electrons in the valence band of p⁺GaN layer of the TJ tunnel into empty states available in the n⁺GaN layer, leaving behind a hole in the p⁺GaN layer. These electrons and holes regenerated at p⁺ and n⁺ side of the TJs are injected into the active region, which in this case is a PN junction. In the case of a multi-quantum well active region LED, the injected hole and electron would recombine radiatively in the QWs. The epitaxial structure (Figure 9.12c) was grown by plasma-assisted MBE on free-standing n-type GaN substrates, and consisted of repeats ($N = 1, 2,$ and 4) of p-GaN/n-GaN junction diodes, and p⁺GaN/GdN nanoisland/n⁺GaN TJs.

Electrical characteristics of 100-$\mu$m² diodes of the cascaded PN junction stack (number of PN junctions $N = 1, 2,$ and 4) are shown in Figure 9.13a and b. The turn-on voltage of the cascaded device increased with the number of PN junctions stacked epitaxially. All the structures showed rectifying behavior, although the device with four junctions has higher leakage current. This effect needs further investigation, and optimization of the growth conditions is expected to mitigate these effects. The series resistance (estimated from the linear portion of the I–V curve) was found to scale with N, and based on this we estimate ~5.7 × 10⁻⁴ $\Omega \cdot$cm² resistivity for each TJ (inset of Figure 9.13b). This resistance would result in a low voltage drop of 57 mV for a current density of 100 A/cm². This report of cascaded III-nitride PN junctions with low series resistance could enable several devices including multijunction solar cells, photodetectors, and multiple active region LEDs.

As our experiment demonstrates the feasibility of cascading active regions, we can use the measured results to model cascaded multiple active region LEDs. To estimate the efficiency of cascaded LED structures, we combine our estimates of tunneling resistance with the experimental data from the commercial LED published elsewhere [99] so that nonidealities of the active region are taken into account. The modeled reference LED emits at 470 nm and consists of a 1.02-mm² mesa area with

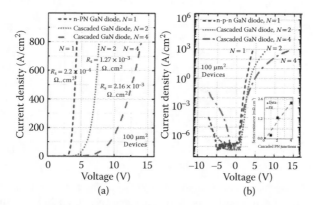

**FIGURE 9.13** The I–V characteristic of the cascaded PN junctions with $N = 1, 3,$ and 4 (a) in linear scale, (b) semi-log scale. The inset to (b) shows the change in series resistance with the number of PN junctions ($N$). (Reproduced with permission from Akyol, F. et al., Tunneling-based carrier regeneration in cascaded GaN light emitting diodes to overcome efficiency droop, *Appl. Phys. Lett.*, 103, 081107, 2013. Copyright 2013, AIP Publishing LLC [9].)

ZnO layer as contact to p-GaN. The differential series resistance ($R_s$) of the reported LED was ~0.02 Ω·cm². We fit the experimental data to the diode equation, which is modified for the case of "$N$" identical LEDs where the series resistance of the TJs in the structure is taken into account [9]. The TJ resistance used for these calculations is the experimental value of $R_{s(TJ)} = 6.4 \times 10^{-4}$ Ω·cm², which was the highest measured resistance value for a single TJ in the cascaded PN junctions discussed earlier. The external quantum efficiency was modeled by curve fitting the experimental data using the ABC model [100]. Combining this data with the I–V characteristics of the reference LED, the wall plug efficiency (WPE) as a function of current for single and cascaded LEDs was obtained. We can now estimate the output power, efficiency, and loss for cascaded LEDs including the active region and TJ losses.

The output power characteristics as a function of input power for cascaded LEDs and the conventional LED are compared (Figure 9.14a). To estimate the optical output power, we assume that a structure with N cascaded stages has $EQE_N = N \times EQE_{(single\ LED)}$. Although absorption loss via free-carriers and InGaN QW layers are expected to increase as "$N$" increases, these are ignored here. As expected, the conventional LED output power increases and saturates below 1 W less than 10 W input power. At the same input power level, relatively lower currents and higher voltages are needed for the cascaded LED structures. As the number of cascaded regions is increased, the saturation is pushed out further, with negligible saturation in output power for the LEDs with $N = 20$ and $N = 50$ cases up to 3.5 W.

These results are more striking when shown as efficiency versus power curves (Figure 9.14b and c). From these curves, it can be seen that for a cascaded structure with $N$ stages, the maximum efficiency point occurs at approximately $N$ times higher power than a single LED. This implies that the maximum device efficiency can be obtained at much higher drive powers even with an LED that exhibits significant droop. Under 10 W input power, modeled conventional LED shows 83% droop and this value decreases as $N$ increases. For the LED with $N = 50$, the droop is 14% at the same input power, which is an enhancement of ~420% in WPE compared to conventional case.

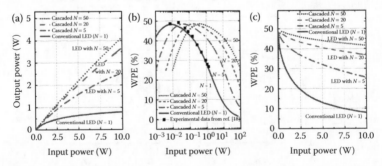

**FIGURE 9.14** The change in (a) output power, (b) wall plug efficiency in logarithmic scale, (c) wall plug efficiency in linear scale of the modeled reference single junction light emitting diode (LED) and cascaded LEDs with $N = 5$, 20, and 50. The experimental data from reference 99 and the modeled fit is also shown in Figure 9.5b. (Reproduced with permission from Akyol, F. et al., Tunneling-based carrier regeneration in cascaded GaN light emitting diodes to overcome efficiency droop, *Appl. Phys. Lett.*, 103, 081107, 2013. Copyright 2013, AIP Publishing LLC [9].)

The cascaded LED structure not only provides an enhancement of external quantum efficiency (EQE) but also suppresses joule heating. The total power dissipated due to series resistance of an LED is $P_R = I^2 R_s$. For a given output power, $R_s \propto N$ and $I \propto 1/N$, and therefore, power dissipated is proportional to $1/N$, which implies that increasing $N$ enables high power operation of LEDs with higher efficiency while reducing overall heat generation. This not only can enhance efficiency but also can enable LED operation at elevated output power, and could eliminate thermal management issues.

Although the proof of concept experiment and modeling for cascaded LEDs provides a solution to the long-standing efficiency droop problem, there are some technological challenges that need to be overcome. First, GaN-based TJs based on MOCVD growth [101,47] are still relatively less efficient than the results described here and reported earlier using MBE growth [43,72]. Another challenge is the activation of buried Mg atoms (p-type dopant) in MOCVD growth, which has been addressed previously using sidewall activation [102]. Finally, reabsorption of photons for LEDs with a large number of active regions, and excessive heating due to un-extracted photons could both reduce the efficacy of these LEDs. Further work is required to address these challenges. The advantages of this cascaded LED design have also been reported recently using detailed modeling [103–105]. Multijunction solar cell with a short circuit current density of 0.28 mA/cm$^2$ and open-circuit voltage of 0.5 V has also been reported recently [106].

## 9.6 SUMMARY

A plot of TJ resistivity as a function of bandgap in different material systems [12,40,43,59,72,107–113], with the results achieved in gallium nitride, is shown in Figure 9.15. Inspite of the large bandgap, using novel heterostructure engineering, a very low tunneling–specific resistance has been achieved in GaN. The first approach involves engineering of polarization-assisted band bending in GaN/InGaN/GaN heterostructures. This was shown to work for both the Ga-polar and N-polar orientations. Using this method, a low-tunneling resistance of $10^{-4}$ $\Omega \cdot$cm$^2$ was measured. Using GdN nanoislands [80] embedded in heavily doped GaN PN junctions, TJs

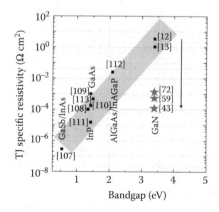

**FIGURE 9.15**  Tunnel junction specific resistivity in different material systems as a function of bandgap. This work represents the lowest reported resistivity for GaN.

along both Ga-polar and N-polar orientations were demonstrated. Weak dependence on crystal polarity makes this approach a viable solution in cases where the polarization field orientation does not align with that of the depletion field. Having realized low resistance TJs using GdN/GaN and InGaN/GaN heterostructures, such TJs were then regrown on a commercial blue LED. A forward voltage drop of 5.3 V at 100 mA is reported in TJLEDs. Proof-of-concept cascading of PN junctions has also been shown, and this enables epitaxial cascading of LEDs for multiple wavelength emission and droop-free single wavelength emission. The use of TJ-cascaded LEDs could lead to unprecedented high power emitters for various applications including lighting, display, and lasers. Development of AlGaN-based TJs could lead to efficient hole injection in UV and DUV emitters, leading to an increase in the efficiency of such devices.

## REFERENCES

1. L. Esaki, "New phenomenon in narrow germanium p-n junctions," *Phys Rev.*, 109: 603, 1958.
2. V.M. Franks, K.F. Hulme, and J.R. Morgan, "An alloy process for making high current density silicon tunnel diode junctions," *Solid-State Electronics*, 8(3): 343–344, 1965.
3. N. Jin, S.-Y. Chung, A. Rice, P.R. Berger, R. Yu, P.E. Thompson, and R. Lake, "151 kA/cm² peak current densities in Si/SiGe resonant interband tunneling diodes for high-power mixed-signal applications," *Appl. Phys. Lett.,* 83(16): 3308–3310, 2003.
4. R. King, D. Law, K. Edmondson, C. Fetzer, G.S. Kinsey, H. Yoon, R. Sherif, and N.H. Karam, "40% efficient metamorphic GaInP/GaInAs/Ge multijunction solar cells," *Appl. Phys. Lett.,* 90(18): 183516–183516-3, 2007.
5. K. Boucart and A.-M. Ionescu, "Double-gate tunnel FET with high-κ gate dielectric," *IEEE Trans. Electron Devices*, 54(7): 1725–1733, 2007.
6. D. Miller, S. Zehr, and J.J. Harris, "GaAs–AlGaAs tunnel junctions for multigap cascade solar cells," *J. Appl. Phys.,* 53(1): 744–748, 1982.
7. G. Verzellesi, D. Saguatti, M. Meneghini, F. Bertazzi, M. Goano, G. Meneghesso, and E. Zanoni, "Efficiency droop in InGaN/GaN blue light-emitting diodes: Physical mechanisms and remedies," *J. Appl. Phys.,* 114(7): 071101–071114, 2013.
8. M. Kneissl, T. Kolbe, C. Chua, V. Kueller, N. Lobo, J. Stellmach, A. Knauer et al., "Advances in group III-nitride-based deep UV light-emitting diode technology," *Semicond. Sci. Technol.,* 26(1): 014036, 2011.
9. F. Akyol, S. Krishnamoorthy, and S. Rajan, "Tunneling-based carrier regeneration in cascaded GaN light emitting diodes to overcome efficiency droop," *Appl. Phys. Lett.,* 103(8): 081107–081107-4, 2013.
10. M.J. Grundmann and U.K. Mishra, "Multi-color light emitting diode using polarization-induced tunnel junctions," *Phys. Status Solidi C,* 4: 2830–2833, 2007.
11. A. Khan, K. Balakrishnan, and T. Katona, "Ultraviolet light-emitting diodes based on group three nitrides," *Nat. Photonics*, 2: 77–84, 2008.
12. M.J. Grundmann, *Polarization-Induced Tunnel Junctions in III-Nitrides for Optoelectronic Applications*, PhD Dissertation, University of California, Santa Barbara, 2007.
13. M. Shatalov, J. Yang, Y. Bilenko, M. Shur, and R. Gaska, "AlGaN deep ultraviolet LEDs with external quantum efficiency over 10%," in *2013 Conference on Lasers and Electro-Optics Pacific Rim (CLEO-PR)*, June 30th-July 4th 2013, IEEE: Kyoto, Japan. 2013.

14. F. Akyol, D. Nath, S. Krishnamoorthy, P. Park, and S. Rajan, "Suppression of electron overflow and efficiency droop in N-polar GaN green light emitting diodes," *Appl. Phys. Lett.*, 100(11): 111118–111118-4, 2012.

15. S. Newman, C. Gallinat, J. Wright, R. Enck, A. Sampath, H. Shen, M. Reed, and M. Wraback, "Wavelength stable, p-side-down green light emitting diodes grown by molecular beam epitaxy," *J. Vacuum Sci. Technol. B*, 31(1): 010601–010603, 2013.

16. J. Verma, J. Simon, V. Protasenko, T. Kosel, H. Grace Xing, and D. Jena, "N-polar III-nitride quantum well light-emitting diodes with polarization-induced doping," *Appl. Phys. Lett.*, 99(17): 171104–171104-3, 2010.

17. Z.Q. Li, M. Lestradet, Y.G. Xiao, and S. Li, "Effects of polarization charge on the photovoltaic properties of InGaN solar cells," *Phys. Status Solidi A*, 208(4): 928–931, 2011.

18. S.-R. Jeon, Y.-H. Song, H.-J. Jang, G.-M. Yang, S.-W. Hwang, and S.-J. Son, "Lateral current spreading in GaN-based light-emitting diodes utilizing tunnel contact junctions," *Appl. Phys. Lett.*, 78(21): 3265–3267, 2001.

19. T. Takeuchi, H. Ghulam, C. Scott, M. Hueschen, R.P. Schneider Jr, C. Kocot, M. Blomqvist et al., "GaN-based light emitting diodes with tunnel junctions," *Jpn J. Appl. Phys.*, 40(8B): L861, 2001.

20. I. Ozden, E. Makarona, A. Nurmikko, T. Takeuchi, and M. Krames, "A dual-wavelength indium gallium nitride quantum well light emitting diode," *Appl. Phys. Lett.*, 79(16): 2532–2534, 2001.

21. M. Diagne, Y. He, H. Zhou, E. Makarona, A. Nurmikko, J. Han, K. Waldrip, J. Figiel, T. Takeuchi, and M. Krames, "Vertical cavity violet light emitting diode incorporating an aluminum gallium nitride distributed Bragg mirror and a tunnel junction," *Appl. Phys. Lett.*, 79(22): 3720–3722, 2001.

22. S.R. Jeon, C.S. Oh, J.-W. Yang, G.M. Yang, and B. Yoo, "GaN tunnel junction as a current aperture in a blue surface-emitting light-emitting diode," *Appl. Phys. Lett.*, 80(11): 1933–1935, 2002.

23. J.J. Wierer, M.R. Krames, J.E. Epler, N.F. Gardner, M.G. Craford, J.R. Wendt, J.A. Simmons, and M.M. Sigalas, "InGaN/GaN quantum-well heterostructure light-emitting diodes employing photonic crystal structures," *Appl. Phys. Lett.*, 84(19): 3885–3887, 2004.

24. F. Bernardini, V. Fiorentini, and D. Vanderbilt, "Spontaneous polarization and piezoelectric constants of III-V nitrides," *Phys. Rev. B*, 56(16): R10024–R10027, 1997.

25. F. Bernardini and V. Fiorentini, "Nonlinear macroscopic polarization in III-V nitride alloys," *Phys. Rev. B*, 64(8): 082507, 2001.

26. F. Bernardini and V. Fiorentini, "Macroscopic polarization and band offsets at nitride heterojunctions," *Phys. Rev. B*, 57(16): R9427–R9430, 1998.

27. V. Fiorentini, F. Bernardini, F. Della Sala, A. Di Carlo, and P. Lugli, "Effects of macroscopic polarization in III-V nitride multiple quantum wells," *Phys. Rev. B*, 60(12): 8849–8858, 1999.

28. C. Wood and D. Jena, *Polarization Effects in Semiconductors: From Ab Initio Theory to Device Applications*, Springer, US, 2007.

29. A. Romanov, T. Baker, S. Nakamura, J. Speck, and A. Romanov, "Strain-induced polarization in wurtzite III-nitride semipolar layers," *J. Appl. Phys.*, 100(2): 023522–023522-10, 2006.

30. L.F. Eastman and U.K. Mishra, "The toughest transistor yet [GaN transistors]," *IEEE Spectrum*, 39(5): 28–33, 2002.

31. O. Ambacher, J. Smart, J. Shealy, N. Weimann, K. Chu, M. Murphy, W. Schaff, L. Eastman et al., "Two-dimensional electron gases induced by spontaneous and piezoelectric polarization charges in N- and Ga-face AlGaN/GaN heterostructures," *J. Appl. Phys.*, 85(6): 3222–3233, 1999.

32. O. Ambacher, B. Foutz, J. Smart, J. Shealy, N. Weimann, K. Chu, M. Murphy et al., "Two dimensional electron gases induced by spontaneous and piezoelectric polarization in undoped and doped AlGaN/GaN heterostructures," *J. Appl. Phys.*, 87(1): 334–344, 2000.

33. J. Ibbetson, P.T. Fini, K.D. Ness, S. DenBaars, J. Speck, and U. Mishra, "Polarization effects, surface states, and the source of electrons in AlGaN/GaN heterostructure field effect transistors," *Appl. Phys. Lett.*, 77(2): 250–252, 2000.

34. I. Smorchkova, C.R. Elsass, J. Ibbetson, R. Vetury, B. Heying, P. Fini, E. Haus, S. DenBaars, J. Speck, and U. Mishra, "Polarization-induced charge and electron mobility in AlGaN/GaN heterostructures grown by plasma-assisted molecular-beam epitaxy," *J. Appl. Phys.*, 86(8): 4520–4526, 1999.

35. L.F. Eastman, V. Tilak, J. Smart, B. Green, E. Chumbes, R. Dimitrov, H. Kim et al. "Undoped AlGaN/GaN HEMTs for microwave power amplification," *IEEE Trans. Electron Devices*, 48(3): 479–485, 2001.

36. D. Nath, Z. Yang, C.-Y. Lee, P. Park, Y.-R. Wu, and S. Rajan, "Unipolar vertical transport in GaN/AlGaN/GaN heterostructures," *Appl. Phys. Lett.*, 103(2): 022102–022104, 2013.

37. G. Gupta, M. Laurent, J. Lu, S. Keller, and U.K. Mishra, "Design of polarization-dipole-induced isotype heterojunction diodes for use in III–N hot electron transistors," *Appl. Phys. Express*, 7(1): 014102, 2014.

38. M. Leroux, N. Grandjean, M. Laugt, J. Massies, B. Gil, P. Lefebvre, and P. Bigenwald, "Quantum confined Stark effect due to built-in internal polarization fields in (Al,Ga)N/GaN quantum wells," *Phys. Rev. B*, 58(20): R13371–R13374, 1998.

39. M. Singh, Y. Zhang, J. Singh, and U. Mishra, "Examination of tunnel junctions in the AlGaN/GaN system: Consequences of polarization charge," *Appl. Phys. Lett.*, 77(12): 1867–1869, 2000.

40. J. Simon, Z. Zhang, K. Goodman, H. Xing, T. Kosel, P. Fay, and D. Jena, "Polarization-induced Zener tunnel junctions in wide-bandgap heterostructures," *Phys. Rev. Lett.*, 103(2): 026801–026804, 2009.

41. C. Wetzel, T. Takeuchi, H. Amano, and I. Akasaki, "Piezoelectric Franz–Keldysh effect in strained GaInN/GaN heterostructures," *J. Appl. Phys.*, 85(7): 3786–3791, 1999.

42. M.F. Schubert, "Interband tunnel junctions for wurtzite III-nitride semiconductors based on heterointerface polarization charges," *Phys. Rev. B*, 81(3): 035303, 2010.

43. S. Krishnamoorthy, F. Akyol, P.S. Park, and S. Rajan, "Low resistance GaN/InGaN/GaN tunnel junctions," *Appl. Phys. Lett.*, 102(11): 113503–113505, 2013.

44. I. Vurgaftman and J.R. Meyer, "Electron bandstructure parameters," in *Nitride Semiconductor Devices: Principles and Simulation*, J. Piprek (editor), Wiley-VCH Verlag, Weinheim, Germany, 2007.

45. E. Kane, "Zener tunneling in semiconductors," *J. Phys. Chem. Solids*, 12(2): 181–188, 1960.

46. M. Grundmann, *Polarization-Induced Tunnel Junctions in III-nitrides for Optoelectronic Applications*, PhD Thesis, University of California, Santa Barbara, 2007.

47. M. Kaga, T. Morita, Y. Kuwano, K. Yamashita, K. Yagi, M. Iwaya, T. Takeuchi, S. Kamiyama, and I. Akasaki, "GaInN-based tunnel junctions in n–p–n light emitting diodes," *Jpn. J. Appl. Phys.*, 52(8): 08JH06, 2013.

48. X. Wang, S.-B. Che, Y. Ishitani, and A. Yoshikawa, "Effect of epitaxial temperature on N-polar InN films grown by molecular beam epitaxy," *J. Appl. Phys.*, 99(7): 073512–073512-5, 2006.

49. K. Xu and A. Yoshikawa, "Effects of film polarities on InN growth by molecular-beam epitaxy," *Appl. Phys. Lett.*, 83(2): 251–253, 2003.

50. R. Singh, D. Doppalapudi, T. Moustakas, and L. Romano, "Phase separation in InGaN thick films and formation of InGaN/GaN double heterostructures in the entire alloy composition," *Appl. Phys. Lett.,* 70(9): 1089–1091, 1997.

51. P. Ruterana, S. Kret, A. Vivet, G. Maciejewski, and P. Dluzewski, "Composition fluctuation in InGaN quantum wells made from molecular beam or metalorganic vapor phase epitaxial layers," *J. Appl. Phys.,* 91(11): 8979–8985, 2002.

52. D. Doppalapudi, S. Basu, K. Ludwig, and T. Moustakas, "Phase separation and ordering in InGaN alloys grown by molecular beam epitaxy," *J. Appl. Phys.,* 84(3): 1389–1395, 1998.

53. L.K. Li, M.J. Jurkovic, W. Wang, J. Van Hove, and P. Chow, "Surface polarity dependence of Mg doping in GaN grown by molecular-beam epitaxy," *Appl. Phys. Lett.,* 76(13): 1740–1742, 2000.

54. V. Ramachandran, R. Feenstra, W. Sarney, L. Salamanca-Riba, J. Northrup, L. Romano and D. Greve, "Inversion of wurtzite GaN(0001) by exposure to magnesium," *Appl. Phys. Lett.,* 75(6): 808–810, 1999.

55. X. Cao, S. Pearton, A. Zhang, G.T. Dang, F. Ren, R. Shul, L. Zhang, R. Hickman, and J. Van Hove, "Electrical effects of plasma damage in p-GaN," *Appl. Phys. Lett.,* 75(17): 2569–2571, 1999.

56. S. Krishnamoorthy, D. Nath, F. Akyol, P.S. Park, M. Esposto, and S. Rajan, "Polarization-engineered GaN/InGaN/GaN tunnel diodes," *Appl. Phys. Lett.,* 97(20): 203502–203503, 2010.

57. S. Krishnamoorthy, P.S. Park, and S. Rajan, "Demonstration of forward inter-band tunneling in GaN by polarization engineering," *Appl. Phys. Lett.,* 99(23): 233504–233504-3, 2011.

58. T. Growden, S. Krishnamoorthy, D. Nath, A. Ramesh, S. Rajan, and P.R. Berger, "Methods for attaining high interband tunneling current in III-Nitrides," in *2012 70th Annual Device Research Conference (DRC),* 20–22 June 2011, IEEE: Santa Barbara, CA, 2012.

59. S. Krishnamoorthy, F. Akyol, and S. Rajan, "InGaN/GaN tunnel junctions for hole injection in GaN light emitting diodes," *Appl. Phys. Lett.,* 105(14): 141104–141104-4, 2014.

60. K. Zhang, H. Liang, Y. Liu, R. Shen, W. Guo, D. Wang, X. Xia et al., "Low Al-composition p-GaN/Mg-doped Al0.25Ga0.75N/n+-GaN polarization-induced backward tunneling junction grown by metal-organic chemical vapor deposition on sapphire substrate," *Sci. Rep.,* 4(6322): 1–7, 2014.

61. S. Rajan, A. Chini, M. Hoi Wong, J.S. Speck, and U.K. Mishra, "N-polar GaN/AlGaN/GaN high electron mobility transistors," *J. Appl. Phys.,* 102(4): 044501–044506, 2007.

62. C.A. Schaake, D.F. Brown, B.L. Swenson, S. Keller, J.S. Speck, and U.K. Mishra, "A donor-like trap at the InGaN/GaN interface with net negative polarization and its possible consequence on internal quantum efficiency," *Semicond. Sci. Technol.,* 28(10): 105021, 2013.

63. A.E. Belyaev, C. Foxon, S. Novikov, O. Makarovsky, L. Eaves, M. Kappers, and C. Humphreys, "Comment on "AlN/GaN double-barrier resonant tunneling diodes grown by rf-plasma-assisted molecular-beam epitaxy" [*Appl. Phys. Lett.,* 81: 1729, 2002], *Appl. Phys. Lett.,* 83(17): 3626–3627, 2003.

64. S. Krishnamoorthy, P.S. Park, and S. Rajan, "III-nitride tunnel diodes with record forward tunnel current density," in *2011 69th Annual Device Research Conference (DRC),* 20–22 June 2011, IEEE: Santa Barbara, CA, 2011.

65. K. Zhang, H. Liang, R. Shen, D. Wang, P. Tao, Y. Liu, X. Xia, Y. Luo, and G. Du, "Negative differential resistance in low Al-composition p-GaN/Mg-doped Al0.15Ga0.85N/n+-GaN hetero-junction grown by metal-organic chemical vapor deposition on sapphire substrate," *Appl. Phys. Lett.,* 104(5): 053507–053507-4, 2014.

66. Y. Zhang, S. Krishnamoorthy, F. Akyol, and S. Rajan, "Design of Ga-polar low resistance polarization engineered GaN/InGaN tunnel junctions," in *Electronic Materials Conference*, Santa Barbara, 2014, June 25–27, 2014, Materials Research Society: Santa Barbara, CA.

67. S. Ahmed, M.R. Melloch, E.S. Harmon, D.T. McInturff, and J.M. Woodall, "Use of nonstoichiometry to form GaAs tunnel junctions," *Appl. Phys. Lett.*, 71: 3667–3669, 1997.

68. C. Kadow, S.B. Fleischer, J.P. Ibbetson, J.E. Bowers, A.C. Gossard, J.W. Dong, and C.J. Palmstrøm, "Self-assembled ErAs islands in GaAs: Growth and subpicosecond carrier dynamics," *Appl. Phys. Lett.*, 75: 3548–3550, 1999.

69. P. Pohl, F.H. Renner, M. Eckardt, A. Schwanhäußer, A. Friedrich, Ö. Yüksekdag, S. Malzer, G.H. Döhler, P. Kiesel, and D.E.A. Driscoll, "Enhanced recombination tunneling in GaAs Pn junctions containing low-temperature-grown-GaAs and ErAs layers," *Appl. Phys. Lett.*, 83: 4035–4037, 2003.

70. J.M.O. Zide, A. Kleiman-Shwarscstein, N.C. Strandwitz, J.D. Zimmerman, T. Steenblock-Smith, A.C. Gossard, A. Forman, A. Ivanovskaya, and G.D. Stucky, "Increased efficiency in multijunction solar cells through the incorporation of semimetallic ErAs nanoparticles into the tunnel junction," *Appl. Phys. Lett.*, 88: 162103–162103-3, 2006.

71. F.L. Bloom, A.C. Young, R.C. Myers, E.R. Brown, A.C. Gossard, and E.G. Gwinn, "Tunneling through MnAs particles at a GaAs p+n+ junction," *J. Vacuum Sci. Technol. B*, 24: 1639–1643, 2006.

72. S. Krishnamoorthy, T.F. Kent, J. Yang, P.S. Park, R.C. Myers, and S. Rajan, "GdN nanoisland-based GaN tunnel junctions," *Nanoletters*, 13(6): 2570–2575, 2013.

73. S. Dhar, O. Brandt, M. Ramsteiner, V.F. Sapega, and K.H. Ploog, "Colossal magnetic moment of Gd in GaN," *Phys. Rev. Lett.*, 94: 037205–037205-4, 2005.

74. W.R.L. Lambrecht, "Electronic structure and optical spectra of the semimetal ScAs and of the indirect-band-gap semiconductors ScN and GdN," *Phys. Rev. B*, 62: 13538–13545, 2000.

75. R. Vidyasagar, S. Kitayama, H. Yoshitomi, T. Kita, T. Sakurai, and H. Ohta, "Study on spin-splitting phenomena in the band structure of GdN," *Appl. Phys. Lett.*, 100(23): 232410–232410-4, 2012.

76. H.J. Trodahl, A.R.H. Preston, J. Zhong, B.J. Ruck, N.M. Strickland, C. Mitra, and W.R.L. Lambrecht, "Ferromagnetic redshift of the optical gap in GdN," *Phys. Rev. B*, 76(8): 085211, 2007.

77. A.R.H. Preston, B.J. Ruck, W.R.L. Lambrecht, L.F.J. Piper, J. Downes, K. Smith, and H.J. Trodahl, "Electronic band structure information of GdN extracted from x-ray absorption and emission spectroscopy," *Appl. Phys. Lett.*, 96(3): 032101–032103, 2010.

78. H. Yoshitomi, S. Kitayama, T. Kita, O. Wada, M. Fujisawa, H. Ohta, and T. Sakurai, "Optical and magnetic properties in epitaxial GdN thin films," *Phys. Rev. B*, 83(15): 155202, 2011.

79. R. Vidyasagar, T. Kita, T. Sakurai, and H. Ohta, "Electronic transitions in GdN band structure," *J. Appl. Phys.*, 115(20): 203717–203717-5, 2014.

80. T.F. Kent, J. Yang, L. Yang, M.J. Mills, and R.C. Myers, "Epitaxial ferromagnetic nanoislands of cubic GdN in hexagonal GaN," *Appl. Phys. Lett.*, 100: 152111–152114, 2012.

81. J. Yang, *PAMBE Growth and Characterization of Superlattice Structures in Nitrides*, Electronic Thesis or Dissertation, Ohio State University, 2013.

82. T. Kent, *III-Nitride Nanostructures for Optoelectronic and Magnetic Functionalities: Growth, Characterization and Engineering*, Electronic Thesis or Dissertation, Ohio State University, 2014.

83. M.A. Scarpulla, C.S. Gallinat, S. Mack, J.S. Speck, and A.C. Gossard, "GdN (111) heteroepitaxy on GaN (0001) by N2 plasma and NH3 molecular beam epitaxy," *J. Cryst. Growth,* 311: 1239–1244, 2009.

84. F. Akyol, D.N. Nath, S. Krishnamoorthy, P.S. Park, and S. Rajan, "Suppression of electron overflow and efficiency droop in N-polar GaN green light emitting diodes," *Appl. Phys. Lett.,* 100: 111118–111118-4, 2012.

85. S. Krishnamoorthy, PhD Dissertation, Ohio State University, 2014.

86. S.-R. Jeon, Y.-H. Song, H.-J. Jang, G.M. Yang, S.W. Hwang, and S.J. Son, "Lateral current spreading in GaN-based light-emitting diodes utilizing tunnel contact junctions," *Appl. Phys. Lett.,* 78: 3265–3267, 2001.

87. T. Takeuchi, G. Hasnain, S. Corzine, M. Hueschen, R.P.S. Jr, C. Kocot, M. Blomqvist, Y. Chang, D. Lefforge, and M.R. Krames, "GaN-based light emitting diodes with tunnel junctions," *Jpn. J. Appl. Phys.,* 40: L861–L863, 2001.

88. J.-H. Lee and J.-H. Lee, "Enhanced output power of InGaN-based light-emitting diodes with AlGaN/GaN two-dimensional electron gas structure," *IEEE Electron Device Lett.,* 31(5): 455–457, 2010.

89. J.-K. Sheu, J.M. Tsai, S. Shei, W. Lai, T. Wen, C.H. Kou, Y. Su, S. Chang, and G. Chi, "Low-operation voltage of InGaN-GaN light-emitting diodes with Si-doped In/sub 0.3/ Ga/sub 0.7/N/GaN short-period superlattice tunneling contact layer," *IEEE Electron Device Lett.,* 22(10): 460–462, 2001.

90. S. Jin An, "Transparent conducting ZnO nanorods for nanoelectrodes as a reverse tunnel junction of GaN light emitting diode applications," *Appl. Phys. Lett.,* 100(22): 223115–223115-5, 2012.

91. G. Koblmuller, R. Chu, A. Raman, U. Mishra, and J. Speck, "High-temperature molecular beam epitaxial growth of AlGaN/GaN on GaN templates with reduced interface impurity levels," *J. Appl. Phys.,* 107(4): 043527–043527-9, 2010.

92. S. Chowdhury, M. Hoi Wong, B. Swenson, and U.K. Mishra, "CAVET on bulk GaN substrates achieved with MBE-regrown AlGaN/GaN layers to suppress dispersion," *IEEE Electron Device Lett.,* 33(1): 41–43, 2012.

93. Y. Shen, G. Mueller, S. Watanabe, N. Gardner, A. Munkholm, and M. Krames, "Auger recombination in InGaN measured by photoluminescence," *Appl. Phys. Lett.,* 91(14): 141101–141101-3, 2007.

94. H.-Y. Ryu, H.-S. Kim, and J.-I. Shim, "Rate equation analysis of efficiency droop in InGaN light-emitting diodes," *Appl. Phys. Lett.,* 95(8): 081114–081114-3, 2009.

95. M.F. Schubert, S. Chhajed, K.-S. Kim, E. Fred Schubert, D.D. Koleske, M.H. Crawford, S.R. Lee, A.J. Fischer, G. Thaler, and M.A. Banas, "Effect of dislocation density on efficiency droop in GaInN/GaN light-emitting diodes," *Appl. Phys. Lett.,* 91(23): 231114–231114-3, 2007.

96. X. Ni, Q. Fan, R. Shimada, U. Ozgur, and H. Morkoc, "Reduction of efficiency droop in InGaN light emitting diodes by coupled quantum wells," *Appl. Phys. Lett.,* 93(17): 171113–171113-3, 2008.

97. C. Wang, C. Ke, C. Lee, S.-P. Chang, W. Chang, J. Li, Z. Li et al., "Hole injection and efficiency droop improvement in InGaN/GaN light-emitting diodes by band-engineered electron blocking layer," *Appl. Phys. Lett.,* 97(26): 261103–261103-3, 2010.

98. J. Iveland, M. Piccardo, L. Martinelli, J. Peretti, J.W. Choi, N. Young, S. Nakamura, J.S. Speck, and C. Weisbuch, "Origin of electrons emitted into vacuum from InGaN light emitting diodes," *Appl. Phys. Lett.,* 105(5): 052103–052104, 2014.

99. A.H. Reading, J.J. Richardson, C.-C. Pan, S. Nakamura, and S.P. DenBaars, "High efficiency white LEDs with single-crystal ZnO current spreading layers deposited by aqueous solution epitaxy," *Opt Express,* 20(101): A13-A19, 2012.

100. G.-B. Lin, D. Meyaard, J. Cho, E. Fred Schubert, H. Shim, and C. Sone, "Analytic model for the efficiency droop in semiconductors with asymmetric carrier-transport properties based on drift-induced reduction of injection efficiency," *Appl. Phys. Lett.,* 100(16): 161106–161106-4, 2012.

101. Z.-H. Zhang, S. Tiam Tan, Z. Kyaw, Y. Ji, W. Liu, Z. Ju, N. Hasanov, X. Wei Sun, and H. Volkan Demir, "InGaN/GaN light-emitting diode with a polarization tunnel junction," *Appl. Phys. Lett.,* 102(19): 193508, 2013.

102. Y. Kuwano, M. Kaga, T. Morita, K. Yamashita, K. Yagi, M. Iwaya, T. Takeuchi, S. Kamiyama, and I. Akasaki, "Lateral hydrogen diffusion at p-GaN layers in nitride-based light emitting diodes with tunnel junctions," *Jpn. J. Appl. Phys.,* 52(8S): 08JK12, 2013.

103. M.-C. Tsai, B. Leung, T.-C. Hsu, and Y.-K. Kuo, "Tandem structure for efficiency improvement in GaN-based light-emitting diodes," *J. Lightwave Technol.,* 32(9): 1801–1806, 2014.

104. J. Piprek, "GaN-based vertical-cavity laser performance improvements using tunnel-junction-cascaded active regions," *Appl. Phys. Lett.,* 105(1): 011116–011116-4, 2014.

105. J. Piprek, "Origin of InGaN/GaN light-emitting diode efficiency improvements using tunnel-junction-cascaded active regions," *Appl. Phys. Lett.,* 104(5): 051118–051118-4, 2014.

106. H. Kurokawa, M. Kaga, T. Goda, M. Iwaya, T. Takeuchi, S. Kamiyama, A. Isamu, and H. Amano, "Multijunction GaInN-based solar cells using a tunnel junction," *Appl. Phys. Express,* 7(3): 034104, 2014.

107. K. Vizbaras, M. Törpe, S. Arafin, and M.-C. Amann, "Ultra-low resistive GaSb/InAs tunnel junctions," *Semicond. Sci. Technol.,* 26: 075021–075021-4, 2011.

108. M.P. Lumb, M.K. Yakes, M. Gonzalez, I. Vurgaftman, C. Bailey, R. Hoheisel, and R. Walters, "Double quantum-well tunnel junctions with high peak tunnel currents and low absorption for InP multi-junction solar cells," *Appl. Phys. Lett.,* 100(21): 213907–213907-4, 2012.

109. H. Nair, A. Crook, and S. Bank, "Enhanced conductivity of tunnel junctions employing semimetallic nanoparticles through variation in growth temperature and deposition," *Appl. Phys. Lett.,* 96(22): 222104–222104-3, 2010.

110. S. Preu, S. Malzer, G.H. Döhler, H. Lu, A.C. Gossard, and L.J. Wang, "Efficient III–V tunneling diodes with ErAs recombination centers," *Semicond. Sci. Technol.,* 25(11): 115004, 2010.

111. N. Suzuki, T. Anan, H. Hatakeyama, and M. Tsuji, "Low resistance tunnel junctions with type-II heterostructures," *Appl. Phys. Lett.,* 88(23): 231103–231103-3, 2006.

112. P. Sharps, N. Li, J. Hills, J. Hou, P.-C. Chang, and A. Baca, "AlGaAs/InGaAlP tunnel junctions for multi-junction solar cells," in *2000 Conference Record of the 28th IEEE Photovoltaic Specialists Conference,* 15–22 September, 2000, IEEE: Anchorage, Alaska 2000.

113. D. Chmielewski, T. Grassman, A. Carlin, J. Carlin, A. Speelman, and S. Ringel, "Metamorphic tunnel junctions for high efficiency III-V/IV multi-junction solar cell technology," in *2013 IEEE 39th Photovoltaic Specialists Conference (PVSC),* 16–21 June 2013, IEEE: Tampa, FL, 2013.

# 10 Trapping and Degradation Mechanisms in GaN-Based HEMTs

*Matteo Meneghini, Gaudenzio Meneghesso, and Enrico Zanoni*

## CONTENTS

## 10.1  INTRODUCTION

Over the last few years, GaN-based transistors have demonstrated to be excellent devices for application in high-frequency systems (radars, communication devices, etc.) and in the power electronics field (power converters, photovoltaic inverters, electric cars, etc.). This is possible thanks to the very unique characteristics of GaN, namely the high electron mobility compared to conventional semiconductors (>2000 cm²/Vs), the high breakdown field (>300 V/μm), and the high thermal conductivity (in excess of 2 W/cmK).

Despite the recent improvements in the growth and fabrication process, the performance and the reliability of GaN-based HEMTs is still limited by a number of mechanisms that must be studied and modeled to make this technology viable for the application market.

More specifically, the dynamic performance of HEMTs is limited by the so-called "current collapse," which is a recoverable decrease in drain current induced by the trapping of carriers within the device structure. Trapping may occur under the gate, thus inducing a shift in the threshold voltage and a variation in the saturation current, or in the access region. In this last case, the most important consequence of trapping is an increase in the on-resistance (dynamic $R_{on}$ increase), which is particularly critical for power devices, since these transistors should have negligible parasitic resistance. The first part of this chapter (Section 10.2) describes the charge trapping mechanisms that limit the dynamic performance of GaN-HEMTs; moreover, we describe and critically compare the different methods that are commonly used for the extraction of the properties of the defects responsible for current collapse (or $R_{on}$ increase): capacitance deep-level transient spectroscopy (C-DLTS), deep-level optical spectroscopy (DLOS), dynamic transconductance measurements ($g_m$-$f$), and drain current transient (DCT) measurements; finally, we describe the properties (activation energy, physical origin) of the defects that are most commonly observed in GaN-based materials and devices.

The second part of the chapter (Section 10.3) deals with the degradation mechanisms that limit the lifetime of state-of-the-art HEMTs: transistor based on GaN may degrade due to several mechanisms. Here, we discuss the degradation processes activated by the high electric field, namely (i) the degradation of the gate junction, due to the exposure to a high off-state bias; (ii) the degradation mechanisms activated by semi-on and on-state operation, and the role of hot electrons in the degradation process; and (iii) the early breakdown processes (punch-through, vertical breakdown, impact ionization), that may severely limit the robustness and the performance of high voltage devices based on GaN.

## 10.2 TRAP-RELATED ISSUES IN GaN-BASED HEMTS

Trapping processes may significantly affect the dynamic performance of GaN-based transistors: one of the most critical issues for RF devices is the current collapse, that is, the recoverable decrease in drain current induced by the exposure to high electric fields. This effect may be due either to a dynamic shift in the threshold voltage (which implies a decrease in the saturation current) or to a dynamic increase in the on-resistance ($R_{on}$). The current collapse (or dynamic $R_{on}$ increase) can be quantitatively evaluated by means of pulsed measurements, carried out starting from various quiescent bias points in the off-state. In most of the cases, a double-pulse setup is adopted: the device is held in a quiescent bias point in the off-state, for example, with a negative gate voltage (e.g., $V_{G,QB} = -6$ V) and a high drain voltage (e.g., $V_{D,QB} = 50$ V). The $I_D$-$V_D$ curves are measured by means of short pulses (e.g., 200 ns–1 μs) starting from this quiescent condition. Several quiescent bias points can be used for this investigation: (a) no bias on the gate and on the drain, that is, $(V_{G,QB}; V_{D,QB}) = (0$ V, $0$ V), which induces a negligible trapping; (b) a negative gate bias, with no bias on the drain (e.g., $(V_{G,QB}; V_{D,QB}) = (-6$ V, $0$ V)), which may imply a significant trapping under the gate; and (c) a negative gate bias, and a high drain bias (e.g., $(V_{G,QB}; V_{D,QB}) = (-6$ V, $50$ V)), which induces a strong trapping both under the gate and in the gate-drain access region.

Figure 10.1a shows an example of pulsed $I_D$-$V_D$ curves measured on a GaN-based transistor starting from various quiescent bias points: the curve measured starting from $(V_{G,QB}; V_{D,QB}) = (0$ V, $0$ V) is the reference measurement, since the device does not show relevant trapping processes in this quiescent condition. On the other hand, the exposure to a negative gate (quiescent) bias and/or to a positive drain (quiescent) bias may induce a significant decrease in drain current, due to charge trapping mechanisms. For the device in Figure 10.1a, charge trapping induces a decrease in the current in the saturation region, with no significant variation in the $R$ on. This result suggests that trapping mostly occurs in the region under the gate, and induces a significant variation in the threshold voltage of the devices, without strong changes in the resistivity of the access regions (see Figure 10.1c). The trapping of charge in the region under the gate can be due to several mechanisms, such as the presence of interface states below the Schottky contact or at the AlGaN/GaN heterostructure [1], the presence of defects within the AlGaN barrier [2], or the existence of traps in the GaN buffer [2,3].

On the other hand, the device in Figure 10.1b shows a different behavior: the exposure to a off-state quiescent bias induces also a significant (and recoverable) increase in the on-resistance; this effect can be explained by considering that for the

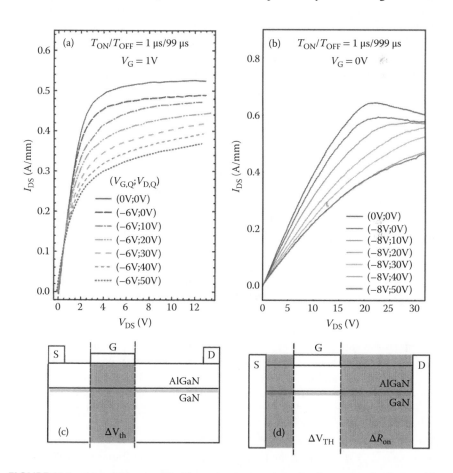

**FIGURE 10.1** (a) and (b) pulsed $I_D$–$V_D$ curves measured starting from several quiescent bias points, in the off-state, on two different samples. (c) Schematic representation of the trapping mechanisms responsible for a dynamic $V_{th}$ shift. (d) Schematic representation of the trapping mechanisms responsible for a dynamic increase in $R_{on}$.

device in Figure 10.1b trapping mostly affects the access regions (see a schematic representation of this process in Figure 10.1d). The trapping of charge in the access region may be due to several processes: (i) surface trapping phenomena, due to a poor passivation of the surface [4], to the trapping of electrons at defects located at the interface between the semiconductor and the passivation layer, or to the injection of electrons from the gate into the passivation layer [5]. (ii) the non-localized trapping of hot electrons in the region between gate and drain [6]; and (iii) the trapping of electrons in the GaN channel or AlGaN barrier layer induced by the off-state electric field [7].

The properties of the defects responsible for the trapping process can be investigated by several methods: C-DLTS measurements, carried out on AlGaN/GaN capacitors, provide a detailed description of the traps located under the gate of the devices [8–10]. C-DLTS is based on the analysis of the capacitance transient induced by the rapid change in the bias voltage, from a "filling bias" ($V_f$, typically above the pinch-off voltage), to a "reverse-bias" ($V_r$, typically lower than the pinch-off voltage) [10]. Consider that we have an electron trap located in the upper half of the bandgap (at the energy $E_C$-$E_T$): when the bias is switched from $V_r$ to $V_f$, the electrons are captured at this defect, due to the variation of the depletion width; when—in $t = 0$ second—the bias is switched back to $V_r$, the trapped electrons are not released instantaneously, since they have to overcome (thermally) the energetic barrier $E_C$-$E_T$. The density of occupied traps (and the junction capacitance) varies exponentially with time, according to

$$N(t) = N_T \, e^{-e_n t} \tag{10.1}$$

where $e_n$ is the (thermal) emission rate of the defect and $N_T$ is the trap concentration. By analyzing the capacitance transients as a function of temperature, it is possible to extrapolate the main properties (activation energy and cross section) of the defects: in fact, the emission rate (for electrons) of a defect is thermally activated, according to the formula

$$e_n = \frac{\sigma_n v_n N_c}{g} \exp\left(-\frac{\Delta E}{kT}\right) \tag{10.2}$$

where $\sigma_n$ is the capture cross section for electrons, $v_n$ is the thermal velocity of the electrons, $N_c$ is the effective density of states in the conduction band, $g$ is the degeneracy of the defect, and $\Delta E$ (equal to $E_C$-$E_T$) is the energetic position of the defect within the bandgap [11]. Moreover, by changing the filling bias it is possible to explore regions close either to the GaN buffer layer or to the AlGaN/GaN interface [9].

State-of-the-art GaN-HEMTs have a very low capacitance (in the order of few pF, i.e., very close to the sensitivity limit of the capacitance meters used for C-DLTS); for this reason, C-DLTS measurements are not carried out on standard HEMTs, but on large-area devices (e.g., FAT-FET structures or transistors with large periphery) with the same epitaxial structure of the HEMTs. This approach provides a detailed characterization of the trap states located in the AlGaN/GaN heterostructure; however, FAT-FETs are significantly different from standard HEMTs in terms of physical dimensions, field plates, and geometry: in addition, also gate leakage (which can be a major source of trapped electrons) and the amplitude of the depleted region can be different in FAT-FETs with respect to HEMTs. As a consequence, in FAT-FETs the trapping/detrapping kinetics and the size of the region affected by trapping can slightly differ from the case of standard transistors; this fact must be taken into account when the properties/location of defects are investigated by capacitance DLTS. Unfortunately, in wide-bandgap materials, some defects may form levels that are too deep to be detected by techniques based on the analysis of the thermal emission processes (such as C-DLTS) [12]; when the defects are too deep to be investigated by conventional C-DLTS, Optical-DLTS (or deep-level optical spectroscopy, DLOS) can be used; this technique probes the energy level of a defect by investigating the photocapacitance transient induced by the exposure to sub-bandgap monochromatic illumination [13]. The transient photocapacitance is proportional to the optical cross section of the defect, which, together with the activation energy, is a specific parameter of a trap [13]. The measurement is repeated at increasing photon energies: the spectral variation of the optical cross section of a defect is evaluated from the time derivative of the photocapacitance transient at $t = 0$ second (i.e., the time where monochromatic light is switched on).

The two methods described above (C-DLTS and DLOS) evaluate the emission rate of a defect by means of time-dependent measurements. Other methods, such as dynamic transconductance ($g_m$-$f$) measurements, can be used to characterize defects in the frequency domain. The conductance of a trap ($G_P/\omega$) can be calculated from the measurement of transconductance versus frequency, according to the formula

$$\frac{G_P(\omega)}{\omega} = -qC_{\text{barrier}}I_D(KT)^{-1}\,\text{Im}\left(\frac{1}{g_m(f)}\right) \tag{10.3}$$

where $\omega = 2\pi f$, $C_{\text{barrier}}$ is the capacitance of the AlGaN barrier, $q$ is the electron charge, $I_D$ is the drain current, $T$ is the temperature and $k$ is the Boltzmann's constant [14–17]. Once $G_P/\omega$ is known, it is possible to extrapolate the main trap parameters by fitting the $G_P/\omega$ curves based on the model

$$\frac{G_P(\omega)}{\omega} = D_t(2\omega\tau)^{-1}\left\{\ln\left[1+(\omega\tau)^2\right]\right\} \tag{10.4}$$

where $\tau$ is the time constant of the trap and $D_t$ is the areal density of the defect. By repeating the measurement at increasing temperatures, it is possible to evaluate the dependence of $\tau$ (related to $e_n$) on temperature (Arrhenius plot) and therefore to extrapolate the activation energy and cross section of the defect. During the dynamic transconductance measurements, the devices are operated in the subthreshold region: in this way it is possible to avoid the shorting effects of the capacitance of the inversion layer [16]. Drain bias is kept relatively low (e.g., 50 mV), to probe the whole gated channel area. The typical frequency range is 1–$10^5$ Hz [16–18]. Based on this technique it is possible to distinguish between interface and buffer defects: the $G_P(\omega)/\omega$ measurements of buffer traps do not show a significant dependence on gate bias, whereas interface states show a significant change in their $G_P(\omega)/\omega$ spectra when gate bias is changed [14,16,17].

The techniques described above (C-DLTS, DLOS, and $g_m$-$f$) provide important results on the properties of traps located under the gate of the devices; however, due to the choice of the bias points (moderate drain bias), or to the layout of the test structures used for the measurements (FAT-FETs and capacitors), they do not provide exhaustive results on the trapping mechanisms that may occur in the gate-drain access region. Moreover, these methods do not provide direct information on the impact of charge trapping on the drain current of the devices, but only indirect information on the variation of junction capacitance or dynamic transconductance induced by the capture of carriers at trap states.

An important method that can be used to directly evaluate the impact of charge trapping on drain current of HEMTs is Drain Current Transient or DCT investigation (also referred to as current-DLTS [2,19,20]). DCT measurements are executed in two phases: during the initial trap-filling phase, the devices are kept for a long time (10–100 seconds) in a quiescent bias point in the off-state (e.g., with negative gate bias, and a high drain bias [$V_{G,F}$;$V_{D,F}$]; during this phase, electrons are trapped under the gate and/or in the gate-drain access region (depending on the adopted bias point). The devices are then switched to the on-state (second phase, transient measurement), and the recovery of the drain current is measured during the whole duration of the de-trapping transient (typically from the 1 μs range to 100–1000 seconds [4,19,20]). In the case of a single electron trap, located at the energy $E_C$-$E_T$ (discrete trap level), during the de-trapping process, drain current shows an exponential increase, whose time constant is equal to $1/e_n$ (where $e_n$ is the emission rate of the trap). Typical DCTs measured on an AlGaN/GaN-HEMT are plotted in Figure 10.2a. Trapping is induced with a filling bias of $(V_{G,F};V_{D,F}) = (-1\ V,\ 27\ V)$. After the device is switched to the measuring point in the on-state (in the saturation region $(V_{G,M};V_{D,M}) = (0\ V,\ 7\ V)$), current shows a significant recovery. The step-like behavior is due to the existence of two different electron emission processes, with different time constants. A simple way to analyze the results of DCT measurements is to plot the derivative of the DCTs ($dI_D/d(\log(t))$, see Figure 10.2b). Each positive peak represents an electron emission process; the time constant of the emission process is correlated to the position of the corresponding peak on the horizontal axis. On the other hand, negative peaks indicate the presence of electron capture processes (which induce a decrease in drain current) and/or of hole emission processes.

Several methods have been proposed to extrapolate the time constant (for emission or capture) of the traps responsible for the DCTs:

1. The fitting of the DCT by a polynomial function, and the subsequent calculation of the time constants from the position of the peaks of the $dI_D/d(\log(t))$ curves on the horizontal axis [19,22]
2. The fitting of the transient data by a sum of a high number (e.g., 100) exponentials with fixed time constant and variable amplitude coefficients [2]

$$I_D(t) = \sum_{i=1}^{100} a_i \exp\left(-\frac{t}{\tau_i}\right) + I_\infty \qquad (10.5)$$

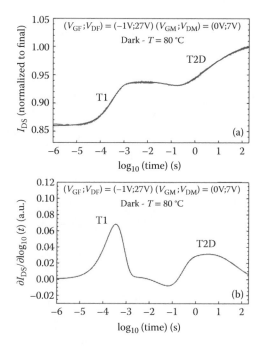

**FIGURE 10.2** (a) Drain current transient measured during the de-trapping phase on a AlGaN/GaN-HEMT. (b) Time constant spectra (derivative of the current transients) corresponding to the drain current transient in (a). (Courtesy of D. Bisi, University of Padova.)

3. The fitting of the data by the sum of $N$ stretched exponential functions (with $N$ typically in the range between 1 and 4, depending on the analyzed current transients) [19]

$$I_D(t) = I_{DS,final} - \sum_{i=1}^{N} A_i \exp\left(-\frac{t}{\tau_i}\right)^{\beta_i} \qquad (10.6)$$

By repeating the measurements at increasing temperature levels (Figure 10.3a) it is possible to extrapolate the Arrhenius plot of the emission process (i.e., the plot of $\ln(T^2{}^*\tau)$ versus $1/kT$), and thus the activation energy and the cross section of the trap. The results of the investigation can depend on the choice of the method used for the extrapolation of the time constant of the de-trapping processes (methods 1–3 above); in the case of a single exponential transient, the three methods provide almost the same results (see the Arrhenius plot of trap T1 in Figure 10.3b). On the other hand, the presence of stretched non-exponential current transients (such as transient T2D in Figure 10.2) leads to spectral dispersion. The derivative spectrum of a stretched exponentials has a broad peak (Figure 10.2b, peak T2D), and the time constants extrapolated by the method (1) above can be significantly different from those extrapolated by fitting the curves based on a stretched exponential

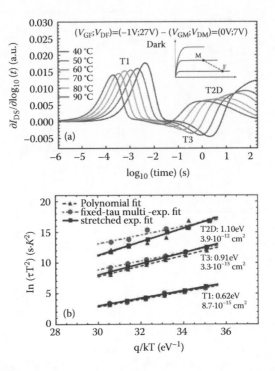

**FIGURE 10.3** (a) Time constant spectra (derivative of the drain current transients) measured at increasing temperature levels on an AlGaN/GaN-HEMT; (b) Arrhenius plot extrapolated from DCT measurements by three methods: polynomial fit and differentiation of the DCT curves, multi-exponential fit, and stretched exponential fit. (Courtesy of D. Bisi, University of Padova.)

function (Figure 10.4). Also the method (2) above is not suitable for the study of stretched exponential transients: in fact, when fitted with the sum of 100 exponentials, a stretched exponential appears as the superposition of several spectral lines.

The presence of stretched exponential and non-exponential transients can be due to several factors: (i) current transients originate from surface states [4]; in this case, the emission of charge may proceed through hopping, rather than via thermal emission; (ii) current transients originate from more than one deep level, with similar emission rates [23]; the superposition of the emission processes generates a broad time-constant spectrum, which can be effectively fitted by a stretched exponential function; (iii) the defects responsible for DCT do not have a discrete energy level, but they form mini-bands [24,25]; (iv) the emission process is assisted by the electric field; due to the non-uniform distribution of the electric field within the device structure, this may result in a spatial variation of the emission rate. In this case, the electrons are emitted faster from the trap levels located in the high field regions [24].

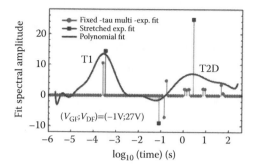

**FIGURE 10.4** Comparison between the methods (1–3) for the extrapolation of the time constant of drain current transients. The trap T1 shows a purely exponential transient, and its time constant can be effectively extrapolated by all the three methods. On the other hand, trap T2D shows a stretched exponential transient, which can be hardly fitted by method (2) (sum of 100 exponentials with fixed tau). (Courtesy of D. Bisi, University of Padova.)

Recent works [20] indicated that for a correct extrapolation of the time constant of the de-trapping processes, it is also important to evaluate the effects of the self-heating of the devices. In fact, neglecting the temperature increase during the DCT measurement may lead to an underestimation of the activation energies of the traps (see Ref. [26] for details). These artifacts can be avoided by accurately measuring the thermal resistance of the devices [27–29], and by using the extrapolated values of the channel temperature (and not just the base-plate temperature) for the construction of the Arrhenius plots.

Several research groups investigated the properties of deep levels in GaN-based materials and devices; as a result, many deep levels were identified. Table 10.1 summarizes the main characteristics (energy level, physical origin) of the deep levels identified in GaN. Figure 10.5 gives a pictorial view of the location of these deep levels in an Arrhenius plot (data were digitized from the reference papers listed in the caption). On the same plot, we report also the curves corresponding to activation energies in the range 0.2–1.2 eV, with a capture cross section of $10^{-15}$ cm$^{-2}$. Remarkably, many of the deep levels cluster in specific regions of the Arrhenius plot, indicating that they have been detected by several research groups, independently from the growth technique used to fabricate the samples, and from the measurement technique. These deep levels may originate from native defects of GaN (such as vacancies, or antisite defects), or to common impurities (e.g., carbon, oxygen) that may be introduced in the samples during the growth process. Table 10.1 provides also information on the deep levels associated to common dopants (such as magnesium, iron, and carbon), which are used to compensate the buffer of GaN-based HEMTs. Table 10.1 can be used as a reference for the interpretation of the results of DCT measurements, and for the identification of the deep levels responsible for current collapse.

**TABLE 10.1**

**Database of the Deep Levels in GaN and Related Materials**

| Reference | Analyzed Samples | Energy Level (eV) | Physical Origin |
|---|---|---|---|
| Umana-Membreno [30], Soh [31], Park [32], Cho [33], Johnstone [34]Cho [35], Choi [36] Arehart [37], Cho [38], Chen [39] | GaN-based devices | EC: 0.09/0.27 | Nitrogen vacancies |
| Chen [39] | n-GaN | EC: 0.12 | Surface |
| Lee [40] | TMGa GaN | EC: 0.14 | Carbon or hydrogen impurities |
| Polyakov [41] | p-GaN | EC: 0.15 | Mg ionization |
| Gassoumi[42], Okino [43] | AlGaN/GaN HEMT | EC: 0.3/0.34 | Possibly AlGaN surface |
| Heitz [44] | Fe doped GaN | EC: 0.34 | $Fe^{3+/2+}$ |
| Umana-Membreno [30] | n-GaN | EC: 0.355 | Mg impurities |
| Soh [31] | Si doped GaN | EC: 0.37/0.4 | Si dopant |
| Caesar[45], Tapajna[22], Lee [40] | Various GaN-based devices | EC: 0.44/0.49 | C/O/H impurities, possibly in nitrogen substitutional position |
| Cho [35], Cho [33], Umana-Membreno [46], Stuchlikova [47] Ashraf [48] Arehart [37], Chung [49], Chen [39], Fang [50] | Various GaN-based devices | EC: 0.5/0.62 | Nitrogen antisites |
| Polyakov [51] | Fe doped GaN | EC: 0.5 | Fe dopant |
| Legodi [52], Arehart [53] | Various GaN-based devices | EC: 0.57 | AlGaN surface |
| Chen [54] | n+p GaN diode | EC: 0.59 | Si dopant |
| Hacke [55], Hierro [56] | Mg doped p-type GaN | EC: 0.6/0.62 | Mg-H complex formation |
| Stuchlikova [47] | AlGaN/GaN HFET | EC: 0.6/0.64 | VGa + Oxygen complex |
| Jhonstone [34] | n-type GaN | EC: 0.613 | Nitrogen vacancies |
| Okino [43] | AlGaN/GaN MIS-HEMT | EC: 0.68 | Surface |
| Silvestri [16] | Fe-doped AlGaN/ GaN HEMT | EC: 0.72 | Fe dopant |
| Asghar [57] | GaN pn diode | EC: 0.76 | Nitrogen interstitials |
| Polyakov [51], Calleja [58] | Fe-doped or U.I.D. GaN | EV + 0.85, 0.94 | Gallium vacancy |
| Fang [50] | n-type GaN on sapphire | EC: 0.89 | Nitrogen interstitials |
| Asghar [57] | GaN pn diode | EC: 0.96 | Gallium vacancy or N interstitials |
| Auret [59], Fang [10], Fang [60] | Various GaN-based devices | EC: 0.95/1.02 | Threading dislocations |

*(Continued)*

**TABLE 10.1 (*Continued*)**

**Database of the Deep Levels in GaN and Related Materials**

| Reference | Analyzed Samples | Energy Level (eV) | Physical Origin |
|---|---|---|---|
| Stuchlikova [47] | AlGaN/GaN HFET | EC: 1.118 | VGa + Oxygen complex |
| Arehart [37], Arehart [61], Arehart [53] Hierro [56] | Various GaN-based devices | EC: 1.28/1.35 | Carbon interstitial defect |
| Sasikumar [62] | AlGaN/GaN HEMT on SiC | EC: 2.3 | Surface |
| Zhang [63] | c/m plane GaN | EC: 2.47/2.49 | VGa + Hydrogen complex |
| Aggerstam [64] | Fe doped GaN | EV + 2.5/3 | Fe dopant |
| Arehart [37], Arehart [61], Hierro [56] | Various GaN-based devices | EC: 2.6/2.64 | VGa or VGa-H or VGa-2H |
| Arehart [65] | AlGaN (Al 30%) | EC: 3.11 | Cation vacancy |
| Sasikumar [66], Arehart [37], Hierro [56] | Various GaN-based devices | EC: 3.2/3.22 | Residual Mg acceptor |
| Zhang [63], Arehart [37], Sasikumar [62], Arehart [61], Henry [67] | Various GaN-based devices | EC: 3.24/3.31 | CN substitutional |
| Sasikumar [62], Arehart [53] | AlGaN/GaN HEMT | EC: 3.7/3.76 | Mg or C substitutional in AlGaN |
| Arehart [65] | AlGaN (Al 30%) | EC: 3.93 | Mg impurities |

*Courtesy of Carlo De Santi and Davide Bisi, University of Padova.*

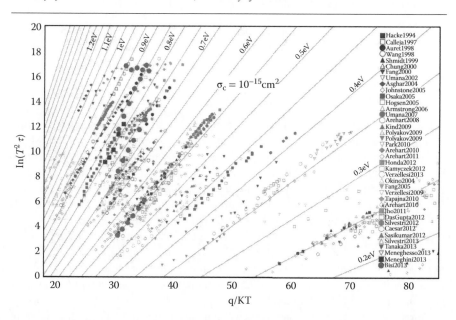

**FIGURE 10.5** Arrhenius plot of the deep levels reported in previous works on GaN-based materials and devices. (Courtesy of D. Bisi, C. De Santi, University of Padova.)

## 10.3  DEGRADATION MECHANISMS OF GaN-BASED HEMTS

Over the last decade, many research groups have extensively studied the degradation of GaN-based HEMTs [18,21, 69–123]; several reliability-limiting mechanisms have been identified, along with the corresponding degradation models. In this section we present an overview of the most important physical processes that limit the reliability of state-of-the-art GaN-HEMTs, by critically comparing experimental data and the results presented in the literature.

More specifically, we focus on the following relevant mechanisms:

1. The degradation of the Schottky junction at the gate, which induces an increase in the gate leakage current. This process has been ascribed to several mechanisms, including converse piezoelectric effect [80,81], electrochemical reactions at the surface [99,100], and defect generation and percolation [102,104].
2. Degradation processes related to hot electrons; these processes may occur when the devices are operated in the semi-on and on-state, and lead to a decrease in drain current (or to an increase in the on-resistance) due to the trapping of electrons in the buffer and/or in the barrier layer.
3. Early breakdown mechanisms that may lead to a gradual or sudden increase in drain current beyond a critical (breakdown) voltage. The existence of high off-state current levels may be related to several processes, such as punch-through [112], surface conduction between gate and drain [115], the breakdown of the gate junction [115], vertical drain-bulk leakage [124], and avalanche processes, initiated by impact ionization [116,117].

These degradation mechanisms are described in detail in the following Section 10.4; a critical review of the data available in the literature is presented throughout the text.

## 10.4  DEGRADATION OF THE SCHOTTKY GATE UNDER REVERSE-BIAS

When GaN-HEMTs are operated in the off-state, the gate-drain Schottky junction is exposed to high reverse bias; for high drain voltages, or for sufficiently long stress times, this can lead to the degradation of the Schottky junction, and to the increase in gate leakage. The reverse-bias degradation of GaN-HEMTs is usually investigated by means of stress tests carried out in reverse-bias conditions (i.e., by applying a reverse-bias to the gate, while keeping drain and source grounded), or in the off-state (i.e., by applying a negative gate bias, and a high drain voltage, with source grounded); Figure 10.6 gives a pictorial representation of these two stress strategies. Since gate-source distance is usually smaller than gate-drain spacing, the first condition results in a significant stress of the gate-source junction; on the other hand, the latter condition is more close to realistic operating conditions, and induces a strong degradation of the gate-drain diode.

In most of the cases, the degradation mechanisms are studied by means of step-stress experiments: the negative gate voltage (in the case of reverse-bias stress) or the positive drain voltage (in the case of off-state stress) is increased repeatedly (e.g., by

5 V every 120 seconds), until failure is reached (see Figure 10.6 for a schematic representation of the step-stress strategy). At each stage of the step-stress experiment, the main electrical parameters of the devices (dc curves, current collapse, transient measurements, electroluminescence signal, etc.) are measured to acquire a complete description of the degradation process.

Several works (see for instance [80,81,87]) demonstrated that, in a step-stress experiment, there is a "critical" voltage beyond which the gate junction starts degrading; this behavior is clearly explained in Figure 10.7a, which reports the results obtained on a representative HEMT submitted to reverse-bias step stress. The (negative) voltage applied to the gate is increased by 5 V every 120 seconds; the gate leakage remains stable up to a stress voltage of −50 V, and shows an abrupt increase after this threshold. The increase in gate leakage is well correlated to the generation of "hot" spots, which can be identified by means of spatially resolved electroluminescence measurements (see representative images in Figure 10.8); these spots indicate the presence of the preferential paths responsible for leakage current conduction; the increase in reverse leakage current occurs in proximity of localized regions, possibly located in proximity of preexisting defective sites [96]. Such defective sites can be identified by atomic force microscopy (AFM) or scanning electron microscopy (SEM), after removing—via chemical etching—the gate metallization from electrically stressed devices [98]. Typical results of this analysis indicate that

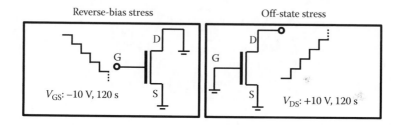

**FIGURE 10.6** Schematic representation of the stress conditions used during reverse-bias stress test (left) and during off-state stress test (right).

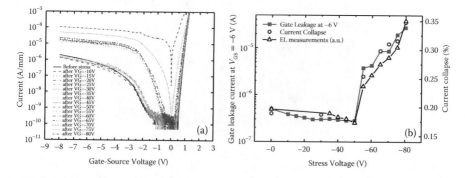

**FIGURE 10.7** (a) Typical current-voltage curves measured on the Schottky diode of an AlGaN/GaN-HEMT submitted to a reverse-bias stress test; (b) variation of gate leakage, electroluminescence signal and current collapse with increasing stress voltage level.

the degradation in the electrical characteristics may be correlated to changes in surface morphology, mostly consisting in the generation of pits and/or linear grooves next to the gate edges. According to Ref. [98], the density and size of the defects depend on stress time and voltage. As a representative example, Figure 10.9 shows typical SEM images collected on two HEMTs stressed at $V_{\text{DGstress}} = 50$ V, for 10 and 1000 minutes. Montes-Bajo et al. [89] demonstrated that there is a one-to-one correspondence between the hot spots detected by electroluminescence (EL) and the surface pits that act at leakage paths (Figure 10.10).

## 10.4.1 CONVERSE PIEZOELECTRIC EFFECT

Earlier works in this field [81,125]suggested that the reverse-bias degradation of HEMTs may be due to converse piezoelectric effect: according to this theory, the exposure to high reverse voltages increases the electric field, and—as a consequence—the tensile stress in the AlGaN barrier (see Figure 10.11a for a schematic representation of this process); beyond a certain (critical) elastic energy, this stress relaxes through the formation of crystallographic defects. Joh et al. [125] developed an effective model for the theoretical calculation of the critical voltage, based on the following steps: (i) calculation of stress, strain and elastic energy as a function of the electric field; (ii) calculation of the electric field at every point in the device, based on

**FIGURE 10.8** EL pattern measured before and after reverse voltage stress on a representative HEMT.

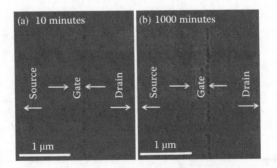

**FIGURE 10.9** SEM images of the pits after stressing at $V_{\text{DGstress}} = 50$ V for (a) 10 minutes and (b) 1000 minutes. (Reprinted with permission from P. Makaram, J. Joh, J. A. del Alamo, T. Palacios, and C. V. Thompson, "Evolution of structural defects associated with electrical degradation in AlGaN/GaN high electron mobility transistors," *Appl. Phys. Lett.* 96, 233509, Copyright 2010, AIP Publishing LLC [98].)

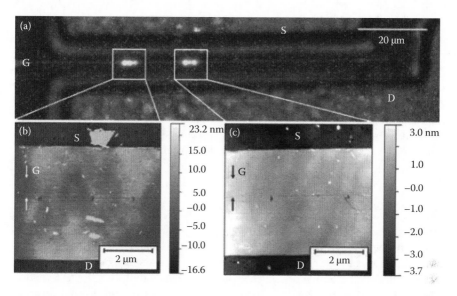

**FIGURE 10.10** Electroluminescence image of a 100-μm-wide HEMT stressed at $V_{DS}$ = 40 V and $V_{GS}$ = −15 V for 4 minutes. (b) and (c) AFM images of the areas inside the squares of (a) after removal of the passivation and contacts. The top and bottom dark bands in the images are the regions where the source and drain contact edges used to be before removal. The thickness and location of the gate contact before removal is indicated with arrows. The two light gray horizontal lines on both sides of the location of the gate contact in (b) are AFM artefacts. (Reprinted with permission from M. Montes Bajo, C. Hodges, M. J. Uren, and M. Kuball, "On the link between electroluminescence, gate current leakage, and surface defects in AlGaN/GaN high electron mobility transistors upon off-state stress," Appl. Phys. Lett., 101, 033508, Copyright 2012, AIP Publishing LLC [89].)

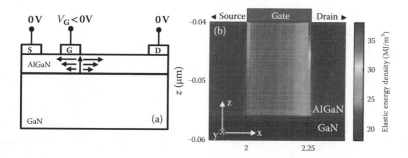

**FIGURE 10.11** (a) Schematic representation of the degradation process according to the converse piezoelectric effect described in Ref. [125]. (b) Elastic energy density in AlGaN and GaN layers of a GaN-HEMT in the off-state at $V_{GS}$ = −5 V and $V_{DS}$ = 33 V. (Reprinted from *Microelectronics Reliability*, Vol 5(6), J. Joh, F. Gao, T. Palacios, J. A. del Alamo, A model for the critical voltage for electrical degradation of GaN high electron mobility transistors, *Microelectronics Reliability*, 767–773, Copyright 2010, with permission from Elsevier [125].)

bi-dimensional electrostatic simulations; (iii) calculation of the elastic energy density at every point in the AlGaN layer; and (iv) extrapolation of the critical voltage based on the comparison between the results obtained in (iii) and the critical elastic energy derived from studies on the epitaxial growth. The results of this analysis are summarized in Figure 10.11b: the elastic energy density peaks close to the gate edge (on the drain side), which is the region where the electric field is maximum; in their example, the elastic energy density increases from 0.31 J/m$^2$ (at zero bias) to 0.46 J/ m$^2$ at the critical voltage. According to Joh and del Alamo these values are similar to the critical elastic energy estimated from material growth data and may lead to the degradation of the devices.

## 10.4.2 Defect Generation and Percolation

Most of the studies quoted above used a step-stress strategy to investigate the degradation of AlGaN/GaN-HEMTs submitted to reverse (or off-state) bias; this strategy provides a quick description of the robustness of a given technology, but it does not give any information on the time dependence of the degradation process. Recently, other research groups [101,104,126] carried out constant voltage experiments to investigate the degradation kinetics of AlGaN/GaN-HEMTs submitted to reverse-bias stress. In most of the cases, devices were stressed at voltages smaller than the "critical voltage" identified by step-stress experiments. A representative example is reported in Figure 10.12: the sample in the figure has a "critical voltage" (estimated by a step-stress experiment) of −40 V (Figure 10.12a); however, this sample can degrade even when it is stressed at very low voltage (e.g., $V_{GS}$ = −15 V), as shown in Figure 10.12b: for long stress time, a sudden, and permanent, increase in gate leakage current is observed.

Several works [68,104,126] pointed out that reverse-bias degradation is a time-dependent process; time-to-breakdown ($t_{bd}$) is significantly dependent on the stress voltage level [127], and degradation may occur even below the "critical voltage" extrapolated from step-stress experiments. For this reason, rather than evaluating the "critical

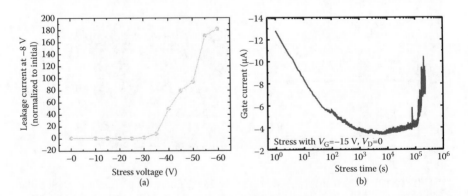

**FIGURE 10.12** (a) Results of a step-stress experiment carried out on an AlGaN/GaN-HEMT; the "critical voltage" of this technology (estimated by a step-stress test) is equal to −40 V. (b) Results of a constant voltage stress test carried out on a sample identical to the one in (a). The stress voltage is $V_{GS}$ = −15 V, and is significantly smaller than the critical voltage.

voltage" of a given technology, it is important to understand how devices behave for long (reverse-bias) stress times, and to study the dependence of time-to-breakdown on the stress voltage level.

The time-dependent increase in leakage current induced by reverse-bias stress has a weak dependence on temperature [127]; the failure data follow a Weibull distribution [127]. This time-dependent degradation was ascribed to a defect generation and percolation process. According to this model, defects are initially formed close to the AlGaN surface; long-term stress results in the propagation of defects within the AlGaN barrier; this eventually results in the generation of preferential paths responsible for leakage current conduction.

### 10.4.3   ELECTROCHEMICAL REACTIONS AT THE SURFACE

The generation of surface pits as a consequence of reverse-bias stress was found to be correlated also to electrochemical reactions occurring at the surface of the devices. By studying the degradation of AlGaN/GaN-HEMTs, Gao et al. [99,100] demonstrated the generation of elongated particles (stringers) on device surface, in proximity of the gate edge (see Figure 10.13b and c). Auger electron spectroscopy (AES) was used to determine the chemical composition of these stringers: the results indicated the presence of gallium, aluminum, and oxygen (Figure 10.13d). The semi-quantitative analysis of the atomic concentration indicated a ratio between the concentration of Al/Ga and the concentration of oxygen equal to 2:3, suggesting that reverse-bias stress induced the generation of $Ga_2O_3$ and $Al_2O_3$. AFM measurements carried out after the removal of these oxide particles revealed the presence of pits, with depth in the range between 25 and 45 nm (Figure 10.13f). These pits are localized at the gate-edge, i.e., in proximity of the regions where the electric field is maximum; the oxidation process is therefore supposed to be accelerated by the electric field. The role of oxygen was confirmed by comparing the behavior of devices stressed in vacuum and in air. In commercial samples, O-related degradation can be prevented by using thick passivation layers and proper device packaging. However, residual oxygen can still be present on the surface of GaN-HEMTs: previous works reported the presence of native oxide with a thickness of few nanometers (1–2 nm) [128]. The generation of oxide may further accelerate the breaking of Al/Ga-N bonds, thus generating nitrogen vacancies or gas; this may result in a further acceleration of the surface oxidation process [99].

## 10.5   HOT-ELECTRON-INDUCED DEGRADATION

The degradation mechanisms described above are triggered and accelerated by the high off-bias applied to the devices. Recent studies (see for instance [18,68]) indicated that AlGaN/GaN-HEMTs may significantly degrade also when they are submitted to semi-on and on-state stress. Under these conditions—if the drain voltage is sufficiently low to prevent the reverse-bias degradation of the gate Schottky junction—the electrical parameters of the devices may degrade due to hot-electron-induced damage. The main consequences of stress can be a shift of the threshold voltage, an increase in the on-resistance, and a reduction of the transconductance.

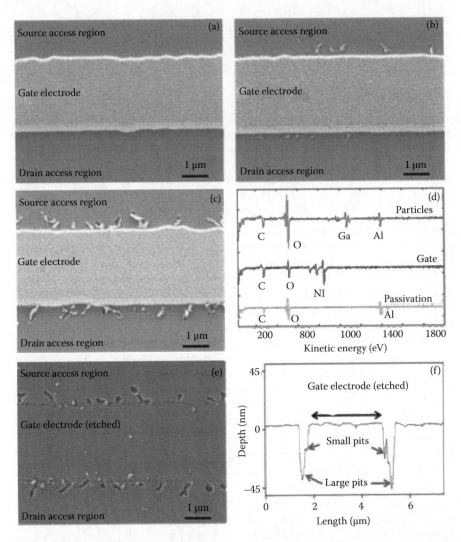

**FIGURE 10.13** Top-view SEM images of an AlGaN/GaN-HEMT stressed at $V_{DS} = 0$ V and $V_{GS} = 40$ V for 60 seconds (a), 600 seconds (b), and 6000 seconds (c). (d) Auger electron spectra results for three different regions of the transistor surface. Panel (e) shows the surface morphology of the device (c) after removal of the gate, and panel (f) shows an AFM depth profile across the gate electrode of the same device. (Reprinted with permission from F. Gao, B. Lu, L. Li, S. Kaun, J. S. Speck, C. V. Thompson, and T. Palacios, "Role of oxygen in the OFF-state degradation of AlGaN/GaN high electron mobility transistors," *Appl. Phys. Lett.* 99, 223506 (2011), Copyright 2011, AIP Publishing LLC [99].)

In GaN devices, the presence of hot electrons is probed by measuring the gate leakage current; this is not possible in GaN-based HEMTs, since gate leakage is dominated by other components, such as tunneling through the AlGaN barrier [129]. For this reason, hot electron effects are usually characterized by means of electroluminescence measurements: when biased in the on- (or semi-on) state, HEMTs may emit a weak luminescence signal, which is proportional to the concentration of electrons in the channel (and therefore to drain current $I_D$), and strongly dependent on the accelerating field. The EL emission is a consequence of the relaxation of highly energetic electrons, which are accelerated by the electric field; light emission may occur either due to inter-valley or intra-valley transitions [130]. In the first case, the electrons are injected to the satellite valleys of the conduction band; light is emitted when the electrons relax back to the Γ valley (the interaction with optical phonons guarantees energy conservation). In the case of intra-valley transitions, the electrons may reach a high energy nonequilibrium condition within the Γ valley itself; photons are emitted when these hot electrons relax to the bottom of the conduction band, via transitions that may involve phonons or ionized impurities. Figure 10.14 reports a typical luminescence pattern measured on a GaN-based HEMT biased in the on-state, and a schematic representation of the mechanisms responsible for light emission in GaN-HEMTs.

Recent works [131–134] investigated—based on Monte Carlo analysis—the energetic distribution of electrons in HEMTs biased in the on-state. Puzyrev et al. [132] simulated AlGaN/GaN-HEMTs with a 25-nm-thick barrier (%Al = 32%) placed on top of 175 nm of GaN. Their simulations indicated that the electric field has its (negative) maximum at the gate edge, on the drain side: with a drain voltage of 10 V, they estimated a longitudinal electric field in excess of $-1.5 \times 10^6$ V/cm, for sufficiently low gate voltages ($V_{GS} = -3$ V). In correspondence of the peak of the electric field (and under the same bias conditions), the average electron energy was estimated to be around 2 eV; in addition, the carrier versus energy distribution has a large tail extending beyond 3 eV.

According to Puzyrev et al. [132], these energy levels exceed the activation energies for the generation/modification of defects within the devices: Puzyrev et al. indicated that dehydrogenation of preexisting defects, such as Ga-vacancies and N-antisites, may play a significant role in the on-state degradation of GaN-HEMTs.

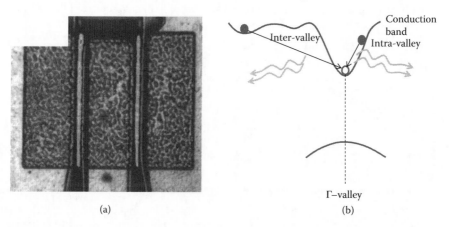

(a)  (b)

**FIGURE 10.14** (a) Typical EL pattern measured in on-state conditions on a GaN-HEMT; (b) schematic representation of the main mechanisms responsible for EL in GaN-based HEMTs, according to the description given in Ref. [130]. (From Gutle, F., Semicond. Sci. Technol. 27, 125003, 2012.)

The dehydrogenation of point defects may explain both the degradation of the transconductance of the devices, and the increase in yellow luminescence [131,135]. The impact of dehydrogenation on the threshold voltage shift is strongly dependent on the nature of the defect: the dehydrogenation of nitrogen antisite defects induces a decrease in the acceptor concentrations, while the dehydrogenation of GaN vacancies leads to an increase in the acceptor concentration, resulting in opposite effects on the threshold voltage shift (see [131] for details).

Meneghini et al. [68,136] demonstrated that on-state stress may induce a significant increase in the on-resistance of AlGaN/GaN-HEMTs (Figure 10.15a), which is well correlated to a decrease in the $EL/I_D$ signal (Figure 10.15b). This last parameter is directly proportional to the electric field in the channel: the results in Figure 10.15 can be interpreted by considering that stress induces the semi-permanent trapping of electrons in the gate-drain access region, thus resulting in an increase in channel resistance and in the reduction of the electric field. By carrying out stress tests in the semi-on and on-state, Meneghini et al. [136] found that there is a direct correlation between the degradation rate and the intensity of the EL signal measured before stress (which is proportional to the number of hot electrons in the channel), see Figure 10.16b and c. This result indicates that hot electrons play an important role in the degradation process; the worsening in the electrical properties of the devices (Figure 10.15a) can be therefore explained by considering that, at high drain voltages, electrons may achieve enough energy to be injected into the AlGaN barrier, at the AlGaN surface or the passivation layer, thus remaining quasi-permanently trapped there. This process results in the increase in the on-resistance, and in the decrease of the electric field. It is noteworthy that an increase in channel temperature does not significantly change the degradation kinetics for hot-electron stress: in Figure 10.16b it is shown that devices stressed at 75°C and 125°C show a degradation rate similar to devices stressed at room temperature. This confirms that degradation rate is mostly determined by the amount of hot electrons in the channel, and does not strongly depend on stress temperature. Silvestri et al. [17] characterized the effect of hot carrier stress by means of dynamic transconductance measurements, to acquire a

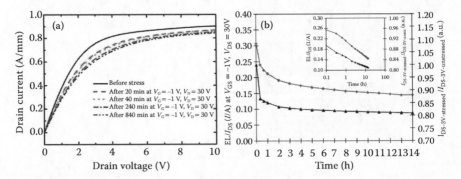

**FIGURE 10.15** (a) $I_D$–$V_D$ curves measured at increasing stress times on a device submitted to on-state operation. (b) Decrease in the on-resistance and $E_L/I_D$ signal measured during a stress test with $V_{DS} = 30$ V, $V_{GS} = -1$ V [136]. (From Meneghini, M et al.,. IEEE Transactions on Device and Materials Reliability, 13, 2, 357–361, 2013.)

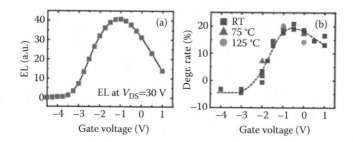

**FIGURE 10.16**   (a) EL signal measured on a similar device before stress, with a drain voltage of 30 V. (b) Dependence of the degradation rate on the gate voltage level applied during stress (for all the devices the drain voltage applied during stress is equal to 30 V) [68]. (From Meneghini, M. et al., *Appl. Phys. Lett.* 100, 233508, 2012.)

more detailed description of the degradation process: according to their conclusions, the damage generated after stress may be located at the AlGaN surface, or—in any case—far from the channel.

## 10.5.1   BREAKDOWN IN GaN-HEMTs

Gallium nitride transistors are expected to play an important role in the next-generation power converters, thanks to the high expected breakdown voltage, which is a direct consequence of the high breakdown field of GaN (>300 V/μm). However, several authors (see for instance [21,112,113,118,124]) demonstrated that the breakdown voltage of GaN-HEMTs can be significantly lower than what is predicted by theoretical calculations: several mechanisms may lead to a gradual or sudden increase in drain current, when the devices are exposed to high drain bias in the off-state. In this section, we review the main mechanisms responsible for off-state current conduction and breakdown.

In most of the cases, the breakdown voltage is defined as the drain voltage level required to reach a given drain current (e.g., 1 mA/mm), with the device biased in the off-state. Several mechanisms may lead to a gradual or abrupt increase in drain current when the drain voltage exceeds a critical level: these mechanisms are discussed in the following.

Figure 10.17 reports the typical drain current versus drain voltage curves measured in the off-state for a GaN-based power MIS-HEMT. To ensure a repeatable and nondestructive characterization, these curves were measured in current controlled mode, by means of a semiconductor parameter analyzer. This approach—which is similar to what proposed by Bahl et al. [137]—prevents a rapid and uncontrolled increase in drain current during the measurements; on the other hand, the conventionally used voltage-controlled characterization may lead to the catastrophic failure of the devices due to the rapid increase in drain current beyond the breakdown voltage. The results in Figure 10.17a indicate that for gate voltages relatively close to the pinch-off voltage (and in the off-state) the subthreshold drain current originates almost completely from drain-source current; this effect is due to the nonoptimized

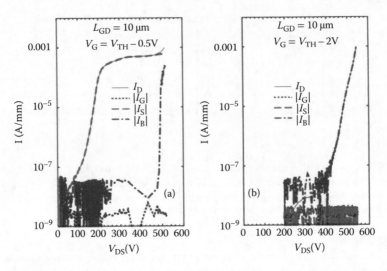

**FIGURE 10.17** Current–voltage measurements measured in the off-state on a GaN-based power MIS-HEMT at two gate voltage levels: (a) $V_{GS} = V_{TH}$-0.5 V and (b) $V_{GS} = V_{TH}$-2 V.

depletion of the channel region, that may result in significant punch-through current components. This process was deeply investigated by Uren et al. [112], based on bi-dimensional simulations of single-heterostructure (SH) HEMTs with a 0.17-μm gate length; the results in Ref. [112] indicate that HEMTs submitted to high drain voltages may show significant short-channel effects, that can be due to punch-through in the GaN buffer. When the device is in the off-state, a measurable subthreshold current may flow deep within the GaN buffer, far from the AlGaN/GaN heterostructure. This effect is particularly prominent for low values of net doping in the buffer (e.g., $1.5 \times 10^{16}$ cm$^{-3}$): under these conditions (and with a gate voltage of −6 V and a drain voltage of 20 V) the conduction band minimum is approximately 200 nm below the AlGaN/GaN heterostructure, and this may favor the flow of significant subthreshold current components. Higher net doping concentrations (e.g., $2 \times 10^{17}$ cm$^{-3}$) can effectively suppress the subthreshold leakage; under these conditions, the channel is pinched-off even when the device is exposed to high drain voltages (see Ref. [112] for more details).

The punch-through current components can be significantly reduced through the adoption of a double-heterostructure (DH) epitaxy: in this case, the AlGaN/GaN heterostructure is grown on top of an AlGaN buffer, whose presence generates a back-barrier that prevents the injection of electrons toward the buffer (see for instance Ref. [113]): the thin channel region behaves as a potential well, that effectively confines the carriers in the GaN layer. Bahat-Treidel et al. [113] simulated the band diagram and electron concentrations in SH and DH HEMTs: the results indicated that in the SH device, although the region under the gate is fully depleted, electrons can travel deep into the buffer, and form a conductive (leaky) path between drain and source. This effect is significantly reduced in DH HEMTs, thanks to the confining barrier of the GaN/buffer heterostructure.

Punch-through current components can be effectively suppressed also by using more negative gate voltages: the comparison between Figure 10.17a and b indicates that by using a gate voltage 2–3 V smaller that the threshold, an almost complete suppression of the drain-source off-state leakage can be achieved (the device in Figure 10.17 is an HEMT with AlGaN backbarrier). Once drain-source leakage is suppressed, the off-state (drain) current can be dominated either by drain-bulk vertical leakage (as in the case of Figure 10.17b) or by the reverse leakage of the gate-drain junction.

Drain-bulk leakage originates from a poor compensation of the buffer, and from the presence of extended defects within the nucleation layer and the buffer. Buffer-related leakage components were extensively studied by Perez-Tomàs et al. [138], who evaluated the behavior of three different substrates: Si (111), c-sapphire ($\alpha$-Al$_2$O$_3$), and GaN (0001). AlGaN/GaN layers were grown by molecular beam epitaxy on top of these substrates to compare the vertical drain-bulk leakage (Figure 10.18): I-V characterization was carried out in a wide temperature range, to extract information on the dominant conduction mechanisms (see Figure 10.19). The results in Ref. [138] demonstrate that (i) at room temperature, the wafer grown on Si-substrate has the largest bulk leakage (0.1–10 $\mu$A for a drain-bulk voltage of $V_{db} = 0$–150 V); the use of free-standing GaN results in drain-bulk leakages in the range $10^{-2}$–w$10^{-1}$ $\mu$A, while, in the moderate voltage range ($V_{db}$<150 V), the vertical leakage of samples with sapphire substrate was found to be negligible. (ii) For all the analyzed substrates, drain-bulk leakage is thermally activated with activation energies in the range 0.10–0.35 eV for the samples grown on free-standing GaN and on silicon. (iii) The vertical current of HEMTs can be reproduced by considering a double conduction mechanism, constituted by a resistive component ($I_R$) and a Poole–Frenkel component ($I_{PF}$). The device with GaN substrate was found to follow

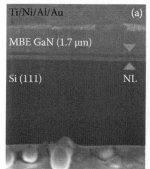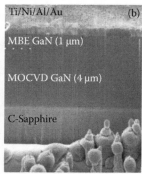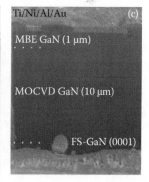

**FIGURE 10.18** SEM micrograph of three different GaN buffer cross sections (a) on silicon, (b) on sapphire, and (c) on FS-GaN. The top Ohmic contact is made of Ti/Ni/Al/Au. The back contact was made directly contacting the different wafers with the chuck of the probe station. NL is the Si/GaN nucleation layer. (Reprinted with permission from A. Pérez-Tomás, A. Fontserè, J. Llobet, M. Placidi, S. Rennesson, N. Baron, S. Chenot, J. C. Moreno, and Y. Cordier. "Analysis of the AlGaN/GaN vertical bulk current on Si, sapphire, and free-standing GaN substrates," *J. Appl. Phys.*, 113, 174501, Copyright 2013, AIP Publishing LLC [138].)

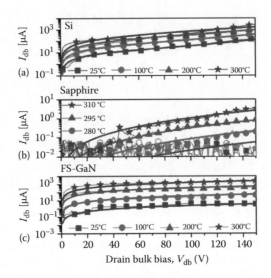

**FIGURE 10.19** AlGaN/GaN/substrate vertical drain-bulk current versus drain-bulk voltage at varying temperatures for the GaN buffer grown (a) on silicon, (b) on sapphire, and (c) on FS-GaN. Note that the leakage current is negligible for the sapphire heterostructure for T lower than 280°C. Solid lines are the simulated current at each T using the resistive and trap assisted current model (Reprinted with permission from A. Pérez-Tomás, A. Fontserè, J. Llobet, M. Placidi, S. Rennesson, N. Baron, S. Chenot, J. C. Moreno, and Y. Cordier. "Analysis of the AlGaN/GaN vertical bulk current on Si, sapphire, and free-standing GaN substrates," *J. Appl. Phys.* 113, 174501, Copyright 2013, AIP Publishing LLC [138].)

only the resistive mechanism, while sapphire substrates follow only Poole–Frenkel trap-assisted conduction. Finally, the samples with Si substrate were found to follow a combination of the two mechanisms. (iv) By comparing the bulk leakage versus temperature curves, Perez Tomàs et al. [138] concluded that at room temperature the most conductive structures are those grown on Si; however, this situation changes for higher temperatures. According to their data, at temperatures higher than 250°C the GaN-on-GaN devices showed the highest drain-bulk currents.

The vertical buffer leakage is strongly influenced by the thickness of the buffer: Rowena et al. [124] investigated the dependence of vertical breakdown voltage on the thickness of the undoped i-GaN layer ($T_{GaN}$) and of the buffer ($T_{buf}$), using silicon as a substrate. Their results indicated that a significant improvement of the breakdown voltage can be obtained by growing i-GaN on a thick buffer. They demonstrated that for a given buffer thickness (1.25 μm), there is only a small increase in breakdown as the thickness of GaN is increased from 0.5 to 1.0 μm. On the other hand, when the thickness of GaN is fixed (e.g., at 0.5 μm), a significant variation in the breakdown voltage can be obtained by increasing the thickness of the buffer from 1.25 to 4 μm. Based on these results, Rowena et al. concluded that a way to improve the vertical breakdown voltage is to grow i-GaN on thick buffers; moreover, the use of thick GaN buffer results in a reduction of the screw and edge dislocation density. By plotting the vertical breakdown voltage as a function of the total epilayer thickness ($T_{GaN} + T_{buf}$), the breakdown field was calculated to be equal to 2.3 MV/cm (the theoretical value is approximately 3 MV/cm).

High breakdown voltages can be obtained also by means of substrate-transfer technology [121,139,140]; since—as discussed below—the use of Si substrate can limit the breakdown voltage of GaN-HEMTs, a significant improvement of the breakdown voltage can be obtained by removing the Si (111) substrate and transferring the epitaxial material on a foreign substrate. Lu et al. [139] demonstrated a breakdown voltage of more than 1500 V with only a 2 μm total epitaxial thickness, by transferring the wafer to a glass substrate. Even better results can be obtained by using substrates with higher thermal conductivity, such as polycrystalline aluminum nitride. Another interesting option was presented by Srivastava et al. [140]: instead of removing the whole substrate, they proposed to enhance the breakdown voltage by locally removing the silicon substrate only under the source-to-drain region. The removal of Si was done by using an $SF_6/C_4F_8$ ICP plasma etching, and an Al hard mask layer. This method, that can be applied also on large wafer sizes; a breakdown voltage of 2200 V was obtained with devices with $L_{GD} = 20$ μm and a buffer thickness of 2 μm. This is significantly higher than the breakdown voltage of 700 V measured before the removal of Si (111) on devices with $L_{GD} > 8$ μm. A similar approach was used also by Herbecq et al. [121].

A different solution was proposed by Umeda et al. [123]: the sheet of electrons at the AlN/Si interface acts as an inversion layer, and can generate a significant leakage current. To reduce the current components associated to this inversion layer, Umeda et al. proposed to use two p+ regions (referred to as "channel stoppers") at the edge of the silicon substrate. In this way, the depletion layer is widened in the Si substrate—for high drain biases—and the total blocking voltage is significantly increased. Based on this approach, Umeda et al. were able to report breakdown voltages of 1340 V with a GaN thickness of 1.4 μm.

High off-state drain current may originate also from the leakage of the reverse-biased gate-drain junction. The breakdown of the gate junction may be related to several factors: as proposed by Tan et al. [115], surface states may generate a significant current conduction, possibly due to hopping of electrons injected from the gate metal to the surface. When the power dissipation of these surface current paths reaches a critical level, breakdown may occur due to thermal runaway [115]. A confirmation of the role of surface states was given by Saito et al. [129]: they suggested that the presence of positive charge (e.g., nitrogen vacancies) at the surface of the AlGaN layer may increase the electric field over the barrier, thus favoring tunneling and impact ionization processes. As a consequence, breakdown voltage decreases with increasing concentration of charged surface defects. Other mechanisms that may lead to high gate-drain leakage are thermionic emission at the Schottky gate [114], Poole–Frenkel emission [141], defect-assisted tunneling [141], and the existence of leakage paths related to the presence of extended defects [96]. Gate leakage current components can be significantly reduced through the use of advanced passivation schemes (e.g., MOCVD or MBE nitride as in [142,143]), or by using the Metal–Insulator–Semiconductor (MIS) approach [144].

The off-state conduction mechanisms described above (punch through, substrate leakage, etc.) induce a gradual increase in drain current, when drain voltage approaches the breakdown. A more abrupt variation in drain current is obtained when avalanche processes are taken into account. Recent works [116,117] proposed that at high drain voltages also impact ionization can play a significant role, and induce a significant increase in breakdown current. Impact ionization is usually simulated by considering that the carrier generation rate (due to impact ionization) is given by

$$G = \frac{\left( \alpha_n |J_n| + \alpha_P |J_P| \right)}{q} \tag{10.7}$$

where $J_n$ and $J_P$ are the electron and hole current densities, respectively, $\alpha_n$ and $\alpha_P$ are the ionization rates for electrons and holes, and $q$ is the electron charge. $\alpha_n$ and $\alpha_P$ can be expressed as

$$\alpha_n = A_n \exp\left( -\frac{B_n}{|E|} \right) \tag{10.8}$$

$$\alpha_P = A_n \exp\left( -\frac{B_P}{|E|} \right) \tag{10.9}$$

Here $E$ is the electric field, while $A_n$, $B_n$, $A_P$, and $B_P$ are coefficients whose values can be deduced from the theoretical work by Bulutay [145]. Hanawa et al. [117] proposed that at high drain bias the holes generated by impact ionization can flow into the buffer, thus being captured by deep donors. This may imply a decrease in the barrier for the injection of electrons to the buffer, and induce an increase in drain current. It is worth noticing that, although predictable at theoretical level, the role of impact ionization is still under investigation.

## CONCLUSIONS

Within this chapter, we have described the main physical mechanisms that limit the performance and the reliability of GaN-based high-electron mobility transistors. In Section 10.2, starting from the definition of current collapse, we have critically compared the methods for the analysis of the properties of the defects in AlGaN/GaN heterostructures, by providing general guidelines for a correct estimation of trap parameters. C-DLTS, DLOS, dynamic transconductance measurements, and DCT investigation have been described and analyzed. Moreover, we have presented a summary of the properties of defects in GaN-based material and devices, which can be useful for correlating the signature of the traps identified by transient measurements with specific defects (either native, or related to foreign impurities).

In Section 10.3, we have discussed the most critical mechanisms that limit the lifetime of GaN-based HEMTs, namely: (i) the degradation of the gate contact, due to the exposure to high off-state bias; the role of converse piezoelectric effect, defect generation and percolation, and electrochemical reactions was extensively discussed based on previous works on this topic; (ii) the degradation mechanisms activated by hot electrons, which can be prominent during semi-on and on-state operation; and (iii) the physical mechanisms that limit the breakdown voltage of GaN-based HEMTs, with focus on punch-through, gate-drain breakdown, vertical leakage, and impact ionization.

The results summarized within this chapter point out that the improvement of the performance and of the reliability of GaN-based transistors requires an integrated approach; on one hand, the growth and processing conditions must be optimized with the aim of reaching high frequency/power operation, and high breakdown voltages. On the other hand, the trapping and degradation mechanisms must be carefully studied, based on specific characterization techniques and stress tests, with the aim of correlating the degradation mechanisms to growth/process parameters, or to the substrate/doping used for the fabrication of the devices. This requires a close collaboration between the industrial/academic laboratories involved in device fabrication, and the research centers dedicated to the study of the parasitic and degradation mechanisms of GaN-based transistors.

## REFERENCES

1. M. Esposto, V. di Lecce, M. Bonaiuti, and A. Chini, "The Influence of Interface States at the Schottky Junction on the Large Signal Behavior of Copper-Gate GaN HEMTs," *J Electron Mater,* 42, 15 (2013).
2. J. Joh and J. A. del Alamo, "A Current-Transient Methodology for Trap Analysis for GaN High Electron Mobility Transistors," *IEEE Transactions on Electron Devices,* 58, 132 (2011).
3. M. Uren, M. Silvestri, M. Casar et al., "Intentionally Carbon-Doped AlGaN/GaN HEMTs: Necessity for Vertical Leakage Paths," *IEEE Electron Device Lett,* 35, 327 (2014).
4. G. Meneghesso, M. Meneghini, D. Bisi et al., "Trapping phenomena in AlGaN/GaN HEMTs: a study based on pulsed and transient measurements," *Semicond Sci Technol,* 28, 074021 (2013).
5. C. Ostermaier, P. Lagger, M. Alomari et al., "Reliability investigation of the degradation of the surface passivation of InAlN/GaN HEMTs using a dual gate structure," *Microelectron Reliab,* 52(9–10), 1812–1815 (2012).
6. C.-Y. Hu and T. Hashizume, "Non-localized trapping effects in AlGaN/GaN heterojunction field-effect transistors subjected to on-state bias stress," *J Appl Phys,* 111, 084504 (2012).
7. K. Tanaka, M. Ishida, T. Ueda, and T. Tanaka, "Effects of deep trapping states at high temperatures on transient performance of AlGaN/GaN heterostructure field-effect transistors," *Japanese J Appl Phys,* 52, 04CF07 (2013).
8. D. V. Lang, "Deep-level transient spectroscopy: a new method to characterize traps in semiconductors," *J Appl Phys,* 45, 3023 (1974).
9. Z.-Q. Fang, B. Claflin, D. C. Look, D. S. Green, and R. Vetury, "Deep traps in AlGaN/GaN heterostructures studied by deep level transient spectroscopy: Effect of carbon concentration in GaN buffer layers," *J Appl Phys,* 108, 063706 (2010).
10. Z.-Q. Fang, D. C. Look, D. H. Kim, and I. Adesida, "Traps in AlGaN GaN SiC heterostructures studied by deep level transient spectroscopy," *Appl Phys Lett,* 87, 182115 (2005).
11. P. M. Mooney, "Defect identification using capacitance spectroscopy," *Semiconduct semimet,* 51B, 93 (1999).

12. A. Armstrong, A. R. Arehart, and S. A. Ringel, "A method to determine deep level profiles in highly compensated, wide band gap semiconductors," *J Appl Phys*, 97, 083529 (2005).

13. A. M. Armstrong, A. A. Allerman, A. G. Baca, and C. A. Sanchez, "Sensitivity of on-resistance and threshold voltage to buffer-related deep level defects in AlGaN/GaN high electron mobility transistors," *Semicond Sci Technol*, 28, 074020 (2013) .

14. H. Haddara, T. Elewa, and S. Cristoloveanu, "Static and dynamic transconductance model for depletion-mode transistors: A new characterization method for silicon-on-insulator materials," *IEEE Electron Device Lett*, 9, 35 (1988).

15. E. H. Nicollian and J. R. Brews, *MOS Physics and Technology*. New York, NY: Wiley, 1982.

16. M. Silvestri, M. J. Uren, and M. Kuball, "Iron-induced deep-level acceptor center in GaN/AlGaN high electron mobility transistors: Energy level and cross section," *Appl Phys Lett*, 102, 073501 (2013).

17. M. Silvestri, M. J. Uren, and M. Kuball, "Dynamic transconductance dispersion characterization of channel hot-carrier stressed 0.25-μm AlGaN/GaN HEMTs," *IEEE Electron Device Lett*, 33, 1550 (2012).

18. A. Sozza, C. Dua, E. Morvan et al., "Evidence of traps creation in GaN/AlGaN/GaN HEMTs after a 3000 hour on-state and off-state hot-electron stress," IEEE International Electron Device Meeting 2005, p. 593.

19. D. Bisi, M. Meneghini, C. de Santi, A. Chini, M. Damman P. Brueckner, et al. "Deep-level characterization in GaN HEMTs—Part I: Advantages and limitations of drain current transient measurements," *IEEE Trans Electron Devices*, 60, 3166 (2013).

20. A. Chini, F. Soci, M. Meneghini, G. Meneghesso, and E. Zanoni, "Deep level characterization in GaN HEMTs—Part II: Experimental and numerical evaluation of self-heating effects on the extraction of traps activation energy," *IEEE Trans Electron Devices*, 60, 3176 (2013).

21. M. Meneghini, D. Bisi, D. Marcon et al., "Trapping and reliability assessment in D-mode GaN-based MIS-HEMTs for power applications," *IEEE Transactions Power Electron*, 29(5), 2199–2207 (2014).

22. M. Tapajna, R. J.T. Simms, Yi Pei, U. K. Mishra, and M. Kuball, "Integrated optical and electrical analysis: Identifying location and properties of traps in AlGaN/GaN HEMTs during electrical stress," *IEEE Electron Device Lett*, 31, 662 (2010).

23. W. R. Thurber, R. A. Forman, and W. E. Phillips, "A novel method to detect nonexponential transients in deep level transient spectroscopy," *J Appl Phys*, 53(11), 7397 (1982).

24. O. Mitrofanov and M. Manfra, "Mechanisms of gate lag in GaN/AlGaN/GaN high electron mobility transistors," *Superlattices and Microstructures*, 34(1–2), 33–53, (2003).

25. P. D. Kirchner, W. J. Schaff, G. N. Maracas, L. F. Eastman, T. I. Chappell, and C. M. Ransom, "The analysis of exponential and nonexponential transients in deep-level transient spectroscopy," *J Appl Phys*, 52(11), 6462 (1981).

26. F. Soci, A. Chini, G. Meneghesso, M. Meneghini, and E. Zanoni, "Influence of device self-heating on trap activation energy extraction," IEEE International Reliability Physics Symposium Proceedings, art. no. 6531988, 3C.6.1, Monterey, CA, 2013.

27. I. Rossetto, M. Meneghini, T. Tomasi, D. Yufeng, G. Meneghesso, and E. Zanoni, "Indirect techniques for channel temperature estimation of HEMT microwave transistors: Comparison and limits," *Microelectron Reliab*, 52, 2093–2097 (2012).

28. S. P. McAlister, "Self heating and the temperature dependence of the dc characteristics of GaN heterostructure field effect transistor," *JVST*, 24(3), 624–628 (2006).

29. J. Joh and J. Del Alamo, "Measurement of channel temperature in GaN high electron mobility transistors," *IEEE Trans Electron Devices*, 56, 2895–2901 (2009).

30. A. Umana-Membreno, J. M. Dell, T. P. Hessler, B. D. Nener, G. Parish, and L. Faraone, "60Co gamma-irradiation-induced defects in n-GaN," *Appl Phys Lett*, 80(23), 4354–4356 (2002).

31. C. B. Soh, S. J. Chua, H. F. Lim, D. Z. Chi, W. Liu, and S. Tripathy, "Identification of deep levels in GaN associated with dislocations," *J Phys Condens Matter*, 16(34), 6305–6315 (2004).

32. Y S. Park, M. Lee, K. Jeon et al., "Deep level transient spectroscopy in plasma-assisted molecular beam epitaxy grown AlGaN/GaN interface and the rapid thermal annealing effect," *Appl PhysLett*, 97, 112110 (2010).

33. H.K. Cho, K.S. Kim, C.-H. Hong, and H.J. Lee, "Electron traps and growth rate of buffer layers in unintentionally doped GaN," *J Crystal Growth*, 223(1–2), 38–42 (2001).

34. D. K. Johnstone, M. Ahoujia, Y. K. Yeo, R. L. Hengehold, and L. Guido, "Deep Centers and their capture barriers in MOCVD-grown GaN," *Mat Res Soc Symp Proc*, 73–83 (2002).

35. H. K. Cho, C. S. Kim, and C.-H. Hong, "Electron capture behaviors of deep level traps in unintentionally doped and intentionally doped n-type GaN," *J Appl Phys*, 94(3), 1485, 1489 (2003).

36. K. J. Choi, H. W. Jang, and J. L. Lee, "Observation of inductively coupled-plasma-induced damage on n-type GaN using deep-level transient spectroscopy," *Appl Phys Lett*, 82, 1233 (2003).

37. R. Arehart, A. Corrion, C. Poblenz et al., "Comparison of deep level incorporation in ammonia and rf-plasma assisted molecular beam epitaxy n-GaN films," *Phys Stat Sol (c)*, 5, 1750 (2008).

38. H. K. Cho, F. A. Khan, I. Adesida, Z. Q. Fang and D. C. Look, "Deep level characteristics in n-GaN with inductively coupled plasma damage," *J. Phys. D*, vol. 41, p. 155314, 2008.

39. S. Chen, U. Honda, T. Shibata et al., "As-grown deep-level defects in n-GaN grown by metal–organic chemical vapor deposition on freestanding GaN," *J Appl Phys*, 112, 053513 (2012).

40. W. I. Lee, T. C. Huang, J. D. Guo, and M. S. Feng, "Effects of column III alkyl sources on deep levels in GaN grown by organometallic vapor phase epitaxy," *Appl Phys Lett*, 67, 1721 (1995).

41. A. Y. Polyakov, N. B. Smirnov, A. V. Govorkov et al., "Comparison of electrical properties and deep traps in p-AlxGa1–xN grown by molecular beam epitaxy and metal organic chemical vapor deposition," *J Appl Phys*, 106, 073706 (2009).

42. M. Gassoumi, B. Grimbert, C. Gaquiere, and H. Maaref, "Evidence of surface states for AlGaN/GaN/SiC HEMTs passivated Si3N4 by CDLTS," *Semiconductors*, 46, 382–385 (2012).

43. T. Okino, M. Ochiai, Y. Ohno, S. Kishimoto, K. Maezawa, and T. Mizutani, "Drain current DLTS of AlGaN-GaN MIS-HEMTs," *IEEE Electron Device Lett*, 25(8), 523–525 (2004).

44. R. Heitz, P. Maxim, L. Eckey, P. Thurian, A. Hoffmann, and I. Broser, "Excited states of Fe3+ in GaN," *Phys Rev B* 55(7), 4382–4387 (1997).

45. M. Caesar, M. Dammann, V. Polyakov et al., "Generation of traps in AlGaN/GaN HEMTs during RF-and DC-stress test," IEEE International Reliability Physics Symposium (IRPS), CD6.1 2012.

46. G. A. Umana-Membreno, G. Parish, N. Fichtenbaum, S. Keller, U. K. Mishra, and B. D. Nener, "Electrically active defects in GaN layers grown with and without Fe-doped buffers by metal-organic chemical vapor deposition," *J Elec Mat*, 37, 569 (2008).

47. L. Stuchlíková, J. Šebok, J. Rybár et al., "Investigation of deep energy levels in heterostructures based on GaN by DLTS," 8th International Conference on Advanced Semiconductor Devices & Microsystems (ASDAM), pp. 135–138, 2010.

48. H. Ashraf, M. Imran Arshad, S. M. Faraz, Q. Wahab, P. R. Hageman, and M. Asghar, "Study of electric field enhanced emission rates of an electron trap in n-type GaN grown by hydride vapor phase epitaxy," *J Appl Phys*, 108(2010), 103708 (2010).

49. H. M. Chung, W. C. Chuang, Y. C. Pan et al., "Electrical characterization of isoelectronic In-doping effects in GaN films grown by metalorganic vapor phase epitaxy," *Appl Phys Lett*, 76(7), 897–899 (2000).

50. -Q. Fang, L. Polenta, J.W. Hemsky, and D. C. Look, "Deep centers in as-grown and electron-irradiated n-GaN," Semiconducting and Insulating Materials Conference, 2000. SIMC-XI. International, pp. 35–42 (2000).

51. A. Y. Polyakov, N. B. Smirnov, A. V. Govorkov, A. A. Shlensky, and S. J. Pearton, "Influence of high-temperature annealing on the properties of Fe doped semi-insulating GaN structures," *J Appl Phys*, 95(10), 5591 (2004).

52. M. J. Legodi, S. S. Hullavarad, S. A. Goodman, M. Hayes, and F. D. Auret, "Defect characterization by DLTS of AlGaN UV Schottky photodetectors," *Phys B*, 308–310, 1189–1192 (2001).

53. A.R. Arehart, A. Sasikumar, G.D. Via et al., "Spatially-discriminating trap characterization methods for HEMTs and their application to RF-stressed AlGaN/GaN HEMTs," *IEEE IEDM* 20.1.1-20.1.4, 2010.

54. X. D. Chen, Y. Huang, S. Fung et al., "Deep level defect in Si-implanted GaN n+-p junction," *Appl Phys Lett*, 82, 3671 (2003).

55. P.Hacke, H. Nakayama, T. Detchprohm, K. Hiramatsu, and N. Sawaki, "Deep levels in the upper band-gap region of lightly Mg-doped GaN," *Appl Phys Lett*, 68(10), 1362–1364 (1996).

56. A. Hierro, S. A. Ringel, M. Hansen, J. S. Speck, U. K. Mishra, and S. P. DenBaars, "Hydrogen passivation of deep levels in n –GaN," *Appl Phys Lett*, 77(10), 1499–1501 (2000).

57. M. Asghar, P. Muret, B. Beaumont, and P. Gibart, "Field dependent transformation of electron traps in GaN p–n diodes grown by metal–organic chemical vapour deposition," *Mat Sci Eng B,* 113 248–252 (2004).

58. E. Calleja, F. J. Sanchez, D. Basak et al., "Yellow luminescence and related deep states in undoped GaN," *Phys Rev B*, 55(7), 4689–4694 (1997).

59. F. D. Auret, S. A. Goodman, F. K. Koschnick, J-M. Spaeth, B. Beaumont, and P. Gibart, "Electrical characterization of two deep electron traps introduced in epitaxially grown n-GaN during He-ion irradiation," *Appl Phys Lett* 73, 3745 (1998).

60. Z.-Q. Fang, G. C. Farlow, B. Claflin, D. C. Look, and D. S. Green, "Effects of electron-irradiation on electrical properties of AlGaN/GaN Schottky barrier diodes," *J Appl Phys*, 105(12), 123704 (2009).

61. A.R. Arehart, T. Homan, M.H. Wong, C. Poblenz, J.S. Speck, and S.A. Ringel, "Impact of N- and Ga-face polarity on the incorporation of deep levels in n-type GaN grown by molecular beam epitaxy," *Appl Phys Lett*, 96(24), 242112 (2010).

62. A. Sasikumar, A. Arehart, S.A Ringel et al., "Direct correlation between specific trap formation and electric stress-induced degradation in MBE-grown AlGaN/GaN HEMTs," *IEEE IRPS*, p. 2C.3.1 (2012).

63. Z. Zhang, C. A. Hurni, A. R. Arehart et al., "Deep traps in nonpolar m-plane GaN grown by ammonia-based molecular beam epitaxy," *Appl Phys Lett*, 100, 052114 (2012).

64. T. Aggerstam, A. Pinos, S. Marcinkevicius, M. Linnarsson, and S. Lourdudoss, "Electron and hole capture cross-sections of Fe acceptors in GaN:Fe epitaxially grown on sapphire," *J Electron Mater*, 36(12), 1621–1624 (2007).

65. A. R. Arehart, A. A. Allerman, and S. A. Ringel, "Electrical characterization of n-type Al0.30Ga0.70N Schottky diodes," *J Appl Phys*, 109(11), 114506 (2011).

66. A. Sasikumar, A. Arehart, S. Kolluri et al., "Access-region defect spectroscopy of DC-stressed N-polar GaN MIS-HEMTs," *IEEE Electron Dev Lett*, 33(5), 658–660 (2012).

67. T. A. Henry, A. Armstrong, K. M. Kelchner, S. Nakamura, S. P. DenBaars, and J. S. Speck, "Assessment of deep level defects in m-plane GaN grown by metalorganic chemical vapor deposition," *Appl Phys Lett*, 100, 082103 (2012).

68. M. Meneghini, A. Stocco, R. Silvestri, G. Meneghesso, and E. Zanoni, "Degradation of AlGaN/GaN high electron mobility transistors related to hot electrons," *Appl Phys Lett*, 100, 233508 (2012).

69. M. Meneghini, A. Stocco, R. Silvestri N. Ronchi, G. Meneghesso, and E. Zanoni, "Impact of Hot Electrons on the Reliability of AlGaN/GaN High Electron Mobility Transistors. In: IEEE IRPS2012, Anaheim, CA, April 15–19, 2012.

70. M. ˇTapajna, N. Killat, J. Moereke et al., "Non-Arrhenius degradation of AlGaN/GaN," *IEEE Electron Device Lett,* 33, 1126 (2012).

71. R. Coffie, Y. Chen, I. P. Smorchkova et al., "Temperature and voltage dependent RF degradation study in AlGaN/GaN HEMTs," Reliability Physics Symposium, pp. 568–569, 2007.

72. P. Saunier, C. Lee, A. Balistreri et al., "Progress in GaN performances and reliability," *Proc IEEE DRC Conf Dig*, 35 (2007).

73. E. Zanoni, G. Meneghesso, G. Verzellesi et al., "A review of failure modes and mechanisms of GaN-based HEMTs," IEDM 2007, pp. 381–384.

74. G. Meneghesso, G. Verzellesi, F. Danesin et al., "Reliability of GaN high-electron-mobility transistors: state of the art and perspectives," *IEEE Trans Dev Mat Rel*, 8(2), 332–343 (2008).

75. G. Meneghesso, M. Meneghini, A. Tazzoli et al., "Reliability issues of Gallium Nitride high electron mobility transistors," *Int J Microwave and Wireless Technol*, 2(01), 39–50 (2010).

76. U.K. Mishra, "Reliability of AlGaN/GaN HEMTs: An overview of the results generated under the ONR DRIFT program," IEEE International Reliability Physics Symposium (IRPS), 2012, vol., no., pp.2C.1.1-2C.1.6, April 15–19, 2012.

77. T. Ohki, M. Kanamura, Y. Kamada et al., "Recent reliability progress of GaN HEMT power amplifiers," *Mater Res Soc Symp Proc* 1432 (2012). doi 10.1557/opl.2012.908.

78. P. Waltereit, J. Kuhn, R. Quay et al., "High efficiency X-band AlGaN/GaN MMICs for space applications with lifetimes above 105 hours," Microwave Integrated Circuit Conference 2012, pp. 123–126.

79. R. Trew, D. Green, and J. Shealy, "AlGaN/GaN HFET reliability," *IEEE Microwave Magazine*, 10(4), 116–127 (2009).

80. J. Joh and J. A. del Alamo, "Mechanisms for electrical degradation of GaN high-electron mobility transistors," IEEE IEDM Tech Dig, (2006).

81. J. Joh, L. Xia, and J. A. del Alamo, "Gate Current Degradation Mechanisms of GaN High Electron Mobility Transistors," Electron Devices Meeting, pp. 385–388 (2007).

82. J. Joh and J. A. del Alamo, "Impact of electrical degradation on trapping characteristics of GaN high electron mobility transistors," IEEE IEDM Tech Dig, 2008.

83. D. Marcon, A. Lorenz, J. Derluyn et al., "GaN-on-Si HEMT stress under high electric field condition," *Phys Stat Sol A*, 6(S2), S1024–S1028 (2009).

84. U. Chowdhury, J. L. Jimenez, C. Lee et al., "TEM observation of crack- and pit-shaped defects in electrically degraded GaN HEMTs," *IEEE Electron Device Lett,* 29, 1098 (2008).

85. S.Y. Park, C. Floresca, U. Chowdhury ET AL., "Physical degradation of GaN HEMT devices under high drain bias reliability testing," *Microelectron Reliab*, 49, 478–483 (2009).

86. T. Kikkawa, K. Makiyama, T. Ohki et al., "High performance and high reliability AlGaN/GaN HEMTs," *Phys Stat Sol A*, 206(6), 1135–1144 (2009).
87. E. Zanoni, F. Danesin, M. Meneghini et al., "Localized damage in AlGaN/GaN HEMTs induced by reverse-bias testing," *IEEE Electron Device Lett*, 30(5), 427 (2009).
88. E.A. Douglas, C.Y. Chang, D.J. Cheney et al., "AlGaN/GaN high electron mobility transistor degradation under on- and off-state stress," Microelectron Reliab, 51, 207 (2010).
89. M. Montes Bajo, C. Hodges, M. J. Uren, and M. Kuball, "On the link between electroluminescence, gate current leakage, and surface defects in AlGaN/GaN high electron mobility transistors upon offstate stress," *Appl Phys Lett*, 101, 033508 (2012).
90. J.A. del Alamo and J. Joh, "GaN HEMT reliability," Microelectron Reliab, 49, 1200–1206 (2009).
91. A. Chini, V. Di Lecce, M. Esposito, G. Meneghesso, and E. Zanoni, "RF Degradation of GaN HEMTs and its correlation with DC stress and I-DLTS measurements," 2009 EUMIC, pp. 132–135, 2009.
92. A. Chini, M. Esposto, G. Meneghesso, and E. Zanoni, "Evaluation of GaN HEMT degradation by means of pulsed I-V, leakage and DLTS measurements," *IEEE Electron Lett*, 45(8), 426–427 (2009).
93. A.R. Arehart, A. Sasikumar, and S. Rajan, "Direct observation of 0.57 eV trap-related RF output power reduction in AlGaN/GaN high electron mobility transistors," *Solid State Electronics,* 80, 19 (2013).
94. C. Hodges, N. Killat, S. W. Kaun, M. H. Wong, F. Gao, and T. Palacios, "Optical investigation of degradation mechanisms in AlGaN/GaN high electron mobility transistors: Generation of non-radiative recombination centers," *Appl Phys Lett*, 97, 223502 (2010).
95. M. Kuball, M. Tapajna, R. J. T. Simms, M. Faqir, and U. K. Mishra, "AlGaN/GaN HEMT device reliability and degradation evolution: Importance of diffusion processes," *Microelectron Reliab,* 51(2), 195–200 (2011).
96. D. A. Cullen, D. J. Smith, A. Passaseo, V. Tasco, A. Stocco, and M. Meneghini, "Electroluminescence and transmission electron microscopy characterization of reverse-biased AlGaN/GaN devices," *IEEE Trans Dev Mat Reliab*, 13, 126 (2013).
97. D. Marcon, J. Viaene, P. Favia et al., "Reliability of AlGaN/GaN HEMTs: Permanent leakage current increase and output current drop," *Microelectron Reliab*, 52(9), 2188–2193 (2012).
98. P. Makaram, J. Joh, J. A. del Alamo, T. Palacios, and C. V. Thompson, "Evolution of structural defects associated with electrical degradation in AlGaN/GaN high electron mobility transistors," *Appl Phys Lett*, 96(23), 233509 (2010).
99. F. Gao, Bin Lu, Libing Li et al., "Role of oxygen in the OFF-state degradation of AlGaN/GaN high electron mobility transistors," *Appl Phys Lett*, 99, 223506 (2011).
100. F. Gao, D. Chen, B. Lu et al., "Impact of moisture and fluorocarbon passivation on the current collapse of AlGaN/GaN HEMTs," *Electron Device Lett*, 33(10), 1378–1380 (2012).
101. M. Meneghini, A. Stocco, M. Bertin et al., "Time-dependent degradation of AlGaN/GaN high electron mobility transistors under reverse bias," *Appl Phys Lett* 100, 033505 (2012).
102. M. Meneghini, A. Stocco, M. Bertin et al., "Electroluminescence analysis of time-dependent reverse-bias degradation of HEMTs: A complete model," Tech Digest IEEE-IEDM 2011, December 5–7, 2011.
103. D. Marcon, X. Kang, J. Viaene et al., "GaN-based HEMTs tested under high temperature storage test," *Microelectron Reliab*, 51(9–11), 1717–1720 (2011).
104. P. Marko, A. Alexewicz, O. Hilt et al., "Random telegraph signal noise in gate current of unstressed and reversebias-bias-stressed AlGaN/GaN high electron mobility transistors," *Appl Phys Lett*, 100, 143507 (2012).

105. P. Marko, M. Meneghini, S. Bychikhin et al., "IV, noise and electroluminescence analysis of stress-induced percolation paths in AlGaN/GaN high electron mobility transistors," *Microelectron Reliab*, 52, 2194 (2012).

106. M. Meneghini, M. Bertin, G. dal Santo et al., "A novel degradation mechanism of AlGaN/GaN/Silicon heterostructures related to the generation of interface traps," 2012 IEEE Electron Device Meeting 13.3.1 (2012).

107. G. Meneghesso, M. Meneghini, A. Stocco et al., "Degradation of AlGaN/GaN HEMT devices: Role of reverse-bias and hot electron stress," *Microelectron Eng*, 109, 257–261 (2013).

108. M. Meneghini, M. Bertin, A. Stocco et al., "Degradation of AlGaN/GaN Schottky diodes on silicon: Role of defects at the AlGaN/GaN interface," *Appl Phys Lett*, 102, 163501 (2013).

109. K. Matsushita, H. Sakurai, S. Teramoto et al., "Influence of SiC Substrate Misorientation on AlGaN/GaN HEMTs Performance," CS MANTECH Conference, May 18–21, 2009, Tampa, FL.

110. Y. Dora, A. Chakraborty, L. McCarthy, S. Keller, S. P. DenBaars, and U. K. Mishra, "High breakdown voltage achieved on AlGaN/GaN HEMTs with integrated slant field plates," *IEEE Electron Device Lett*, 27, 713 (2006).

111. R. Chu, A. Corrion, M. Chen et al., "1200-V normally off GaN-on-Si field-effect transistors with low dynamic on -resistance," *Electron Dev Lett*, 32, 632–634 (2011).

112. M. J. Uren, K. J. Nash, R. S. Balmer et al., "Punchthrough in short-channel AlGaN/GaN HFETs," *IEEE Trans Electron Devices*, 53(2), 395–398 (2006).

113. E. Bahat-Treidel, O. Hilt, F. Brunner, J. Würfl, and G. Tränkle, "Punchthrough-voltage enhancement of AlGaN/GaN HEMTs using AlGaN double-heterojunction confinement," *IEEE Transactions on Electron Devices*, 55, 3354 (2008).

114. O. Hilt, P. Kotara, F. Brunner, A. Knauer, R. Zhytnytska, and J. Würfl, "Improved Vertical Isolation for Normally-Off High Voltage GaN-HFETs on n-SiC Substrates," *IEEE Transactions on Electron Devices*, 60, 3084 (2013).

115. W. S. Tan, P. A. Houston, P. J. Parbrook, D. A. Wood, G. Hill, and C. R. Whitehouse, "Gate leakage effects and breakdown voltage in metalorganic vapor phase epitaxy AlGaN/GaN heterostructure field-effect transistors," *Appl Phys Lett*, 80, 3207 (2002).

116. M. Wang and K. J. Chen, "Off-state breakdown characterization in AlGaN/GaN HEMT using drain injection technique," *IEEE Transactions on Electron Devices*, 57, 1492 (2010).

117. H. Hanawa, H. Onodera, A. Nakajima, and K. Horio, "Numerical Analysis of breakdown voltage enhancement in AlGaN/GaN HEMTs with a high-k passivation layer," *IEEE Transactions on Electron Devices*, 61(3), 769–775 (2014).

118. M. Meneghini, A. Zanandrea, F. Rampazzo et al., "Electrical and electroluminescence characteristics of AlGaN/GaN high electron mobility transistors operated in sustainable breakdown conditions", *Jap JAppl Phys*, 52, 08JN17 (2013).

119. S. R. Bahl, M. Van Hove, X. Kang, D. Marcon, M. Zahid, and S. Decoutere, "New Source-side Breakdown Mechanism in AlGaN/GaN Insulated-Gate HEMTs," Proceedings of the 25th International Symposium on Power Semiconductor Devices and ICs ISPSD 2013.

120. F. Medjdoub, M. Zegaoui, B. Grimbert, N. Rolland, and P. A. Rolland, "Effects of AlGaN back barrier on AlN/GaN-on-silicon high-electron-mobility transistors," *Appl Phys Express*, 4, 124101 (2011).

121. N. Herbecq, I. Roch-Jeune, N. Rolland et al., "1900V, 1.6mΩcm2 AlN/GaN-on-Si power devices realized by local substrate removal," *Appl Phys Express*, 7, 034103 (2014).

122. M. Meneghini, G. Cibin, M. Bertin et al., "Off-state Degradation of AlGaN/GaN power HEMTs: Experimetnal demonstration of time-dependent drain-source breakdown," *IEEE Trans Electron Devices*, Vol 61, no. 6, pp. 1987-1992, 2014.

123. H. Umeda, A. Suzuki, Y. Anda et al.," Locking-Voltage Boosting Technology for GaN Transistors," Electron Devices Meeting (IEDM), 2010 IEEE International, 20.5.1, 20.5.4 (2010).

124. I. B. Rowena, Susai L. Selvaraj, and T. Egawa, "Buffer Thickness Contribution to Suppress Vertical," *IEEE Electron Device Lett,* 32, 1534 (2011).

125. J. Joh, F. Gao, T. Palacios, and J. A. del Alamo, "A model for the critical voltage for electrical degradation of GaN high electron mobility transistors," *Microelectron Reliab,* 50, 767–773 (2010).

126. D. Marcon, G. Meneghesso, T.L. Wu et al., "Reliability analysis of permanent degradations on AlGaN/GaN HEMTs," *IEEE Trans Electron Devices,* 60 (10), 3132–3141 (2013), ISSN: 0018-9383.

127. D. Marcon T. Kauerauf, F. Medjdoub et al., "A Comprehensive Reliability Investigation of the Voltage-, Temperature- and Device eometry-Dependence of the Gate Degradation on state-of-the-art GaN-on-Si HEMTs," Proc. IEEE IEDM 2010, 20.3.1.

128. S. Singhal, J. C. Robers, P. Rajogopal et al., "GaN-on-Si failure mechanisms and reliability improvements," *IEEE Int Reliab Phys Symp Proc,* 95, 2006.

129. W. Saito, M. Kuraguchi, Y. Takada, K. Tsuda, I. Omura, and T. Ogura, "Influence of surface defect charge at AlGaN–GaN-HEMT upon schottky gate leakage current and breakdown voltage," *IEEE Transactions on Electron Devices,* 52, 259 (2005).

130. F. Gutle, V. M. Polyakov, M. Baeumler, F. Benkhelifa, S. Muller,and M. Dammann, "Radiative inter-valley transitions as a dominant emission mechanism in AlGaN/GaN high electron mobility transistors," *Semicond Sci Technol,* 27, 125003 (2012).

131. Y. S. Puzyrev, T. Roy, M. Beck et al., "Dehydrogenation of defects and hot-electron degradation in GaN high-electron-mobility transistors," *J Appl Phys,* 109, 034501 (2011).

132. Y. Puzyrev S. Mukherjee, Jin Chen, T. Roy, M. Silvestri, R. D. Schrimpf, D. M. Fleetwood et al. Gate Bias Dependence of Defect-Mediated Hot-Carrier Degradation in GaN HEMTs," *IEEE Trans Electron Devices,* 61(5), 1316, 1320 (2014).

133. Y. S. Puzyrev, B. R. Tuttle, R. D. Schrimpf, D. M. Fleetwood, and S. T. Pantelides, "Theory of hot-carrier-induced phenomena in GaN high-electron-mobility transistors," *Appl Phys Lett,* 96, 053505 (2010).

134. S. Mukherjee, Y. Puzyrev, J. Hinckley, R. D. Schrimpf, D. M. Fleetwood, J. Singh, and S. T. Pantelides et al., "Role of bias conditions in the hot carrier degradation of AlGaN/GaN high electron mobility transistors," *Phys Status Solidi C,* 10(5), 794–798 (2013).

135. C. H. Lin, T. A. Merz, D. R. Doutt, M. J. Hetzer, Joh, Jungwoo, A. del Alamo et al., "Nanoscale mapping of temperature and defect evolution inside operating AlGaN/GaN high electron mobility transistors," *Appl Phys Lett,* 95, 033510 (2009).

136. M. Meneghini, G. Meneghesso, and E. Zanoni, "Analysis of the Reliability of AlGaN/GaN HEMTs Submitted to On-State Stress Based on Electroluminescence Investigation," *IEEE Trans Device Mater Reliab,* 13(2), 357–361 (2013).

137. S. R. Bahl and J. A. del Alamo, "A new drain-current injection technique for the measurement of breakdown voltage in FETs," *IEEE Transactions on Electron Devices,* 40(8), 1558–1560 (1993).

138. A. Pérez-Tomás, A. Fontserè, J. Llobet et al., "Analysis of the AlGaN/GaN vertical bulk current on Si, sapphire, and free-standing GaN substrates," *Jl Appl Phys,* 113, 17450 (2013).

139. B. Lu and T. Palacios, "High breakdown (>1500 V) AlGaN/GaN HEMTs by substrate-transfer technology," *IEEE Trans Electron Devices,* 31, 951 (2010).

140. P. Srivastava et al., "Record breakdown voltage (2200 V) of GaN DHFETs on Si with 2-μm buffer thickness by local substrate removal," *IEEE Electron Device Lett,* 32, 30 (2011).

141. S. Turuvekere, N. Karumuri, A. A. Rahman, A. Bhattacharya, A. DasGupta, and N. DasGupta, "Gate leakage mechanisms in AlGaN/GaN and AlInN/GaN HEMTs: Comparison and modeling," *IEEE Trans Electron Devices*, 60, 3157 (2013).

142. B. Heying, I.P. Smorchkova, R. Coffie et al., "In situ SiN passivation of AlGaN/GaN HEMTs by molecular beam epitaxy," *Electronics Lett,* 43, 14 (2007).

143. M. Germain et al., "In-situ passivation combined with GaN buffer optimization for extremely low current dispersion and low gate leakage in Si3N4/AlGaN/GaN HEMT devices on Si (111)," *Phys. Stat Sol (C)*, 5, 2010 (2008).

144. M. Van Hove, S. Boulay, S. R. Bahl et al., "CMOS process-compatible high-power low-leakage AlGaN/GaN MISHEMT in silicon," *IEEE Electron Device Lett*, 33(5), 667 (2012).

145. C. Bulutay, "Electron initiated impact ionization in AlGaN alloys," *Semicond Sci Technol*, 17(10), 59–62 (2002).

# Index